Nucleic Acids and Protein Synthesis in Plants

NATO ADVANCED STUDY INSTITUTES SERIES

A series of edited volumes comprising multifaceted studies of contemporary scientific issues by some of the best scientific minds in the world, assembled in cooperation with NATO Scientific Affairs Division.

Series A: Life Sciences

Recent Volumes in this Series

The series is published by an international board of publishers in conjunction with NATO Scientific Affairs Division

A	Life Sciences	Plenum Publishing Corporation
B	Physics	New York and London
C	Mathematical and Physical Sciences	D. Reidel Publishing Company Dordrecht and Boston
D	Behavioral and Social Sciences	Sijthoff International Publishing Company Leiden
E	Applied Sciences	Noordhoff International Publishing Leiden

Nucleic Acids and Protein Synthesis in Plants

Edited by

L. Bogorad
Harvard University
Cambridge, Massachusetts

and

J. H. Weil
Université Louis Pasteur
Strasbourg, France

PLENUM PRESS • **NEW YORK AND LONDON**
Published in cooperation with NATO Scientific Affairs Division

Library of Congress Cataloging in Publication Data

Nato Advanced Study Institute on Nucleic Acids and Protein Synthesis in Plants, Strasbourg, 1976.
Nucleic acids and protein synthesis in plants.

(Nato advanced study institutes series: Series A, Life science; v. 12)
"Lectures presented at the NATO Advanced Study Institutes on Nucleic Acids and Protein Synthesis in Plants, sponsored by NATO, FEBS, EMBO, at a colloquium sponsored by CNRS, held in Strasbourg, France, July 15-24, 1976."
Includes index.
1. Protein biosynthesis–Congresses. 2. Nucleic acid synthesis–Congresses. 3. Botanical chemistry–Congresses. I. Bogorad, Lawrence, 1921-   II. Weil, Jacques Henry. III. North Atlantic Treaty Organization. IV. Title. V. Series.
QK898.P8N37 1976   581.8'732   77-23267
ISBN 978-1-4684-2777-6   ISBN 978-1-4684-2775-2 (eBook)
DOI 10.1007/978-1-4684-2775-2

Lectures presented at the NATO Advanced Study Institute on
Nucleic Acids and Protein Synthesis in Plants sponsored by NATO,
FEBS, EMBO, at a Colloquium sponsored by CNRS, held in
Strasbourg, France, July 15-24, 1976

© 1977 Plenum Press, New York
Softcover reprint of the hardcover 1st edition 1977
A Division of Plenum Publishing Corporation
227 West 17th Street, New York, N.Y. 10011

NUCLEIC ACIDS AND PROTEIN SYNTHESIS IN PLANTS

an Advanced Course sponsored by
NATO, FEBS, EMBO
a colloquium sponsored by CNRS
Strasbourg, July 15-24, 1976

———————————— *Organized by* ————————————

L. BOGORAD

The Biological Laboratories
Harvard University
16 Divinity Avenue
Cambridge, Massachusetts

J.H. WEIL

Institut de Biologie Moléculaire et Cellulaire
Université Louis Pasteur
15, rue Descartes
67084 Strasbourg Cedex, France

———————————— *Organizing Committee* ————————————

Institut de Biologie Moléculaire et Cellulaire
Université Louis Pasteur
15, rue Descartes
67084 Strasbourg Cedex, France

G. BURKARD	P. IMBAULT	L. OSORIO
J. BLUM-CANADAY	G. JEANNIN	J. RAMIASA
E. DUBOIS-FISCHEL	A. KLEIN	V. SARANTOGLOU
P. GUILLEMAUT	S. MONTASSER	A. STEINMETZ
	A. M. OHLENBUSCH	

Secretary: **M. J. KRAUS**

Preface

During the summer of 1974 we discussed the state of molecular biology and biochemical developmental biology in plants on a few occasions in Paris and in Strasbourg. The number of laboratories engaged in such research is minute compared with those studying comparable problems in animal and bacterial systems, but by then much interesting work had been done and a great momentum was building. It seemed to us that the summer of 1976 would be a good time to review these areas of plant biology for students as well as advanced workers. We outlined a program for a course to colleagues both in Europe and the United States and asked a few potential lecturers if they would be interested. The response was not just positive; it was overwhelmingly enthusiastic. Those who had some acquaintance with Alsace, and especially with Strasbourg, invariably told us that they had two reasons for being enthusiastic about participating - the subject and the proposed site.

The lectures published here* reflect the diversity of current research in plant molecular biology and biochemical developmental biology. Each lecture gives us a glimpse of the depth of questions being asked, and sometimes answered, in segments of this field of investigation. This research is directed at fundamental biological problems, but answers to these questions will provide knowledge essential for bringing about major changes in the way the world's agricultural enterprise can be improved.

---

*An International Colloquium on Nucleic Acids and Protein Synthesis in Plants sponsored by the Centre National de la Recherche Scientifique was held in July, 1976, in Strasbourg parallel with this Advanced Course. The 100 or so brief contributions to the Colloquium are being published by Centre National de la Recherche Scientifique (15, quai Anatole France, 75700 Paris, France). Readers of this volume may also be interested in the proceedings of an international conference on "The Genetics and the Biogenesis of Chloroplasts and Mitochondria" held in Munich later in the summer of 1976; this volume is edited by Professor Th. Bücher and is to be published by Elsevier Press.

It is unfortunate that this volume cannot bring to the reader the sense of discovery that came to participants as chapters of modern experimental plant biology were described, explained, and discussed enthusiastically. As organizers as well as participants we are grateful to the lecturers for clear expositions both in their formal presentations and in response to questions. We are grateful to our fellow participants in the Advanced Study Institute for their contributions to the lively discussions.

Instructors and students in this advanced course represented a substantial fraction of all of the laboratories of the world working on molecular biology and biochemical developmental biology in plants. The generous financial support we received permitted participation not only by Europeans but also by lecturers and students from the United States and elsewhere in the world.

We are grateful to the North Atlantic Treaty Organization, the Federation of European Biochemical Societies, the European Molecular Biology Organization, the Centre National de la Recherche Scientifique, and the National Science Foundation of the United States for financial support which made this advanced course possible. Its success was in large measure also the result of careful planning and a great deal of work by the Organizing Committee at the Institute of Molecular and Cellular Biology at the University Louis Pasteur in Strasbourg. We are also indebted to officials responsible for housing participants at the University Louis Pasteur for their efforts, which virtually completely relieved the participants of concern for food and shelter, thus adding to the thinking time available to each of us.

Lawrence Bogorad

Jacques H. Weil

Contents

DNA SYNTHESIS DURING MICROSPOROGENESIS

Herbert Stern

Department of Biology, B-022
University of California, San Diego
La Jolla, California 92093

INTRODUCTION

Meiosis is a major and critical event in microsporogenesis. In the course of that event, the processes of pairing and crossing-over activate elements of chromosome organization that are probably not required for vegetative growth and reproduction. Even where mitotic crossing over does occur, its frequency is rare presumably because a supportive organization is lacking. An apparently loose form of pairing between homologous chromosomes has been reported for somatic cells of various plant species (4), but such "somatic association" is both morphologically and functionally distinct from meiotic pairing and its accompanying exchange. The latter are virtually unique to meiocytes. Although the regulation of chromosome reproduction and transcription imposes broad requirements on chromosome organization, it is doubtful that such organization encompasses the special needs of chromosomes in meiosis. The major thrust of this presentation is to demonstrate that there are indeed components of chromosome organization which, thus far at least, have been revealed only during meiosis. Two of these components are expressed in the sequence organization of DNA and thus are fixed characteristics of the linear differentiation of the chromosome (4).

Despite the simplicity of textbook diagrams, the alignment and pairing of chromosomes, let alone their crossing-over, is a complex process requiring major structural accommodations. Paired or "synapsed" chromosomes normally reflect the linear ordering of their constituent genes, a relationship that has been most thoroughly examined and documented in cytogenetic studies of

1

Zea mays (16).   Inversions, translocations, insertions and
deletions of chromosome segments are all reflected in the geometry
of chromosome pairing during meiosis.   Nevertheless, it is most
improbable that the alignment of homologs directly involves even
as much as 0.5% of the genome (20, 21).   The process of synapsis
occurs at a stage (zygotene) when chromosomes have already undergone an appreciable degree of compaction; in *Lilium*, the length
of pairing chromosomes is about 10^{-4} that of the DNA (Preben Holm,
private communication).   Homologous segments, even when present in
two different chromosomes as result of translocation, align suc-
cessfully without involving the full span of their DNA sequences.
Moreover, the fidelity of alignment, whether it be entirely or
only partially homologous, can itself be regulated by specific
genes or accessory chromosomes (3, 17).   The highly sensitive
regulation of pairing homology is of major importance in the evolution of established allopolyploids such as *Triticum aestivum*.
Thus, not only must homologous alignment be effected by a limited
number of DNA sites that are strategically positioned along the
length of the chromosomes, but the <u>process</u> of that alignment must
be mediated by specific gene products that affect both the
stringency and extent of matching between homologous chromosome
segments.

     Crossing-over in higher plants is dependent on prior chromo-
some synapsis and is subject to constraints that are superimposed
on those set by the molecular mechanisms mediating DNA strand
recombination.   The frequency with which crossing-over occurs
between two loci in a chromosome is not purely a function of the
physical distance between them.   Phylogenetic differences in
frequency is the rule rather than the exception.   Flowering plants
have a relatively large intergenic distance for a particular
crossover frequency compared with fungi (21).   The average intergenic distance per crossover in *Lilium*, for example, has been
calculated to be about 10^4 times that in *Saccharomyces cerevisiae*
(21), and many similar examples could be chosen from the biological
kingdom to illustrate the general point that the larger the size of
the genome, the lower the frequency of crossing-over for a given
physical distance in a chromosome.   Meiosis in higher plants must
be so organized as to limit the frequency of exchanges per unit of
DNA without, however, compromising the regularity of its occurrence.   Moreover, frequency of crossing-over per unit length of
DNA is not uniform for all chromosomes within a species or even
for all regions within a chromosome (14).   Heterochromatic segments of chromosomes show very little, if any, crossing-over.   In
some species, crossing-over may be regularly excluded from a particular pair of chromosome arms, and in many, if not most species,
a gradient in the frequency of crossing-over is present along
chromosome arms.   Although the distances between crossovers may
appear small when viewed through a light microscope, they are

greater by several orders of magnitude than an entire chromosome of *Saccharomyces cerevisiae*.

It is useful, though slightly provincial, to point out that in development of cytogenetics, the plant kingdom has made possible a variety of insights into meiosis, insights which would have been much more difficult to obtain from animal sources. The meiotic behavior of chromosomes in polyploid situations and the attendant fine tuning of homologous pairing, the influence of accessory chromosomes on the pairing process, the direct cytogenetic analysis of meiotic phenomena, and the biochemical events in meiosis have all had their primary origins in studies of plant systems. The description of DNA metabolism is thus by necessity, rather than choice, derived almost entirely from plant studies. A weakness of the description is that it is derived mainly from liliaceous plants; a broader taxonomic base would be highly desirable. However, no other group of plants has yet been found that provides as favorable a degree of meiotic synchrony and of meiotic duration for the biochemical analysis of meiosis.

In keeping with the broad purpose of the course--an examination of nucleic acid and protein synthesis in plants--this article will address itself to DNA synthesis in pollen mother cells of *Lilium*. Three categories of DNA synthesis occur, each with its own set of distinctive features, and each occurring at a specific interval during the life of the pollen mother cell. The first of these intervals occupies the premeiotic S-phase and, as in mitotic cell cycles, constitutes the time of chromosome reproduction. The second interval covers the stage of chromosome pairing (zygotene) during which 0.3-0.4% of the genome undergoes a delayed replication. Completion of pairing and of replication is followed by a marked shift to a pattern of DNA metabolism that is highly characteristic of pachytene, the stage commonly associated with crossing-over. The semiconservative replication of zygotene is succeeded at pachytene by an intense interval of repair-replication activity. The entire succession of events extends for 6-8 days and it is the temporal separation of the individual events that has made possible the kinds of analyses reported here.

A.   THE PREMEIOTIC S-PHASE

Although the number of organisms examined is small, it would appear that for any given species, the S-phase preceding meiosis is of longer duration than the S-phase in any mitotic cycle (2). Just as S-phases following fertilization appear to be inordinately short, those preceding meiosis appear to be inordinately long, of the order of 3-4 or even more times the duration of S-phases in somatic cells. In *Lilium* the S-phase of microsporocytes is

48 hours whereas that of root tip cells is 8-12 (6). The significance of S-phase attenuation is unclear, but its widespread
occurrence points to a special, even if still unidentified,
meiotic function. The boundaries of the S-phase have never been
as precisely defined in meiocytes as they have been in mitotic
cells. With very few exceptions, a G-2 interval is prominently
present between completion of chromosome reproduction and the
beginning of mitosis. In meiocytes, on the other hand, designa-
tion of a G-2 interval may be no more than a formalism. In some
species a separation between S-phase and the leptotene-zygotene
stages has been difficult to establish, especially in situations
where DNA amplification occurs (5). While the data on DNA
replication can be fitted into an orderly pattern by making a
strong distinction between syntheses related to pairing or
crossing-over and those related to subsequent developments in
microsporo- or gametogenesis, the precise endpoint of the
attenuated premeiotic S-phase remains difficult to define.

Our poor understanding of the mechanisms regulating DNA
replication in eukaryotes limits the approaches available for
probing the long duration of the premeiotic S-phase. There is,
nevertheless, convincing evidence that the molecular rate of DNA
replication--or fork migration--is approximately the same for all
tissues despite major differences in S-phase duration (1, 2).
Based on the assumption that compact chromatin obstructs the
necessary interactions between replicases and initiation sites for
replication, the proposal has been made that nuclei regulate the
duration of S-phase by regulating the extent of chromosome compaction (1). It is therefore of interest that striking patterns
of transient heterochromatinization (or chromatin compaction) have
been observed in microsporocytes prior to and during the premeiotic
S-phase. Oono and Hotta (15) have fractionated the chromatin of
Lilium microsporocytes into diffuse and compact components and
found that prior to leptotene, microsporocytes have only 1/3 of
their total chromatin in the diffuse form. Moreover, they
determined that virtually all of the DNA synthesis during the
first part of the S-phase was confined to this diffuse fraction.
Two-thirds of the nuclear DNA was delayed in its replication until
after the first 20 hours of S-phase. Comparable results have been
obtained by Holm (6) using autoradiography. Superficially at
least, there appears to be a simple relationship between the degree
of chromatin compaction and the occurrence of DNA synthesis in
microsporocytes. If that relationship proves to be correctly
interpreted, the data would add experimental support to the general
proposition that chromatin compaction serves to regulate the
chromosomal rate of DNA replication (1). If so, the questions of
how compaction itself is regulated and what the functional role
of S-phase attenuation might be still remain to be answered!

B. Z-DNA REPLICATION

In *Lilium*, as indicated earlier, the alignment of homologous chromosomes coincides with the delayed replication of about 0.3-0.4% of the nuclear DNA. The functional importance of this replication is made evident by a failure of pairing to proceed if Z-DNA synthesis is inhibited (13). Although the delayed replication at zygotene might be considered a product of S-phase attenuation, no insights can be gained from the classification. S-phase and Z-DNA replication are separated by the leptotene stage during which time chromosomes undergo considerable morphological modification. Pairing and Z-DNA synthesis begin after completion of leptotene. Despite the very many studies of cell cycles, no evidence has yet been recorded for a delayed replication of DNA during mitotic prophase. In this respect, the occurrence of Z-DNA replication is unusual and its functional role intriguing. Its temporal coincidence with chromosome pairing, and the arrest of pairing upon selective inhibition of DNA synthesis, invite the speculation that DNA replication is essential to the process of pairing. Various models can be and have been proposed to relate the two events but they remain purely speculative and will not be described here (19). The important point is that a specific DNA component, distributed within and among all the chromosomes, has been programmed to replicate at a distinctive time, one that coincides with the homologous pairing of chromosomes.

Several properties of Z-DNA add to the probability of its distinctive role in meiotic DNA metabolism. Although late replicating, it does not appear to be a component of heterochromatin as is evident from the more or less uniform distribution of Z-DNA label among all chromosomes when examined at metaphase (12). Oono and Hotta (15) have found a much larger proportion of Z-DNA than of any other DNA component to be present in the diffuse fraction of chromatin. Its GC content is about 50% in contrast with the bulk of nuclear DNA whose GC content is about 40% (7). It occurs in segments of an average size between 5×10^3-10^4 base pairs (10). If the segments were distributed uniformly within the genome, which is doubtful, the distances between segments would be about 3×10^6 base pairs, and the number of segments per chromatid would be approximately 2-3,000 (9, 10). The disposition of these segments along the length of the chromosome is a point of major interest but data on this question are still being obtained.

When DNA is labeled at zygotene and analyzed on a CsCl gradient, the label is confined to the heavy end of the gradient and has an isopycnic profile characteristic of a set of satellite DNA's (7). Its isopycnic behavior and its late replication might lead to the prediction that Z-DNA is a collection of highly repeated sequences, a property that has been considered to be

desirable for effecting chromosome synapsis. That this is not at
all the case is demonstrated by C_0t analysis. Regardless of
whether the analysis is performed by following the Z-label in the
presence of total DNA or by using Z-DNA which has been purified
from a CsCl gradient the reassociation data show unequivocally
that Z-DNA belongs to the category of unique sequences in *Lilium*
(9). The precise number of copies has not been determined, but it
is of the order of 1-5 per haploid genome and, as such, Z-DNA is
probably as unique as any of the other unique components in the
Lilium genome.

The absence of multiply repeated sequences in Z-DNA gives it,
at least theoretically, a capacity to lend precision to chromosome
alignment, although it detracts from the capacity to facilitate
alignment, a property more readily provided by repeated sequences.
The serious problem posed by its sequence characteristics is not,
however, related to the mechanics of a pairing function, but
rather to its satellite behavior in a CsCl gradient. If, as is
commonly the case, DNA breakage in the course of extraction were
largely random, one would not expect an almost total banding of
labeled Z-DNA in CsCl solution as a satellite cluster, under
conditions where the weight average of the Z-DNA segments is
between $0.5-1 \times 10^4$ base pairs, and that of total DNA is about
$1.5-3 \times 10^4$ base pairs. The explanation of satellite behavior
turns out to be straightforward but it nevertheless provides
further evidence that certain elements of DNA sequence organization are specifically related to meiotic functions of chromosomes.

When DNA containing labeled zygotene sequences is prepared
from protoplasts under conditions of minimal shear and spun in a
neutral glycerol gradient, the label cosediments with total
nuclear DNA which has a mean molecular size well in excess of 10^6
base pairs (10). If Z-DNA were present as free segments of 10^3-
10^4 base pairs, it would have traced a distinctive sedimentation
profile. This not being the case, it would appear that the zygotene label is linked to adjacent DNA stretches much as any of the
other sequences in the genome. A striking contrast is provided
however by gentle extraction of DNA in denaturing medium. Under
such conditions, Z-DNA sediments as a discrete band with a much
lower S value than total DNA, thus indicating the presence of
single strand gaps at the ends of the labeled Z-DNA segments.
The existing gaps render double-stranded DNA highly susceptible
to mechanical shearing in the gapped regions. That such gaps can
account for the isopycnic behavior of Z-DNA when prepared by
standard procedures is demonstrable by mild stirring of a high
molecular weight lysate. The stirring, like denaturation,
releases the Z-DNA as a relatively low molecular weight fraction.
The satellite behavior of Z-DNA in CsCl gradients is thus a
product of its incomplete replication, the gaps or nicks bracketing

the Z-DNA sequences providing sites that are readily sheared when
DNA is prepared by standard procedures.

The requirement of shearing to release labeled Z-DNA duplexes
implies that the "old" DNA strands are intact even though their
newly replicated complements are not. If so, it is to be expected
that the gaps in the new strands would be healed at some stage of
microsporogenesis. If healing did not occur, half the chromosomes
would show breaks at the first microspore mitosis. That healing
does occur before termination of meiosis can be demonstrated
experimentally as can also the absence of interruptions bracketing
the Z-DNA stretches in the "old" strands. The time of gap closure
is demonstrated by labeling microsporocytes during zygotene,
transferring them to non-radioactive medium, and harvesting them
at successive intervals of meiosis (1). DNA prepared from each
group of cells by standard procedures is analyzed by equilibrium
centrifugation in a CsCl gradient. Two principal types of radioactivity profiles are observed. Preparations from cells harvested
prior to the metaphase of meiosis display a typical satellite
Z-DNA band. By contrast, preparations from cells beyond metaphase
display little or no satellite in their radioactivity profiles.
Thus, labeled Z-DNA segments which behave as discrete fragments
during meiotic prophase no longer do so beyond meiotic metaphase.
The gaps bracketing the Z-DNA stretches are filled as the cells
progress through meiosis following the completion of prophase.
In a strict sense, meiotic chromosome replication is not complete
until many days after termination of the S-phase when the metaphase
stage is reached.

To demonstrate that the old DNA strand remains intact throughout meiosis requires a more elaborate procedure than isopycnic
centrifugation. Radioactivity cannot be used to trace the behavior
of old Z-DNA strands because they cannot be differentially labeled.
The position of Z-DNA in a gradient can however be traced by
hybridization to previously labeled and purified Z-DNA. Nuclear
DNA prepared from cells at different meiotic stages is centrifuged
to equilibrium in a CsCl solution (10). The position of Z-DNA
sequences in the isopycnic profiles is determined by collecting
fractions and testing their capacity to hybridize to purified
Z-DNA. If the DNA is prepared by standard procedures, the profiles
of hybridization are essentially the same as the radioactivity
profiles of labeled Z-DNA prepared from cells harvested at different meiotic stages. Extracts of DNA from premetaphase cells
hybridize Z-DNA sequences in the satellite region of the gradient,
whereas those from postmetaphase cells hybridize Z-DNA across the
mainband region. However, different hybridization profiles are
obtained with DNA prepared by gently lysing zygotene or pachytene
protoplasts in an alkaline medium. Under these conditions
shearing is avoided and relatively few breaks are introduced into

the single DNA strands during the *in vitro* manipulations. The
hybridization capacity of the denatured DNA thus prepared from
zygotene and pachytene cells is approximately evenly distributed
between satellite and mainband regions. Such a distribution would
result from a situation in which half the Z-DNA sequences, being
present in the uninterrupted old strands, would sediment with total
DNA whereas newly replicated Z-DNA, being unligated, would sediment
in the satellite regions. As expected, a partitioning between
satellite and mainband regions is not observed in similar preparations of DNA from postmetaphase cells.

A significant aspect of these studies on Z-DNA lies in the
evidence that a specific set of DNA sequences is not only delayed
in its replication for about 3 days but is further delayed in its
ligation to adjacent DNA sequences for an additional 5 days. As
mentioned earlier, no direct experimental evidence yet exists for
the mechanisms by which Z-DNA functions in one or more meiotic
processes. The circumstantial evidence, however, is very strong
and as such it points in a particular direction. Chromosomes
require matching sites to achieve homolog alignment under conditions of partial compaction. They probably also need devices to
resolve sister chromatids in the course of, or subseuqent to,
chiasma formation. Matching needs could be met by a transient
availability of intercalary stretches of single-stranded DNA made
possible by coordinating delayed replication with synapsis. The
close association of sister chromatids during meiosis and their
ultimate separation could be achieved by having the old DNA
strands retain their original duplex association in those regions
where gaps are still present in the otherwise replicated complementary strands. Whether these needs are actually met in this way
is still very much an open question, but it is a question which
will be possible to answer experimentally.

C. PACHYTENE DNA SYNTHESIS

A striking feature of meiotic DNA metabolism is the sharp
switch in character of DNA synthesis when cells, having completed
synapsis, leave zygotene and enter the pachytene stage (8). Semiconservative replication ceases and repair-replication begins.
Zygotene synthesis is sensitive to hydroxyurea; pachytene is not
(7). If the strands replicated during premeiotic S-phase are made
heavy with BdUrd and the cells radio-labeled during pachytene, the
label is found in both strands, unlike the Z-DNA label which is
found only in one. The data therefore compel the conclusion that
following completion of chromosome pairing, cells enter into an
interval of relatively intense repair-replication activity. Just
how this activity relates to crossing-over is a matter of some
conjecture, but it is at least demonstrable that in those

conditions where chiasmata are not formed, as in an achiasmatic
hybrid, or in colchicine induced achiasmatics, repair-replication
activity does not occur. For this and other bits of circumstantial evidence, it is reasonable to suppose that pachytene
metabolism is related to crossing-over (21).

The regulation of pachytene repair-replication activity is of
considerable interest inasmuch as it is confined to the one
interval in meiotic prophase. The confinement is not due to a
lack of repair-replication capacity at other meiotic stages. Cells
irradiated with UV or X-rays at any stage of meiosis, except for
the metaphase interval, respond with repair-replication (21).
Several enzymes which presumably are required for repair-replication,
specifically polymerase, polynucleotide kinase, and ligase are
present at nearly the same levels of activity during intervals prior
to pachytene (11). One critical regulatory factor appears to be the
activation of an endonuclease, or perhaps even its synthesis (11).
Endonuclease activity varies markedly during meiosis. Its activity
becomes detectable *in vitro* toward the end of zygotene and reaches
a peak of activity in early pachytene; beyond pachytene, activity
is barely detectable. The enzyme appears to be unique to the
meiocytes. Although endonuclease activity has been identified
in somatic cells, it can be distinguished from the meiotic endonuclease by various properties and can also be separated from it.
The meiotic endonuclease is specific for double-stranded DNA and
cleaves in a 3'phosphoryl:5'OH pattern.

The *in vivo* action of this nuclease is demonstrable by sizing
nuclear DNA at different stages of meiosis (8). DNA released from
protoplasts by lysis in either neutral or alkaline medium is
analyzed on a glycerol gradient to obtain a profile of molecular
sizes for each meiotic stage. Neutral preparations show no change
in sedimentation profile throughout the course of meiosis. The
frequency of double-stranded breaks, whatever their number, remains
approximately constant. The behavior of the single-stranded DNA
is dramatically different. The sedimentation profile of pachytene
preparations is consistently bimodal with peaks at about 105S and
63S. The bimodality is not observed in lysates from cells at
stages before and after pachytene; all such lysates of single-
stranded DNA show only one peak at about 105S. A high frequency
of single-strand breaks is clearly a characteristic of pachytene.
The appearance of single-stranded nicks closely parallels that of
endonuclease activity. This is accompanied by a major increase in
the ratio of 5'OH to 5'phosphoryl termini as would be expected if
the nicks were formed by the meiotic endonuclease. It may be
concluded that an essential feature of the meiotic program is to
introduce a substantial number of single-stranded nicks into the
DNA at about the time that the chromosomes have completed synapsis.
Presumably, such nicking sets the stage for crossing-over although

the frequency of nicks is several orders of magnitude higher than
the frequency of exchanges. The mechanism whereby 10^3-10^4 nicks
are translated into a crossover remains to be studied.

A significant point to be clarified in connection with the
programmed nicking is whether such nicking is randomly placed or
whether it is confined to specific chromosomal regions. This has
been partly answered by C_0t analysis (9, 18). The reassociation
kinetics of pachytene replicated DNA ("P-DNA") is distinctly
different from the genome as a whole and from zygotene-labeled
DNA. The striking characteristic of P-DNA is the high proportion,
70-80%, that reanneals at a C_0t value below 10^3moles.sec.liter^{-1}.
Repair-replication, and hence nicking, occurs preferentially in a
population of repeated sequences which have been estimated as
being reiterated about 2000 times per genome. The exact length of
these sequences has not been determined but they are probably of
the order of 100-170 base pairs. Their distribution appears to be
broad. If pachytene-labeled DNA is centrifuged to equilibrium in
a CsCl gradient and the gradient divided into heavy, medium, and
light fractions, the C_0t curves of each of the three fractions
are superimposable. By the criterion of buoyant density, at least,
the selectively labeled repeated sequences are more or less evenly
distributed within the genome.

That the short repeat sequences are distinctively associated
with the meiotic repair-replication process can be demonstrated by
irradiating meiocytes with UV or X-ray and analyzing the subsequently labeled DNA for its C_0t characteristics. In all cases,
radiation-induced repair-replication tracks the C_0t profile of
total DNA regardless of the stage of meiosis at which the cells
are irradiated. Radiation-induced repair-replication appears to
be randomly distributed; repair-replication that is endogenously
activated during pachytene is a special condition of meiotic
organization. The mechansim effecting the selectivity has not yet
been clarified. Changes in chromosomal proteins which would
render the pachytene sites much more susceptible to damage are
unlikely. If such were the case, irradiation at pachytene would
reinforce and not obscure the pachytene pattern. It would appear
rather that the synthesis of DNA during pachytene is programmed to
occur in specific regions of the DNA, and that such regions have
a special function in meiosis.

D.   GENERAL FEATURES OF DNA METABOLISM DURING MEIOTIC PROPHASE

On the broad issue of how the different categories of DNA
synthesis relate to one another and to the meiotic process, only
rather general comments can be made. Certain basic pieces of
information are still lacking and these are not easy to obtain in

Lilium. That Z-DNA belongs to the unique end of the C_ot profile
is certain, but whether its replication during synapsis reflects
the requirement for special DNA sites or whether unique sequences
in general have a special role in synapsis is a question that
cannot be answered without a much more detailed knowledge of the
Lilium genome. For plants in general, and for the *Liliaceae* in
particular, reiterated DNA sequences constitute an unusually large
fraction of the genome. How much of that genome is subject to the
processes of synapsis and crossing-over is undetermined. The
relationships of P-DNA and Z-DNA would be much easier to work out
in a genome that was more sparing in DNA content. It would be
difficult, nevertheless, to ignore the temporal correlations in
Lilium between the two categories of synthesis and their corresponding cytological events. The basis for inferring a functional
relationship is unequivocal.

More specific comments can be made on the narrower issue of
the distinctiveness of the two categories of DNA synthesis. It
has already been indicated that Z-DNA does not participate to any
measurable extent in pachytene repair-replication. Such behavior
raises the question of how Z-DNA escapes ligation during an
interval when all other gaps appear to be healed. That Z-DNA gaps
are repairable even during zygotene or pachytene has been demonstrated in isolated nuclei (10). If supplemented with appropriate
enzymes, such nuclei can repair all the gaps adjacent to Z-DNA.
Under *in vivo* conditions, Z-DNA gaps must be protected against
enzyme action. In nuclei treated with DNase, most of the DNA is
hydrolyzed to acid soluble components. However, no more than 50%
of Z-DNA is thus hydrolyzed during an interval when more than 90%
or more of the bulk DNA is degraded (21). The basis for this
resistance is unclear. It does not apply to other hydrolytic
enzymes. Staphylococcal nuclease digests Z-DNA in chromatin as
effectively and to the same extent as it does bulk DNA. On the
other hand, P-DNA chromatin appears to be more susceptible to the
action of this nuclease. After staphylococcal nuclease digestion,
P-DNA radioactivity is consistently found in 70-100 base pair
fragments if the digestion is performed at the pachytene stage.
By contrast, DNA labeling induced by radiation damage behaves like
total DNA when the chromatin is digested with nuclease. Thus,
superimposed on the specific DNA sequences that function in meiosis
are localized differences in chromatin organization associated with
these sequences. The extent of the differences is unknown but it
is sufficient to result in contrasting susceptibilities of the DNA
to enzyme digestion.

The complexity of meiosis should not serve to obscure the
simple pointers to DNA behavior that studies of the phenomenon
have thus far provided. A great deal more will have to be learned
about chromosome organization before incisive observations on

meiotic regulation can be made. It is nevertheless clear that
studies of meiocytes themselves provide information on chromosome
organization, the most important of which is that specific segments of DNA play a particular role in chromosome behavior during
meiosis. Whether these segments function uniquely in meiosis
remains to be determined, but it is already clear that highly
specific sets of segments rather than a randomly selected set are
active in the course of chromosome pairing and crossing-over.
DNA sequence specificities and associated differences in chromatin
characteristics reflect in a still undefined way an aspect of
intrachromosomal organization that is essential to the achievement
of recombination in large sized genomes.

REFERENCES

1. Blumenthal, A. B., H. J. Kriegstein and D. S. Hogness. The
    units of DNA replication in *Drosophila melanogaster* chromo-
    somes. Cold Spring Harbor Symp. on Quant. Biol. 38, 205, 1973.

2. Callan, H. G. DNA replication in chromosomes of eukaryotes.
    Cold Spring Harbor Symp. in Quant. Biol. 38, 195, 1973.

3. Evans, G. M. and A. J. Macefield. The effect of B chromosomes
    on homoeologous pairing in species hybrids. 1. *Lolium
    temulentum* x *L. perenne*. Chromosoma 41, 63, 1973.

4. Feldman, M. Regulation of somatic association and meiotic
    pairing in common wheat. Proc. 3rd Inter. Wheat Genetics
    Symposium, pp 31-40, 1968.

5. Ghatnekar, R., A. Lima-de-Faria, S. Rubin and K. Menander.
    Development of human male meiosis *in vitro*. Hereditas 78,
    265, 1974.

6. Holm, P. (ms in preparation; Carlsberg Laboratories,
    Copenhagen, Denmark).

7. Hotta, Y. and H. Stern. Analysis of DNA synthesis during
    meiotic prophase in *Lilium*. J. Mol. Biol. 55, 337, 1971.

8. Hotta, Y. and H. Stern. DNA scission and repair during
    pachytene in *Lilium*. Chromosoma 46, 279, 1974.

9. Hotta, Y. and H. Stern. Zygotene and pachytene-labeled
    sequences in the meiotic organization of chromosomes. In
    The Eukaryote Chromosome, edited by W. J. Peacock and
    R. D. Brock, pp 283-300, Canberra: Austr. Natl. Univ.
    Press, 1975.

10. Hotta, Y. and H. Stern. Persistent discontinuities in late replicating DNA during meiosis in *Lilium*. Chromosoma 55, 171, 1976.

11. Howell, S. H. and H. Stern. The appearance of DNA breakage and repair activities in the synchronous meiotic cycle of *Lilium*. J. Mol. Biol. 55, 357, 1971.

12. Ito, M. and Y. Hotta. Radioautography of incorporated ^{3}H-thymidine and its metabolism during meiotic prophase in microsporocytes of *Lilium*. Chromosoma 43, 391, 1973.

13. Ito, M., Y. Hotta and H. Stern. Studies of meiosis *in vitro*. II. Effect of inhibiting DNA synthesis during meiotic prophase on chromosome structure and behavior. Devel. Biol. 16, 54, 1967.

14. Jones, G. H. Correlated components of chiasma variation and the control of chiasma distribution in rye. Heredity 32, 375, 1974.

15. Oono, K. and Y. Hotta. The relation of chromosome compaction to DNA synthesis during meiosis in *Lilium*. (ms. submitted)

16. Rhoades, M. M. Studies on the cytological basis of crossing-over. In *Replication and Recombination of Genetic Material*, edited by W. J. Peacock and R. D. Brock, pp 229-241, Canberra: Australian Academy of Science, 1968.

17. Riley, R. Cytogenetics of chromosome pairing in wheat. Genetics 78, 193, 1974.

18. Smyth, D. R. and H. Stern. Repeated DNA synthesized during pachytene in *Lilium henryi*. Nature New Biol. 245, 94, 1973.

19. Stern, H. and Y. Hotta. DNA metabolism during pachytene in relation to crossing over. Genetics 78, 227, 1974.

20. Stern, H. and Y. Hotta. Biochemical controls of meiosis. Ann. Rev. Genetics 7, 37, 1974.

21. Stern, H. and Y. Hotta. Biochemistry of meiosis. Proc. Roy. Soc. (London) 1976 (in press).

STRUCTURE OF CHLOROPLAST DNA

K.K. Tewari, R. Kolodner, Nathan M. Chu,
and Robert R. Meeker

Department of Molecular Biology and Biochemistry
University of California, Irvine
Irvine, California 92664

INTRODUCTION

We had previously reported the isolation of circular chloroplast (ct) DNA molecules from pea leaves.[6] Circular pea ctDNA was found to have a molecular weight of 90×10^6 with no evidence of inter- or intramolecular heterogeneity.[1] Recently we have extensively studied the size and structure of ctDNAs from pea, bean, spinach, lettuce, corn, and oats.[2] As much as 89% of the ctDNAs from these higher plants has been obtained in circular form. The DNA preparations were also found to contain circular and catenated dimers of the circular monomer. The molecular sizes of these circular ctDNA molecules relative to internal standards has been found to range from 85×10^6 to 97×10^6. The molecular size of these ctDNAs was also determined by renaturation kinetics and found to range from 82×10^6 to 93×10^6. The excellent agreement between the molecular weights of these circular DNA molecules obtained by electron microscopy and the molecular weight of the unique sequences of these ctDNAs determined by renaturation kinetics suggests that the sequence of a circular ctDNA molecule represents the entire information content of the ctDNA. The molecular weight of pea Form I ctDNA monomer has been determined by equilibrium sedimentation and found to be 89.1×10^6, which is close to the values determined by electron microscopy and renaturation kinetics. The sedimentation properties of the native and denatured forms of the open and closed circular ctDNAs from pea, lettuce, and spinach have also been studied by analytical ultracentrifugation.[3]

We have further analyzed the genetic content of the pea ctDNA and the gross arrangement of its base sequences by carrying out de-

naturation mapping studies.[4] The denaturation mapping technique is
able to locate AT rich regions consistently in individual molecules
by mounting the molecules for electron microscopy under levels of
partial denaturation. Using this technique, pea ctDNA molecules
have been found to be homogeneous, i.e. molecules of different base
sequences do not exist in the pea ctDNA. Furthermore, it has been
possible to locate areas of base sequence heterogeneity by this
method which could not be detected by thermal denaturation studies
or by renaturation kinetics experiments. No evidence of repeating
sequences in the pea ctDNA has been found by the denaturation mapping experiments. We have also investigated the structure of pea
ctDNA circular dimers by constructing denaturation maps of such
dimers and comparing them to the denaturation map of circular monomers. The data have shown that theses circular dimers are formed
by "head to tail" fusion of two circular monomers.

   We have recently investigated the effect of alkali and a mixture of RNases A and T1 on the structure of the closed circular
ctDNAs from pea, lettuce, and spinach plants.[5] The results have
shown that these DNAs contain covalently linked ribonucleotides.
The kinetics of the alkali nicking of these ctDNAs has shown that
each ctDNA contains only one population of molecules. By comparing
the rates of alkali nicking of these ctDNAs with the rate of alkali
hydrolysis of RNA under identical conditions, we have estimated the
number of ribonucleotides that are present in each of these ctDNAs.
The pea and spinach ctDNAs contain a maximum of 18 + 2 ribonucleotides, while lettuce ctDNA contains a maximum of 12 + 2 ribonucleotides. We have further analyzed the structure of pea ctDNA by
electron microscopic methods. These studies with alkali-fragmented
DNA and reannealed alkali-fragmented DNA have also shown that pea
ctDNA contains 19 alkali-labile sites at specific locations. A map
containing the relative location of each site has been constructed.
These results suggest that individual ribonucleotides are located
at the alkali-labile sites in the pea ctDNA.

   ctDNA has been shown to contain genes for the 70S ribosomal
RNA of chloroplasts.[6] Detailed hybridization studies using pea
ctDNA and ct-rRNA isolated from 70S ribosomes from pea leaves have
shown that there are two ct-rRNA genes in the ctDNA.[7] We have
studied the ct-rRNA genes in the ctDNAs from bean, lettuce, spinach,
corn, and oats. The experiments have been carried out by hybridizing the ctDNAs from the above plants with ^{32}P-labeled ct-rRNA
from pea chloroplasts. The amount of hybridization in each case
accounted for approximately two ct-rRNA gene equivalents in those
DNAs. Competition hybridization and thermal stability experiments
involving homologous and heterologous ct-rRNA-DNA combinations have
shown that within the limits of this technique the base sequences
of these genes are very similar in all the higher plants studied.[8]
These results are strikingly different from those reported in other

systems. For example, the number of cytoplasmic rRNA genes in the
nuclear DNA (nDNA) varies considerably among the different angiosperm species. Simialrly, distantly related bacterial species have
been shown to contain rRNA genes which show only partial sequence
homology.

Pea ctDNA has also been found to hybridize with purified tRNAs
of chloroplast. The level of hybridization obtained suggests that
there are 30-40 tRNA genes in ctDNA. The fraction of DNA hybridizing with tRNAs bands at a lighter density in CsCl density gradients
compared to the fraction of DNA hybridizing with ribosomal RNA
(rRNA). The ct-tRNAs have been found to have no complementary base
sequences with tRNAs from bacteria, yeast, and animals. Also, cytoplasmic tRNAs from pea has not been found to compete with ct-rRNAs.
We have aminoacylated tRNAs with individual amino acids using aminoacyl tRNA synthetases from chloroplasts. The isolated aminoacyl
tRNA has been hybridized with ctDNA. The data have shown the presence of leucine, arginine, and phenylalanine tRNA genes in ctDNA.
Similarly, other aminoacyl tRNAs are being studied.

The saturation hybridizations with the in vivo labeled RNAs
from chloroplast have shown almost complete transcription of ctDNA
in vivo. The preliminary results show that mRNA transcripts of
ctDNA occur with and without poly A at their 3' end.

RESULTS AND DISCUSSION

Denaturation and Renaturation Properties of ctDNA

The purity of ctDNAs from different plants was routinely analyzed by their Cot½ values and equilibrium cesium chloride (CsCl)
density gradient centrifugation after denaturation and renaturation
in an analytical ultracentrifuge. Using T4 DNA as a standard, the
ctDNAs were observed to have distinctly different T_m values. The
most striking difference was between pea ctDNA, which had a T_m that
was $1^{O}C$ below the T_m of T4 DNA, and corn ctDNA whose T_m was $1.9^{O}C$
above the T_m of T4 DNA. These differences (see Table I) in the T_m
values of the ctDNAs were independent of the rate of temperature
rise that was used in these experiments; rates ranging from $1^{O}C$
per min to $1^{O}C$ per 10 min were used.

In order to study the possible existence of heterogeneity in
the melting patterns of the ctDNAs, their melting profiles were
differentiated. The higher plant ctDNAs melted more broadly than
T4 DNA. The pea, lettuce, and spinach ctDNAs melted as a single
component showing only one inflection point. The corn ctDNA melted
considerably more broadly than the rest of the ctDNAs which might

suggest the presence of some heterogeneity. This possible heterogeneity is not as obvious as the results obtained with the ctDNAs
from Euglena, Chlamydomonas. and Chlorella which melt heterogene-
ously and show several components.

The renaturation kinetics of ctDNAs was studied using frag-
ments of 1.5×10^5 and 14×10^5 daltons at different salt concentrations.
The renaturation reaction of the ctDNAs showed second order kine-
tics regardless of the fragment size or salt concentrations. The
linearity of the plot and the good extrapolation to 1.0 indicated
the absence of any rapidly renaturing components within the limits
of detection in the spectrophotometer.

The molecular weight of the unique sequence of the ctDNAs from
pea, lettuce, spinach, beans, corn, and oats were calculated using
the ratio of the Cot½ values of the ctDNAs to T4 DNA or pea ctDNA.
The average values that were determined for each ctDNA under the
different salt conditions and with different fragment sizes are
shown in Table I. The molecular weights of ctDNAs were found to
range from 82×10^6 to 93×10^6. The data suggest that corn and oat
ctDNA are smaller than pea, spinach, bean, and lettuce ctDNAs.

The Structure of Circular ctDNA

When the chloroplast lysate was centrifuged in a CsCl-ethidium
bromide (EtBr) gradient, the ctDNA was recovered in two fluorescent
bands. The ctDNA from up to six such gradients was pooled and recentrifuged. As much as 30% of the total ctDNA could be recovered
in the lower band of the CsCl-EtBr density gradients. In a large
number of experiments with oats, corn, peas, lettuce, spinach, and
beans, an average of 20% of the total ctDNA was recovered in the
lower band, 70% of the ctDNA was recovered in the upper band, and
10% of the ctDNA was found in the area between the two bands.
When the lower band ctDNA was examined in the electron microscope,
more than 90% of the DNA molecules were present as supertwisted
molecules (Fig. 1b,c). Most of the remaining molecules were relaxed circles (Fig. 1a). The ctDNA molecules that were found in
the region of the gradient between the upper and lower bands mostly
consisted of supertwisted circular DNA molecules. In an average
preparation, 60% of the upper band ctDNA consisted of relaxed circular molecules while the rest of the DNA molecules were linear.
No linear molecules that were longer than the circular molecules
were observed. In general, 70% of the total ctDNA from any of the
plants examined was recovered as supertwisted or relaxed circular
DNA molecules. After nicking the supertwisted ctDNA molecules,
there were three species of circular DNA molecules present; single
length circular molecules, double length circular molecules, and
catenated dimers which appeared to consist of two topologically

interlocked single length circular molecules. The frequency of
these species on the lower band ctDNA from pea, spinach, lettuce,
corn, and oats ranged from 3 to 4%.

The Size of the Circular ctDNA

The closed circular ctDNA molecules were nicked with γ-rays
and their size was determined by measuring their lengths relative
to the length of ØX RFII DNA, which was used as an internal standard in all experiments. The distribution of the lenghts of the
circular pea, lettuce, spinach, corn, and oat ctDNA molecules (in
ØX units) is presented in Fig. 2. The results showed that each
ctDNA contained both single length and double length circular
molecules. Circular molecules of other sizes were not observed.
The sizes of the single length circular ctDNA molecules in ØX units
are presented in Table I. The data show that there are significant
differences between the sizes of all of the ctDNAs that were ex-
amined. The molecular size of ctDNAs ranged from 85×10^6 to 97×10^6.

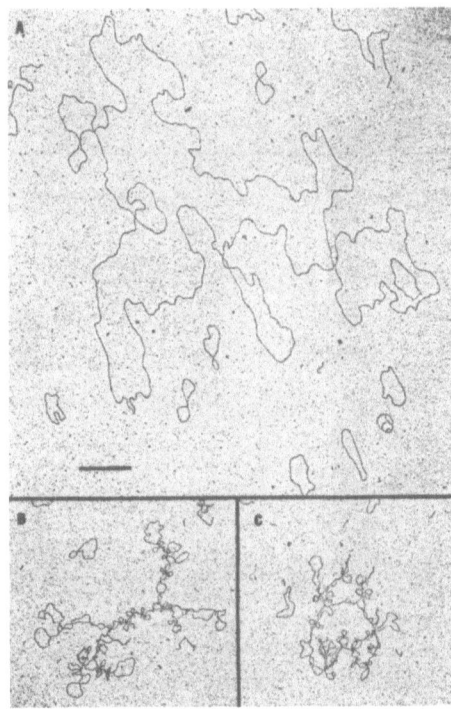

Fig. 1. Super-twisted and relaxed circular ctDNA molecules. (C and
B) Super-twisted circular oat ctDNA molecules; (A) relaxed circular
lettuce ctDNA molecule. The small circular DNA molecules are
ØX RF II DNA. The bar indicates 1 μm.

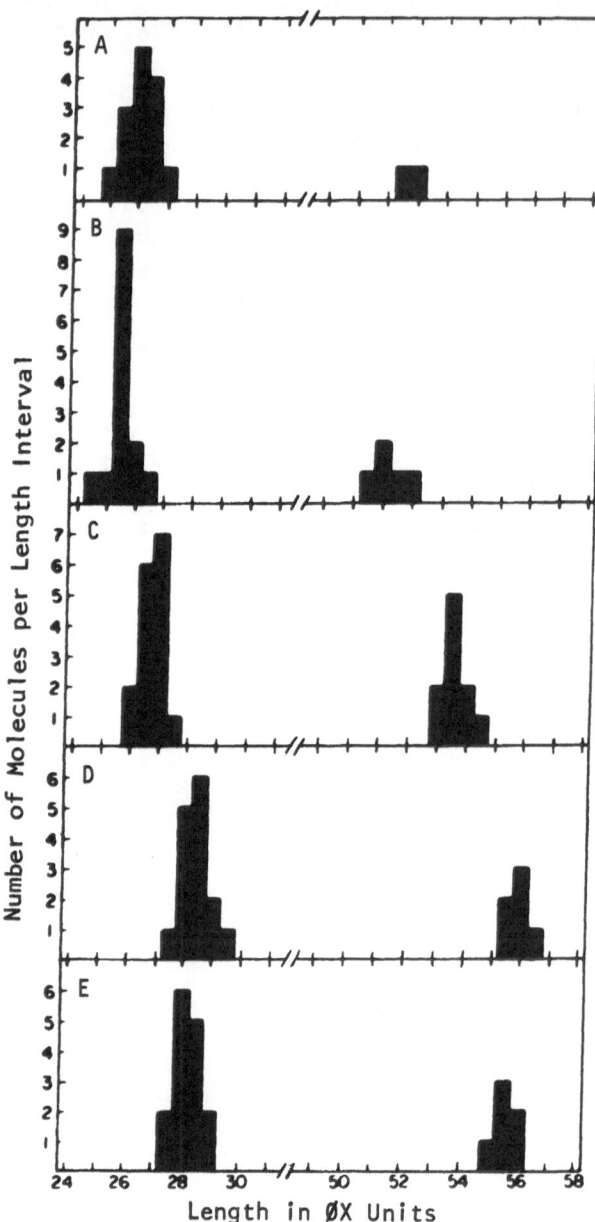

Fig. 2. Length distribution of the circular ctDNA molecules.
(A) Oat ctDNA; (B) corn ctDNA; (C) pea ctDNA; (D) lettuce ctDNA;
(E) spinach ctDNA. The lengths are expressed in units of ØX length.
The large classes of circular DNA molecules are two times the size
of the small class of circular DNA molecules.

TABLE I

The Molecular Size of ctDNAs

DNA	Molecular Size (M)				
	$\emptyset$X units	M x 10^{-6}a	M x 10^{-6}b	M x 10^{-6}c	M x 10^{-6}d
Pea	26.7+0.1	88.4+0.3	89.1	86.9+2.1	89.1
Lettuce	29.0+0.1	95.8+0.3	96.7+0.5	90.9+2.0	93.1+2.0
Spinach	28.5+0.2	94.3+0.6	95.1+1.0	87.9+2.7	90.1+2.7
Corn	25.6+0.2	84.7+0.6	85.4+1.2	83.5+2.8	85.6+2.9
Oats	25.9+0.1	85.7+0.3	86.4+0.6	81.7+3.0	83.7+3.0

a. Calculated using an average M for $\emptyset$X RF monomer DNA of 3.31 x 10^6.
b. Calculated assuming an M of 89.1 x 10^6 for pea ctDNA.
c. Calculated from renaturation rates. The molecular weight of T4 DNA has been taken to be 106 x 10^6.
d. Calculated from renaturation rates. The molecular weight of pea ctDNA has been taken to be 89.1 x 10^6.

TABLE II

The Values of $S^o_{20,w,Na}$+ of ctDNA

Configuration	Pea	Spinach	Lettuce
	S	S	S
Form I	89.1	97.1	94.9
(neutral)	S.D.+0.4	S.D.+0.6	S.D.+o.3
Form II	58.3	60.6	60.8
(neutral)	S.D.+0.3	S.D.+0.4	S.D.+0.1
Form II	55.57	57.7	57.8
(alkaline)	S.D.+0.08	S.D.+0.5	S.D.+0.5
Form IV	240	264	264.1
(alkaline)	S.D.+0.7	S.D.+3.0	S.D.+1.5
Form IV	146.8	---	170
(neutral)	S.D.+1.1	---	S.D.+10.0

Sedimentation Properties of ctDNA

The pea Form I ctDNA obtained from the lower band of a CsCl-
EtBr density gradient, was sedimented in neutral 3M CsCl. Only one
component sedimenting at about 86S was observed. When this DNA was
irradiated with 1,000 rads of γrays, 50% of the DNA was converted
to a slower moving form sedimenting at about 55S. Higher doses of
γrays converted more of the faster moving Form I DNA to the slower
moving nicked circular Form II DNA, but no other forms were produced. When the pea Form I ctDNA was sedimented on 3M CsCl containing 0.2M NaOH, all of the DNA was present as the fast moving
Form IV, sedimenting at 240S. The sedimentation values of Form I,
II, and IV from pea, spinach, and lettuce are given in Table II.
The molecular weight of pea ctDNA by equilibrium sedimentation has
been found to be 89.1×10^6. This is in excellent agreement with the
value for pea ctDNA of 88.4×10^6 determined by electron microscopy
using ΦX RF II DNA as internal standard. We have derived an em-
pirical relationship between S and molecular weight for open circular DNA molecules.[3] Using this relationship and the value of
$S^0_{20,w}$ for open circular lettuce and spinach ctDNAs, their molecular
weights have been calculated to be 98.2×10^6 and 97.2×10^6 respectively. These values are in good agreement with 96.7×10^6 for lettuce ctDNA and 95.1×10^6 for spinach ctDNA which were calculated by
electron microscopic measurements.

Denaturation Maps of the Circular Monomer

The circular ctDNA molecules were spread onto a hypophase of
45% formamide.[4] At this concentration, 70% of all of the circular
molecules were partially denatured, containing from one to seven
denatured regions. The average amount of the denaturation was 2.5%.

All of the interpretable circular molecules containing two or
more denatured regions were photographed, measured as described,
and were arbitrarily linearized. These linearized molecules were
arranged in a map starting with the most denatured molecules first,
so that the greatest number of denatured regions matched between
the molecules. After the molecules were matched by this criterion,
they were further matched with respect to the length of each individual denatured region. This denaturation map is given in Fig. 3.
In order to assess the accuracy of this method, denaturation maps
were also produced from the same molecules using the least denatured
molecules first, and arranging them as described above. A denaturation map was also constructed by picking the molecules randomly.
All of the denaturation maps that were produced were identical.
This map contained six major denatured regions, two minor denatured regions, and two places where only one molecule was denatured. A region corresponding to 34% of the pea ctDNA molecule

contained no denatured regions. The circular pea ctDNA molecules were spread onto a hypophase of 47% formamide. All of the circular molecules were partially denatured, and contained from 11 to 33 denatured regions. The average amount of denaturation was 22%. A typical molecule from this spreading is presented in Fig. 4. A denaturation map was constructed from these molecules by the methods described above. Because of a large number of denatured regions in these molecules, the map was constructed by using molecules showing intermediate levels of denaturation first. A map was also produced by randomly picking the molecules. All of the maps were identical. This map contained 31 distinct denatured regions as well as six distinct regions showing no denaturation. It was possible to locate all of the denatured regions produced at the 2 5% level of denaturation.

The circular pea ctDNA was spread onto a hypophase of 51% formamide. All of the circular molecules were partially denatured and contained from 18 to 43 denatured regions. The average amount of denaturation was 44%. A denaturation map was constructed using these molecules as described above. There were only two distinct undenatured regions which matched two undenatured regions in the map at 22% denaturation. The five denatured regions between these two undenatured regions were also present in the denaturation map at 22% denaturation.

Denaturation Maps of the Circular Dimers

In our preparations, 3% of the pea ctDNA molecules were present as circular dimers. Our success in constructing denaturation maps of the circular monomers has enabled us to investigate the integration of the monomer units into the circular dimer. In a head to tail circular dimer the two monomer units will be in tandem repeat. Therefore, in a partially denatured circular dimer, every denatured region will have a corresponding denatured region one monomer length away. In a head to head circular dimer the two monomer units will be integrated in reverse repeat. Therefore, in a partially denatured circular dimer, it will be possible to locate two points of symmetry on the molecule. The correspnding denatured regions will then be equidistant on either side of these points. The structure of the circular dimer was examined by constructing individual denaturation maps of 10 circular dimers from three spreadings. The maximum number of matchings for each of the two types of integration was determined as described. There were more matched denatured regions when the data were analyzed by head to tail test than head to head test. Thus, the circular dimers are integrated in a head to tail fashion. This conclusion has been confirmed by constructing a denaturation map of the circular dimer.

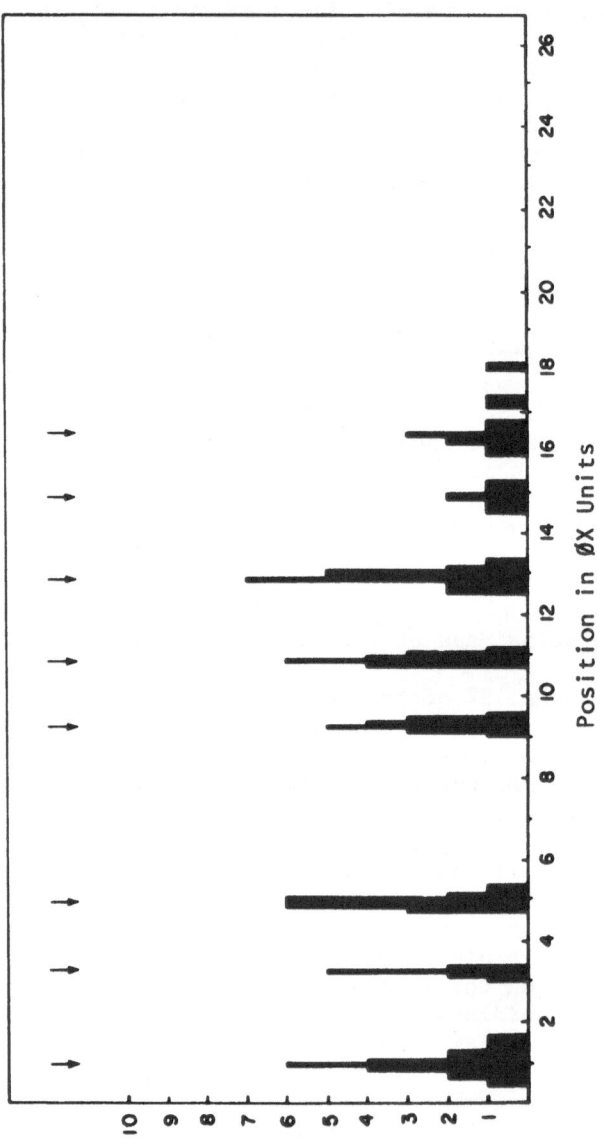

Fig. 3. Denaturation map of the pea ctDNA. Average amount of denaturation was 2.5%. Map represents circular DNA molecules, and has been linearized for display purposes. ➔ designates denatured regions found in more than one molecule. The six denatured regions on the left side of the figure have been referred to as "major" denatured regions in the text.

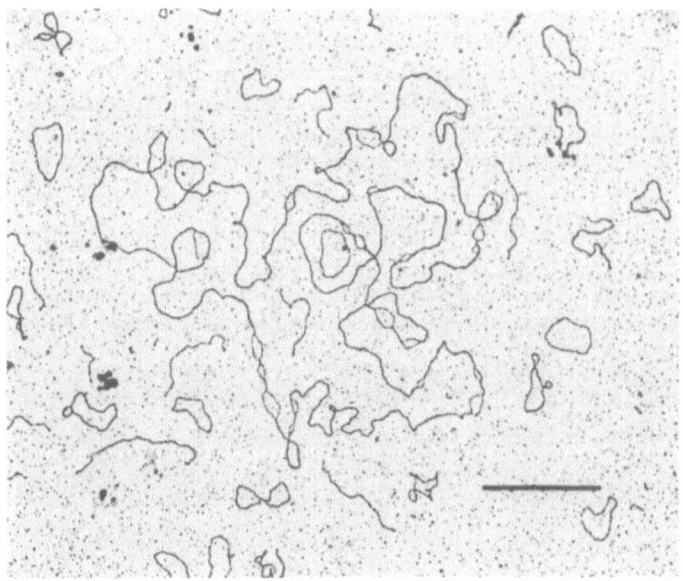

Fig. 4. Partially denatured pea ctDNA molecule showing 22 denatured regions. Hypophase contained 47% formamide in 1 x TE.
Spreading solution contained 77% formamide in 10 x TE. Small circular molecules are SS and DS ∅X DNA. Bar indicates 1 μm.

Alkali Lability of ctDNA

        Covalently closed circular pea ctDNA was centrifuged through
3M CsCl/0.2M NaOH/0.01M EDTA, and the centrifugation cell was
scanned at 6-min intervals. The DNA zone that was present in each
scan is given in Fig. 5. The observed decrease in the size zones
could not be accounted for by the effect of radial dilution. By
plotting the log of the per cent of the DNA remaining against time
for spinach ctDNA, the half-life of spinach Form I ctDNA was 10 + 1
min. Similar alkaline sedimentation experiments were performed
with both ∅X 174 RFI and G4 RFI monomers and dimers. There was no
detectable loss of either of these DNAs during alkaline sedimentation. Similarly, there was no loss of DNA when closed circular
ctDNA was centrifuged through neutral sedimentation solvent. Thus,
the disappearance of closed circular spinach ctDNA from the Form IV
zone during sedimentation through the alkaline sedimentation solvent is due to a specific effect of alkali on the DNA. The loss
of spinach ctDNA from the Form IV peak is due to alkali-induced
single strand breaks which convert the rapidly sedimenting Form IV
configuration (264S) to the more slowly sedimenting denatured single
stranded form of spinach ctDNA (57.7S). The closed circular ctDNAs

from pea and lettuce plants were also analyzed by the above methods.
Under identical conditions, pea Form IV ctDNA disappeared with a
half-life of 10 + 1 min, while lettuce Form IV ctDNA disappeared
with a half-life of 15 + 1 min. The kinetics of degradation of
pea, lettuce, and spinach Form I ctDNAs was first order and it has
been possible to observe the degradation of each ctDNA for a minimum of four half-lives and, in all cases, the kinetic data were
first order, showing only one component.

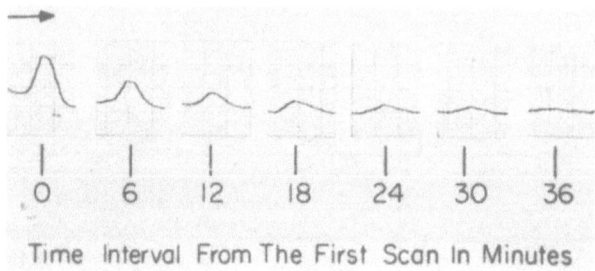

Time Interval From The First Scan In Minutes

Fig. 5. Alkali breakdown of pea ctDNA. Pea Form I ctDNA was centrifuged in the alkaline sedimentation solvent at 16,000 rpm. The
Form IV peak from scans taken at 6-min intervals during this experiment is shown to illustrate the loss of DNA from the Form IV
zone. No other zones were present during this experiment, or in
other experiments of this type. The zero time peak presented here
was scanned at 4 min after layering, and the 36-min peak represents
4% of the applied DNA. The experiment presented here represents a
period of 4.2 half-lives. Sedimentation is from left to right.

Ribonuclease Lability of ctDNA

The alkali sensitivity of ctDNA could be due to the presence
of covalently inserted ribonucleotides. This was tested by incubating ctDNA with RNases A and T1 under conditions where these enzymes will digest RNA in an RNA-DNA duplex as well as double stranded
RNA.[4] The pea Form I ctDNA was incubated with a mixture of RNase
A and RNase T1 for increasing lengths of time. The covalently
closed circular pea ctDNA (89.1S) was successively converted to the
open circular form (58.3S). A 3-hour incubation with RNase A and
T1 quantitatively converted pea, lettuce, and spinach Form I ctDNA
to the open circular form, but did not produce any unit length
linear (50S) or smaller molecules. G4 RFI monomers and dimers were
also incubated with RNase A and T1 under identical conditions.
These molecules were not nicked in these experiments.

Number of Ribonucleotides in ctDNA

To determine the number of ribonucleotides present in the ctDNA, we have compared the rate of alkaline hydrolysis of ctDNA to the rate of alkaline hydrolysis of E. coli [^{32}P] RNA under identical conditions. The kinetic data of hydrolysis of the [^{32}P] RNA were first order, and the RNA had a half-life of 180 min. This half-life represents the rate of breakage of a single RNA-RNA phosphodiester bond. The rate of nicking pea and spinach ctDNA was 18 times as fast as the rate of breaking a single RNA-RNA phosphodiester bond. Therefore, pea and spinach Form I ctDNA nick at rates that would be expected if they each contained a maximum of 18 ± 2 ribonucleotides. Similarly, the lettuce Form I ctDNA was nicked 12 times as fast as an RNA phosphodiester bond, which corresponds to the presence of 12 ± 2 ribonucleotides in the lettuce ctDNA. The ctDNAs we have examined are nicked at rates that are 100 to 150 times faster than the rates of nicking of viral and E. coli DNAs, which do not contain covalently inserted ribonucleotides.

Structure of Nicked ctDNA

The ctDNAs could contain the ribonucleotides located at one or several sites in either one or both strands of the DNA. To investigate this, nicked ctDNA was studied by sedimentation analysis. Each of the three Form I ctDNAs was incubated for 3 hours in 0.2M NaOH at 20°C, and was then sedimented in alkaline 3M CsCl. In each case, 20 to 40% of the DNA sedimented at the position of intact single strands, while the rest of the DNA sedimented more slowly as a broad zone. This indicated that the alkaline labile sites could be located at multiple positions in both strands. When the Form I ctDNAs were incubated for 3 hours with RNases A and T1, a similar sedimentation pattern resulted. Using centrifugation techniques, it would require large quantities of ctDNA to determine if RNase treatment of alkaline hydrolysis of Form I ctDNA produces specific size classes of fragments. Therefore, this problem was studied by electron microscopy. Pea Form I ctDNA was incubated with 0.2M NaOH at 20° for 16 hours (96 half-lives) and was mounted for electron microscopy by the formamide technique to visualize single stranded DNA.⁵ The length distribution of the fragment sizes produced by this treatment is presented in Fig. 6. A large number of fragment size classes were observed. The largest fragment was 12.4 ∅X units long. The length of intact pea ctDNA is 26.7 ∅X units. These results indicated that the alkali labile sites in pea ctDNA were located in both strands of the DNA, since we did not observe unit length single stranded circular molecules.

Map of the Nicks in Pea ctDNA

The previous experiments had indicated that pea ctDNA might
contain covalently inserted ribonucleotides located at specific
sites. In order to investigate this further, the pea Form I ctDNA
was incubated in 0.2M NaOH for 3 hours (18 half-lives), which will
nick more than 99% of the molecules, but will not digest all of the
alkali labile sites. These fragments were then partially reannealed
to produce molecules that generally had single stranded tails and
internal duplex sections. This proceedure will generate a number of
overlapping molecules from which a map can be constructed containing
the positions of the nicks relative to each other. This map should
be circular with a monomer repeat length of 26.7 ⌀X units (the length
of pea ctDNA), if the alkali labile sites are at specific sites.

A map of the relative positions of the alkali labile sites in
pea ctDNA is presented in Fig. 7. Nineteen single strand breaks
were located at distinct sites. All of the molecules were consistent with this map. It should be noted that at Sites 6 and 16, two
single strand breaks were mapped on oppostie strands at the same
position. When two nicks mapping on opposite strands at the same
site were found on a hybrid molecule, one end of the molecule
appeared to be fully duplex. The repeat length of the map was 26.7
⌀X units.

Strand Location of the Alkali labile Sites of Pea ctDNA

If the map presented in Fig. 7 accurately represents the locations of the alkaline labile sites, it should be possible to make
unambiguous strand assignments for each of the nicks. The two nicks
that define a single strand tail of a reannealed molecule are located
on opposite strands. Using this criteria, it was possible to make a
list of the nicks that were located on opposite strands from each
other. The nicks at positions 1,5,8,11,12,14, and 15 (Fig. 7, Strand
A) were located on one strand, while the nicks at positions 2,3,4,7,
9,10,13, and 17 (Fig. 7, Strand B) were located on the other strand.
This method located nicks on both strands at positions 6 and 16,
which agrees with the previous finding that there was one nick located on each strand at these two positions.

The significance of the individually inserted ribonucleotides
in Form I ctDNA is not understood at this time.. The ribonucleotides
could arise from a nonstringent DNA polymerase, but in that case
they would not be located at specific positions in the ctDNA molecule. It is possible that the ribonucleotides are remnants of RNA
primers which are now generally believed to initiate DNA replication. In pea ctDNA, replication is initiated at two sites located
on opposite strands of the DNA molecule. Replication is bidirec-
tional and at least 50% of the ctDNA is synthesized bidirectionally.

If the ribonucleotides resulted from incomplete excision of the primers that are used for the initiation of pea ctDNA replication, we should expect to observe only two alkali labile sites. If they were remnants of primers for "Okazaki fragments," we would expect to observe on the order of 40 to 50 alkali labile sites. In addition, they would be located more uniformly than the alkali labile sites we have observed. It should be pointed out that in all systems in which RNA primers for DNA replication have been studied, the primer is completely excised in the mature DNA except under abnormal conditions. We consider it possible that these ribonucleotides have some function in ctDNA. These sites could be involved in transcription, recombination, or some other process that requires specific recognition sites. It will require further experimentation to test these possibilities.

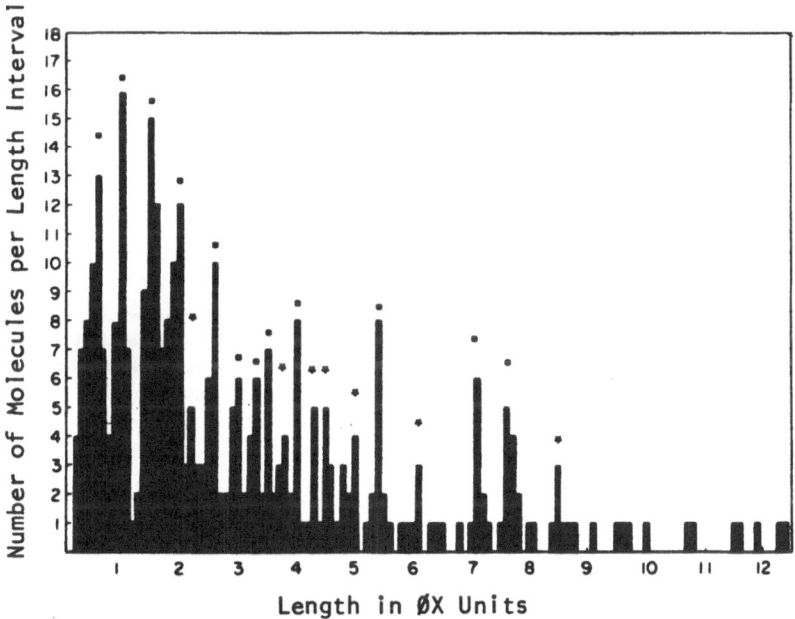

Fig. 6. Length distribution of fragments produced by the alkali hydrolysis of pea Form I ctDNA. Pea Form I ctDNA (10 µg/ml) was incubated at 20° in 0.2M NaOH 0.04M EDTA for 16 hours (96 half-lives) and was neutralized with 1.8M Tris-HCl 0.2M Tris. This DNA was spread with single stranded ØX 174 DNA by the formamide technique; the spreading solution contained 50% formamide, and the hypophase contained 20% formamide. Fields were selected and photographed randomly, and all of the molecules on a negative were measured. ■, size classes that match the fragment lengths predicted by the map that is presented in Fig. 7. *, size classes that match fragment lengths resulting from incomplete digestion of pea Form I ctDNA, as predicted by the map presented in Fig. 7.

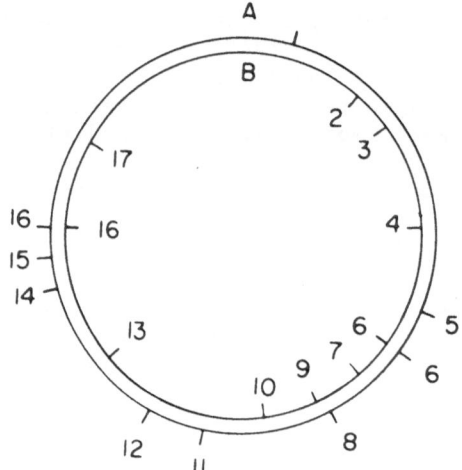

Fig. 7. Map of the positions of the alkali labile sites in pea ctDNA. Strand assignments for each site are those described in the text.

Inverted Repeat Sequences in ctDNA

The ctDNA from spinach, lettuce, and corn has been found to contain one long stretch of inverted repeat sequences. The repeat length of this sequence corresponds to a molecular size of 15×10^6. The size of the repeat length is almost the same in all three ctDNAs. Denaturation mapping has shown that repeat sequences occur in the same position in all the circular molecules of ctDNA. No inverted repeating sequences have been found to be present in pea ctDNA.

TABLE III

ctDNA source	Molecular size $\times 10^{-6}$		
	Single strand region	Double strand region	SS region
Lettuce	12.0	15	53.5
Spinach	11.4	15	52.8
Corn	11	15	43.7

The supercircled ctDNAs were nicked by γrays to obtain about 50% molecules as open circles. 1 μl of 0.2M NaOH–20 M EDTA was

added to 0.05 µg of ctDNA in 9 µls. After 2 min at room temperature, 2 µl of 1.8M Tris HCl-2M Tris base was added and the DNA was
mounted for electron microscopy in 50% formamide.[4]

Information Content in ctDNA

Ribosomal RNA genes in ctDNAs. The ct-rRNA from pea leaves
was hybridized with 1.0 µg of ctDNA from spinach, lettuce, bean, and
corn. The saturation curves for such experiments are presented in
Fig. 8. It is seen from the figure that the level of hybridization
increases with increasing RNA to DNA ratio (R/D) and reaches a maximum when this ratio is 5.0. It may be noted that most of the
hybridization has taken place when R/D is only 2.0. These experiments indicate that all the ct-rRNA sites on the ctDNA are saturated
and the ct-rRNA preparations are not heterogeneous. In a control
experiment ct-rRNA from pea has been hybridized with pea ctDNA.
The rate of hybridization of this homologous system follows closely
that of the heterologous system involving spinach ctDNA. Hybridization reaches maximum of 4.22% when R/D is 5.0. Under similar
conditions ctDNAs from bean and lettuce hybridize to 4.14% and
4.46% respectively. Experiments with ctDNA from corn leaves, a
monocotyledonous plant, mimic the saturation curve obtained with
pea ctDNA, with a maximum hybridization of 3.96%. Thus, the data
show that spinach, lettuce, bean, and corn ctDNAs hybridize between
3.96% and 4.46% with ^{32}P-labeled pea ct-rRNA. This range of hybridization with the different ctDNAs is of the same order as that obtained in hybridization with pea ctDNA. The molecular weights of
the ctDNAs from pea, spinach, bean, corn, and oats have been found
to be 89×10^6, 97×10^6, 92×10^6, 85×10^6, and 86×10^6, respectively.
Thus, the hybridization obtained with pea ct-rRNA indicates that the
amount of single stranded ctDNA complementary to ct-rRNA ranges
from 3.5 to 4.1×10^6 daltons. This level of complementarity shows
that there are two gene equivalents of ct-rRNA in the ctDNAs
of higher plants, assuming a molecular weight of 1.8×10^6 for total
ct-rRNA.

Specificity of ct-rRNA·DNA Hybridizations

In the above experiments, the number of ct-rRNA gene equivalents in the ctDNAs of higher plants was studied using ^{32}P-labeled
pea ct-rRNA. In order to find out the specificity of such heterologous hybridizations, we have carried out competition experiments
and thermal stability analysis of the DNA·rRNA hybrids. In the
competition hybridization experiments of Fig. 9, it is clearly seen
that heterologous rRNA is equally effective in competition compared
to homologous rRNA. These competition experiments were confirmed
using homologous and heterologous 23S and 16S rRNA. The competition experiments demonstrate that base sequences of ct-rRNA from

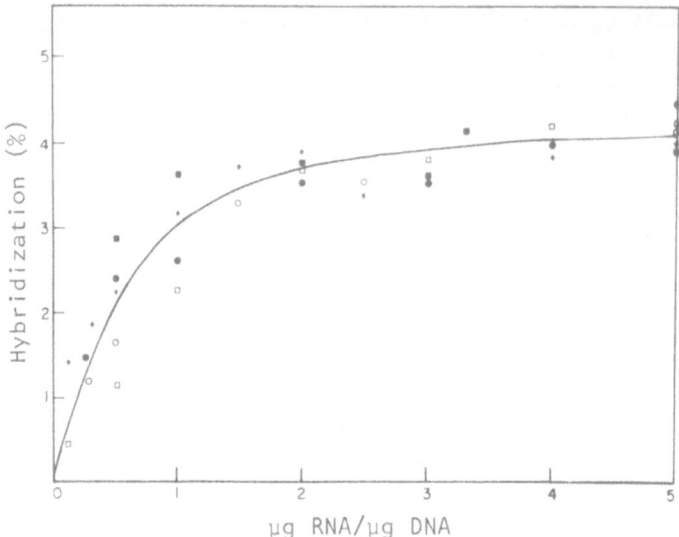

Fig. 8. Hybridization of ^{32}P-labeled pea ctDNAs with different plants.

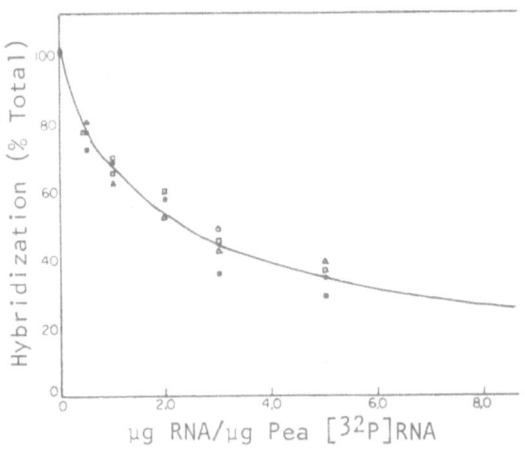

Fig. 9. Competition hybridization experiments. ctDNAs from different plants were hybridized with ^{32}P-labeled pea ct-rRNA in the presence of increasing amounts of unlabeled ct-rRNA from different plants.

the various plants are similar to each other. These conclusions
were born out by the thermal denaturation studies on the heterologous DNA-rRNA hybrids. The observations on the conservation of
two ct-rRNA genes in the plants which have diverged through millions
of years and have been frequently cultivated through all civiliza-
tions is unique in molecular evolution.

tRNAs Genes in ctDNA

tRNAs from pea chloroplasts were purified and labeled by in
vitro iodination. Hybridization of ^{125}I-labeled tRNA with ctDNA
showed that $1.23 \pm 0.35\%$ of the ctDNA was complementary to tRNAs (Fig.
10). This level of hybridization will account for 30-40 tRNAs genes
in ctDNA. The sequence specificity of ctDNA-tRNA hybrid has been confirmed by competition hybridization and thermal stability studies on
DNA-tRNA hybrids. Competition experiments have shown that: 1) unlabeled ct-tRNA competes with labeled tRNA at the theoretically expected level, 2) pea cytoplasmic tRNAs do not compete significantly
with ^{125}I-ct-tRNA, and 3) the bacterial (E. coli), yeast and calf
liver tRNAs do not compete with ^{125}I-ct-tRNA (Fig. 11). These competition experiments clearly indicate that the base sequences of
ct-tRNA genes are unique. Further analysis of the ctDNA by fractionation in a CsCl density gradient and hybridization of the various fractions with tRNAs and rRNA has shown that the rRNA and tRNAs
genes are not closely localized in the DNA molecule (Fig. 12). What
is the nature of the 30-40 tRNAs genes in ctDNA? Are they composed
of repeating units of one or two individual aminoacyl tRNAs or are
there copies of tRNA for each of the 20 different amino acids? In
order to answer this question we have aminoacylated ct-tRNAs with
^{3}H-labeled leucine, phenylalanine or arginine. The individually
aminoacylated tRNA was hybridized with ctDNA. A saturation curve
with leucyl tRNA is shown in Fig. 13. All of the three aminoacyl
tRNAs were found to hybridize with ctDNA. Our preliminary work
along these lines suggests that probably all or nearly all the 20
individual amino acids have tRNAs coded for in the ctDNA.

Complete Transcription of ctDNA

It is apparent from the above data that only about 5×10^6 daltons of ctDNA is utilized in the coding of rRNA and tRNAs. The remaining 45×10^6 daltons of ctDNA is available for the formation of
mRNA transcripts. Are all of the base sequences of ctDNA transcribed
in chloroplasts? The previous experiments have shown that about 20%
of the ctDNA was complementary to the RNA synthesized in vitro by
the isolated tobacco chloroplasts.[10] These data indicated that the
ctDNA is probably expressed in messenger type RNA. We have now
carried out saturation hybridization experiments with the in vivo
labeled total pea chloroplast RNA and pea ctDNA. It has been possi-

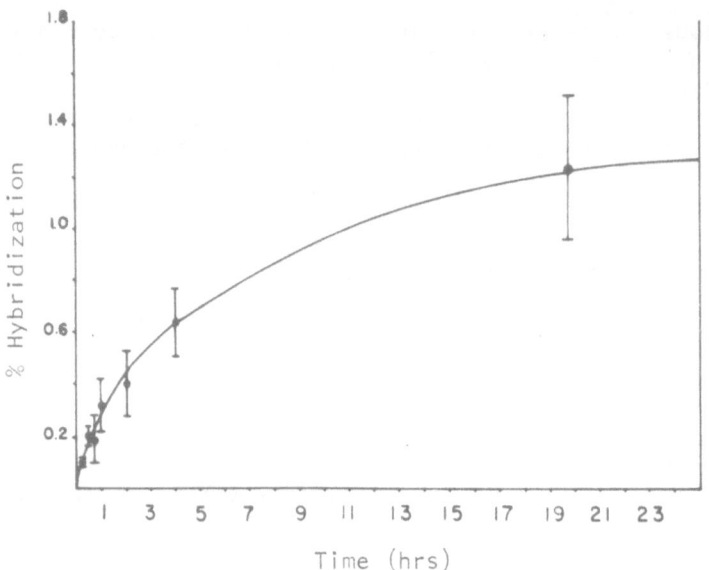

Fig. 10. Hybridization of ctDNA with in vitro ^{125}I labeled tRNA.
1 µg of ct DNA and 2 µg of labeled tRNAs were used in the experiment.

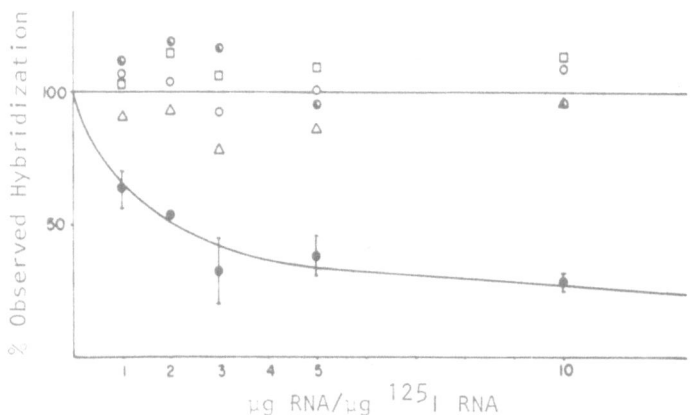

Fig. 11. Competition hybridization experiments. Pea ctDNA was hybridized with 2 µg of ^{125}I-ct-tRNA in the presence of increasing
amounts of pea ct-tRNAs (●—●), pea cytoplasmic tRNAs (○—○), E. coli
tRNAs (◐—◐), yeast tRNAs (△—△), and calf liver tRNAs (□—□).

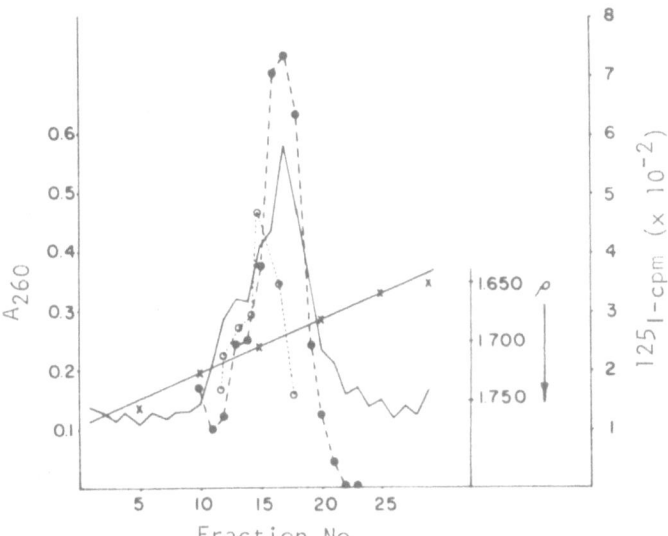

Fig. 12. Hybridization of different buoyant density fractions of ctDNA with rRNA and tRNAs. ctDNA was sheared to 2.8x10^6 daltons, centrifuged to equilibrium in a CsCl density gradient, and the different fractions were hybridized to rRNA and tRNAs. (——) ctDNA A$_{260}$, (●—●) ^{125}I-ct-tRNAs-ctDNA hybrid counts, (O—O) ^{125}I-ct-rRNA-ctDNA hybrid counts, and (—✕—) density.

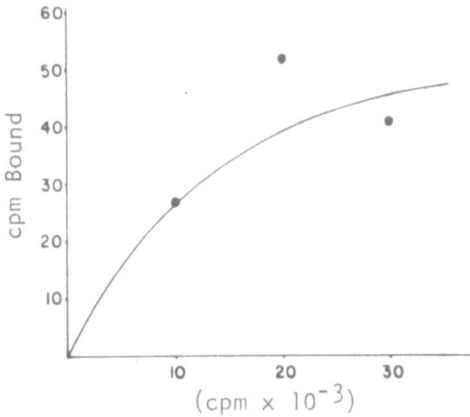

Fig. 13. Saturation hybridization of ctDNA with ^{3}H leu-ct-tRNA. 5 µg of ctDNA was hybridized with 10,000, 20,000, and 30,000 cpm/ml of ^{3}H leu-ct-tRNA.

ble to saturate about 40-50% of the base sequences of ctDNA at R/D
ratios of 200 to 300. These data show that RNAs equivalent to one
complete strand of ctDNA are present in chloroplasts. The in vivo
labeled RNA was fractionated on a poly U sepharose column or on poly
U cellulose filters. About 5% of the total RNA was found to bind
with the poly U in the presence of 25% formamide-0.7M NaCl-50mM Tris-
10mM EDTA, pH 7.5. The bound RNA (presumably containing poly A)
was eluted from the column by passing 90% formamide-10mM EDTA-10mM
KPO_4, pH 7.5. Non poly A and poly A containing RNAs were found to
hybridize with 30-35% and 10-20% of the ctDNA, respectively. These
preliminary experiments indicate the presence of both poly A and
non poly A containing RNA transcripts of ctDNA in chloroplasts.
Current efforts are directed towards further characterization of
these mRNA transcripts.

The experiments are in progress to construct a physical map of
ctDNA by using restriction enqonucleases. It will be then possible
to localize ribonucleotides,[11] rRNA genes, tRNAs genes, messenger RNA
and initiation and termination sites of DNA replication in the physical map of ctDNA.[12] This approach will, hopefully, help us in understanding the replication and differentiation of chloroplasts.

REFERENCES

1. Kolodner, R. and Tewari, K. K. (1972) J. Biol. Chem., 247, 6355.
2. Kolodner, R. and Tewari, K. K. (1975) Biochim. Biophys. Acta,
    402, 372.
3. Kolodner, R., Tewari, K. K. and Warner, R. C. (1976) Biochim.
    Biophys. Acta (in press).
4. Kolodner, R. and Tewari, K. K. (1975) J. Biol. Chem. 250, 4888.
5. Kolodner, R. and Tewari, K. K. (1975) J. Biol. Chem. 250, 7020.
6. Tewari, K. K. and Wildman, S. G. (1968) Proc. Nat. Acad. Sci.,
    USA.
7. Thomas, J. R. and Tewari, K. K. (1974) Biochim. Biophys. Acta
    361, 73.
8. Thomas, J. R. and Tewari, K. K. (1974) Proc. Nat. Acad. Sci.
    71, 3147.
9 Meeker, R., Thomas, J. R., and Tewari, K. K. (1976) Plant Physiol.
    in press.
10 Tewari, K. K. (1971) Ann. Rev. Plant Physiol. 22, 141.
11 Kolodner, R., Warner, R. C., and Tewari, K. K. (1975) J. Biol.
    Chem. 250, 7020.
12. Kolodner, R. and Tewari, K. K. (1975) Nature 256, 708.

PURIFICATION AND CHARACTERIZATION OF SOYBEAN
DNA-DEPENDENT RNA POLYMERASES AND THE MODULATION
OF THEIR ACTIVITIES DURING DEVELOPMENT

Tom J. Guilfoyle and Joe L. Key

Departments of Botany and Biochemistry
University of Georgia
Athens, Georgia 30602

INTRODUCTION

Following the report of Roeder and Rutter, (1, 2) describing
multiple forms of RNA polymerase in animal cells, studies on the
structure and function of the eukaryotic RNA polymerases have
progressed rapidly. Three major classes of nuclear RNA polymerase have been described: I or A, II or B, and III or C (1-5).
The nomenclature used throughout this paper is that adopted from
Roeder and Rutter (1) which is derived from the elution pattern
of the RNA polymerases from DEAE Sephadex. Each class has dis-
tinguishing characteristics (i.e., sensitivity to the fungal toxin,
α-amanitin, elution patterns from cation and anion exchange resins,
salt optima, and template preferences) as well as distinct subunit
structures (5-9). In addition, each of the three classes can be
subdivided into 2 or 3 species which are resolved by gel electrophoresis under both denaturing and nondenaturing conditions (7-9).

RNA polymerase I is localized within the nucleolus (1, 10),
transcribes the genes coding for 18s and 28s rRNA (11), and is refractory to α-amanitin (3, 10, 12). In contrast to animal RNA polymerase I enzymes, yeast RNA polymerase I is 50% inhibited at approximately 600 μg/ml α-amanitin (13). RNA polymerase II is localized
within the nucleoplasm (1, 10), synthesizes HnRNA and presumably
mRNA (14), and is 50% inhibited at approximately 0.03 μg/ml α-
amanitin in animals (15-17). Yeast RNA polymerase II is about 100-
fold less sensitive to the fungal toxin (13, 18). RNA polymerase
III is localized in the nucleoplasm and possibly the cytoplasm (9,
19, 20), transcribes the genes coding for tRNA and 5s RNA (21, 22)
and is 50% inhibited at approximately 20 μg/ml α-amanitin in

vertebrates (16, 23, 24). Again the yeast enzyme III is about 100-
fold more resistant to α-amanitin (13). The class III enzymes have
the unique property of co-eluting with RNA polymerase I at low salt
on DEAE cellulose (24, 25) while eluting at high salt on DEAE Sephadex (9, 24, 25).

Higher plants have been shown to possess multiple forms of
RNA polymerase activity (26-30), but the literature dealing with
these enzymes is not extensive and in some instances is rather
confusing (31, 32). RNA polymerases I and II have been reported
from several higher plants (26-30, 33), and these enzymes appear
to have properties similar to other eukaryotic RNA polymerases.
RNA polymerase II enzymes from several higher plant tissues have
been purified to high specific activity (34-36), but definitive
subunit structures are not available. The detection of RNA polymerase III in higher plants has been reported in only a couple of
instances (37,38). Failure to detect the class III enzyme probably
relates to its low level in most plant tissues and the purification
and assay procedures employed (9, 16, 23, 28).

As an extension of our studies on RNA metabolism associated
with normal and auxin-induced growth in the soybean (39, 40, 41),
we have initiated studies to analyze the mechanisms operative in
the regulation of RNA synthesis in this system (42, 43). The work
to date has focused primarily on the purification and characterization of the soybean RNA polymerases (44, 45). This approach reflects our conviction that the regulation of these enzymes will
not be understood without a thorough knowledge of the enzymes per
se. Further it seems highly unlikely that any advances in the
understanding of the regulation of the RNA polymerases will be
gleaned from in vitro studies using crude enzyme preparations, undefined templates, and heterogeneous preparations of "factors". In
this paper, we summarize the purification, characterization, and
subunit structure analyses of soybean RNA polymerases I, IIa and
IIb. In addition, the activities of RNA polymerase I and II have
been monitored during normal and auxin-induced growth transitions
in the soybean hypocotyl using both isolated nuclei and enzymes
which have been solubilized from these nuclei and subsequently
fractionated on DEAE cellulose.

RESULTS

1. Properties of the Soybean RNA Polymerases

Early in this work, we demonstrated that soybean chromatin
which was prepared by standard methods at pH 8 (46) contained a high
level of RNA polymerase activity which was refractory to α-amanitin,

but only a very low level of α-amanitin-sensitive activity was expressed in these chromatin preparations (46, 47). A somewhat higher level of α-amanitin-sensitive RNA polymerase activity was associated with chromatin which was isolated at pH 6 (46). Most of the α-amanitin-sensitive activity was recovered from the post-chromatin supernatant while no α-amanitin-insensitive activity was detected in this fraction. The RNA polymerase activities associated with the chromatin and post-chromatin supernatant were solubilized and fractionated on DEAE cellulose (46). The RNA polymerase activity solubilized from pH 8 chromatin fractionated as a single peak (Fig. 1A), eluting at 0.12 M ammonium sulfate, and was refractory to α-amanitin. The RNA polymerase activity recovered from the supernatant eluted as a single peak at 0.22 M ammonium sulfate (Fig. 1B) and was completely inhibited by low levels of α-amanitin. These data suggested that a class I RNA polymerase activity was associated with the chromatin while most of the class II RNA polymerase activity was free as a "soluble" enzyme or was released from nuclei during cell breakage and chromatin preparation.

In an attempt to localize both types of RNA polymerase activity which would presumably be associated with nuclei in vivo, methods were adapted for the purification of soybean nuclei (48, 49). Since salt concentration and α-amanitin have been shown to affect the expression of RNA polymerases I and II in animal nuclei (50-52), the RNA polymerase activity of soybean nuclei was assayed over a range of ammonium sulfate concentrations in the presence and absence of 4 μg/ml α-amanitin (Fig. 2B). The α-amanitin-insensitive activity was optimal at relatively low salt concentrations (50-100 mM ammonium sulfate) while the α-amanitin-sensitive activity was optimal at much higher salt concentrations (200-300 mM ammonium sulfate). The salt optima of RNA polymerases I and II expressed in nuclei are a property of the chromatin which they transcribe rather than a property of the enzymes themselves (17; See section 3c). When nucleoli were isolated (53) and assayed for RNA polymerase activity, only α-amanitin-insensitive activity was detected, and this activity was optimal at 50-100 mM ammonium sulfate (Fig. 2A).

The RNA polymerase activities were solubilized from purified nuclei and nucleoli by stirring in a 0.5 M ammonium sulfate buffer according to the methods of Guilfoyle et. al., (42). The solubilized proteins were fractionated on DEAE cellulose columns using a linear 0.05 to 0.5 M ammonium sulfate gradient. Each fraction was assayed for RNA polymerase activity in the presence and absence of 4 μg/ml α-amanitin. The DEAE cellulose profile from nuclei (Fig. 3B) showed two peaks of RNA polymerase activity eluting at 0.12 and 0.22 M ammonium sulfate. Peak I activity was refractory to α-amanitin while peak II activity was completely inhibited by low levels of the toxin. The relative levels of α-amanitin-insensitive and -sensitive

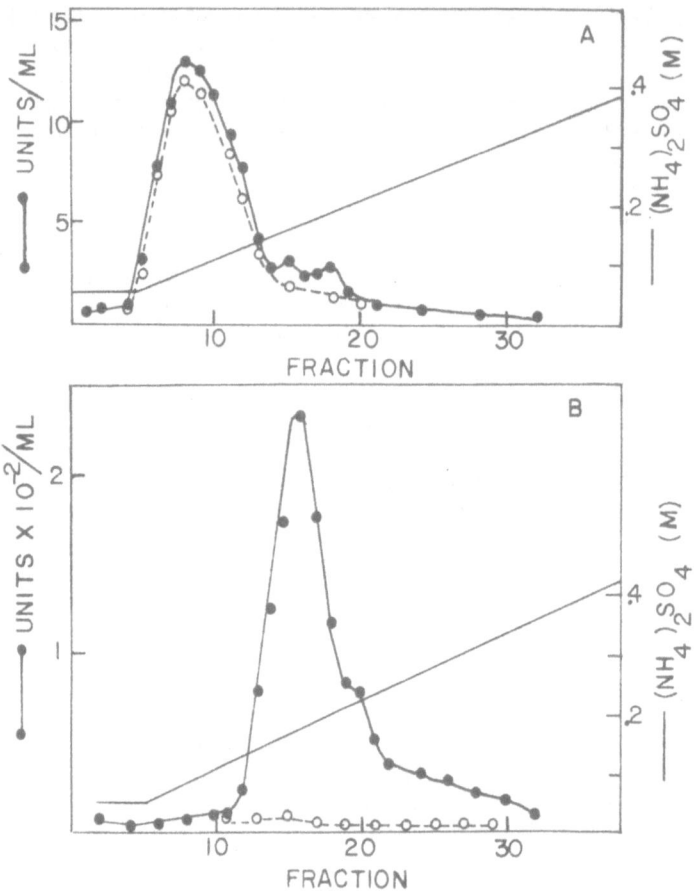

Fig.1. DEAE cellulose profiles of RNA polymerase activity (A) solubilized from chromatin prepared at pH 8.0 and (B) recovered from
soybean post-chromatin supernatant. Procedures for solubilization,
DEAE cellulose chromatography, and assay conditions have been described elsewhere (46). (●———●), minus α-amanitin; O– – –O) plus
4 µg/ml α-amanitin.

activities observed in Fig. 3B (peaks I and II, respectively) were
similar to the relative activities expressed in isolated nuclei as
assayed in Fig. 2B. The RNA polymerase activity solubilized from
nucleoli fractionated as a single peak on DEAE cellulose (Fig. 3A)
corresponding to the α-amanitin-insensitive activity observed with
isolated nucleoli (Fig. 2A). The DEAE cellulose peak of the solubilized nucleolar enzyme elutes at 0.12 M ammonium sulfate and is
refractory to α-amanitin.

As an additional approach in establishing the identity of the

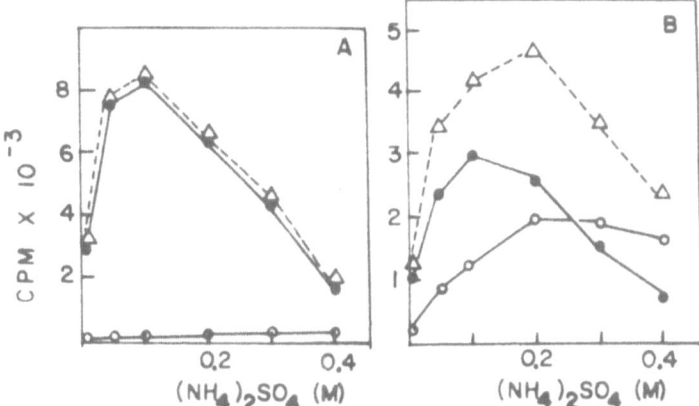

Fig.2. Incorporation of (^{3}H)-UMP into RNA by isolated (A) nucleoli and (B) nuclei at various ammonium sulfate concentrations in the presence and absence of 4 μg/ml α-amanitin. Assay conditions have been described (48). (△----△), total activity; (●———●), α-amanitin-insensitive activity; (○———○), α-amanitin-sensitive-activity.

two major RNA polymerase activities associated with soybean nuclei, the proportion of the rRNA in the product of nuclei-directed RNA synthesis was determined under conditions which maximally accentuate differences in the expression of the two RNA polymerase activities (i.e., 50 mM ammonium sulfate which favors the expression of RNA polymerase I and 300 mM ammonium sulfate which favors the expression of RNA polymerase II; see Fig. 2B). The RNA products from these assays were subjected to competition RNA-DNA hybridization analysis (54) using the methods of Reeder and Roeder(11). Data from these experiments are summarized in Table 1. When nuclei-directed RNA synthesis occurred under conditions which are optimal for the α-amanitin-insensitive activity, (i.e., 50 mM ammonium sulfate) about 35% of the RNA product was rRNA in the absence of the toxin while 60-70% of the RNA product was rRNA when the reaction mixture included α-amanitin. When the reaction was conducted under conditions optimal for the expression of the α-amanitin-sensitive activity (i.e., 300 mM ammonium sulfate) only 5-10% of the product was rRNA. Thus, the α-amanitin-insensitive enzyme, which chromatographs as RNA polymerase I on DEAE cellulose and which is localized primarily within the nucleolus, synthesizes as its major product rRNA in isolated nuclei. Since these competition hybridization experiments measure only mature 18s and 25s rRNA species, the actual level of rDNA transcription by the α-amanitin-insensitive enzyme under optimal conditions is greater than the 60-70% detected by our methods (54).

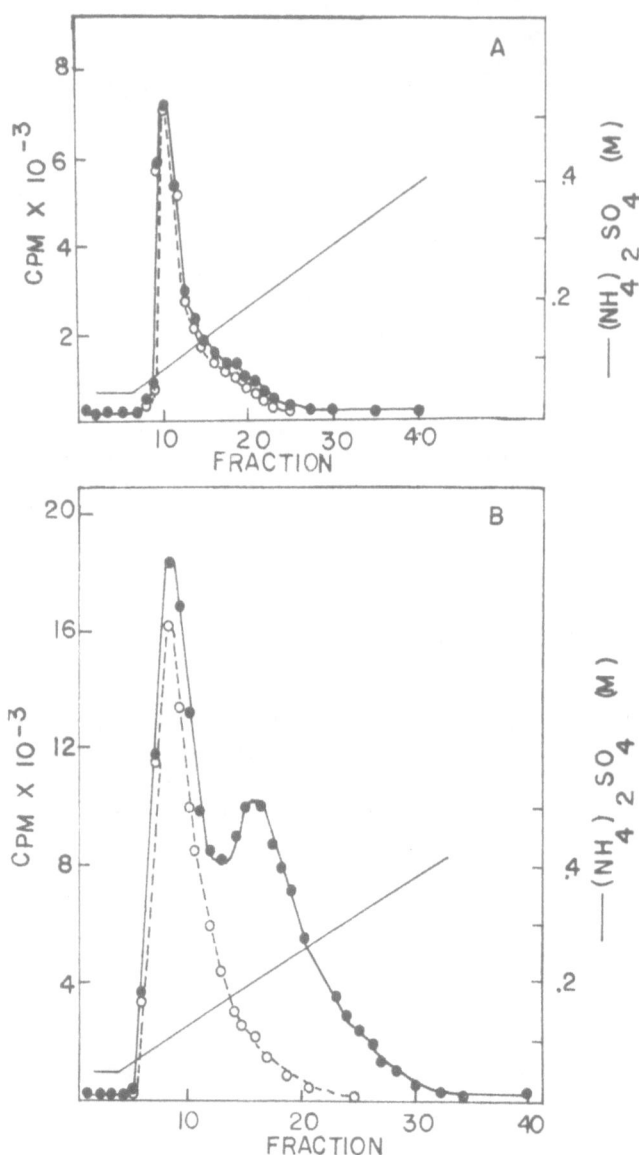

<u>Fig.3.</u> DEAE cellulose profiles of RNA polymerase activity solu-
bilized from isolated (A) nucleoli and (B) nuclei. Procedures
for solubilization, DEAE cellulose chromatography, and assay
conditions have been described (48). (●━━━●), minus α-amanitin;
(○----○), plus 4 μg/ml α-amanitin.

Table 1. Ribosomal RNA as a Major Transcription Product
of the α-Amanitin-Insensitive RNA Polymerase in Nuclei

Conditions*	Experiment	% rRNA
50 mM $(NH_4)_2SO_4$	1	34
	2	36
50 mM $(NH_4)_2SO_4$+α-amanitin	1	71
	2	58
300 mM $(NH_4)_2SO_4$	1	8
	2	5

* Nuclei were incubated as described (54) under conditions which
maximally accentuate differences in the α-amanitin-sensitive
and -insensitive RNA polymerase activities. The (^{3}H) RNA product was subjected to competition RNA-DNA hybridization analyses
using the methods of Reeder and Roeder (11).

2. Purification and Characterization of Soybean RNA Polymerases

   a. RNA polymerase I. Studies of plant RNA polymerase I enzymes have been limited because of the reported instability of the
enzymes following solubilization and because of low enzyme activity
(26-30, 36). In the present study pH 8 chromatin from auxin-treated
soybean hypocotyl was used as the source of RNA polymerase I since
the level of activity is up to 20-fold higher than in control tissue.
Stabilization of the enzyme required the adoption of procedures which
avoid the use of Mg^{++} and low salt conditions which favor aggregation
and precipitation of the solubilized enzyme (44). In addition, high
concentrations of enzyme protein were maintained throughout the purification procedures.

   Chromatin was isolated from hypocotyl tissue and solubilized
in a TGDEP buffer (50 mM Tris-HCl, 30% glycerol, 10 mM dithiothreitol, 0.1 mM EDTA, and 0.5 mM phenylmethylsulfonylfluoride) containing
0.5 M ammonium sulfate (44). The fraction of chromatin which was
not solubilized by this procedure was pelleted by ultracentrifugation. Dilution to low ionic strength prior to centrifugation as
described for the solubilization for some animal enzymes (1, 2) results in precipitation of large and variable amounts of soybean RNA
polymerase I.

A summary of the purification procedure is given in Fig. 4A.
The solubilized chromatin proteins were first fractionated by Agarose A-1.5m gel filtration at high ionic strength (0.5 M ammonium
sulfate). Failure to develop the column under high ionic strength
conditions results in little or no purification because of the ag-
gregative properties of the enzyme preparation (44). Under high
salt conditions, this step yields a 10-fold purification and
greatly reduces the tendency of the enzyme to aggregate at low salt.
DEAE cellulose, CM-Sephadex, and phosophocellulose chromatography
steps were achieved in a short time span in order to assure the
maintenance of high levels of RNA polymerase I activity. Utilizing
linear gradients from 0.05 to 0.5 M ammonium sulfate, the RNA polymerase I activity eluted as a single peak from DEAE cellulose at
0.12 M ammonium sulfate and from CM-Sephadex at 0.1 M ammonium sulfate. In order to have a concentrated fraction of RNA polymerase I
protein for sucrose gradient centrifugation, the enzyme was step
eluted from phosphocellulose with 0.3 M ammonium sulfate following
a 0.15 M ammonium sulfate wash of the column. The inclusion of
0.3 M ammonium sulfate in the sucrose gradients prevented aggregation of the enzyme and resulted in high recovery (80-90%) of enzyme activity after 60 hours of centrifugation. Purified RNA polymerase I was dialyzed against TDEP buffer containing 50% glycerol
and 0.3 M ammonium sulfate and stored at -80 C without loss of activity for at least 6 months.

About a 300-fold purification was achieved from the chromatin and about a 20,000-fold purification based on total tissue
protein. The purified enzyme had a specific activity of 200-300
nmoles UMP incorporated/mg protein/ 30 min at 28 C using denatured
DNA as template. The specific activity was about 10-fold higher
using poly (dC) as template. The specific activity using denatured
DNA is comparable to that of other purified RNA polymerase I enzymes (7, 55-58).

The sucrose gradient purified enzyme was subjected to SDS
gel electrophoresis, resulting in the gel pattern shown in Fig. 5.
Evidence to date suggests that seven of the protein bands (labeled
a-g) are components of RNA polymerase I. The molecular weights
of the putative subunits are: a - 183,000, b - 136,000, c - 50,000,
d - 46,000, e - 40,000, f - 33,000, and g - 28,000. While a and b
are present in stoichiometric amounts, the smaller components vary
from a molar ratio of about 0.5 to 2 relative to a and b. This lack
of stoichiometry of the soybean RNA polymerase I putative subunits
is not unique to this system, and possible reasons for this lack
of stoichiometry have been discussed (44, 58). The putative subunit structure of soybean RNA polymerase I is strikingly similar to
a highly purified yeast RNA polymerase I (provided by Drs. G. Hager
and W. Rutter) (Fig. 6). A purity for the soybean enzyme of about
90% is indicated by the SDS gel profiles and the high specific activity of the enzyme. It may be that some of the "contaminating"

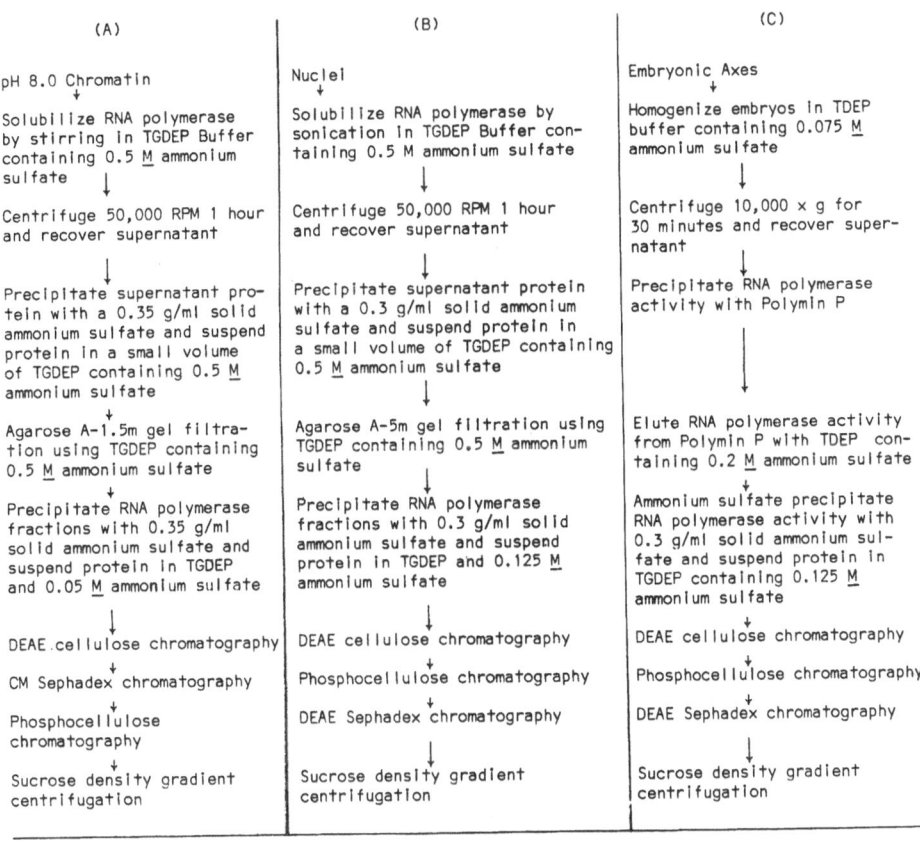

(A)

pH 8.0 Chromatin
↓
Solubilize RNA polymerase by stirring in TGDEP Buffer containing 0.5 M ammonium sulfate
↓
Centrifuge 50,000 RPM 1 hour and recover supernatant
↓
Precipitate supernatant protein with a 0.35 g/ml solid ammonium sulfate and suspend protein in a small volume of TGDEP containing 0.5 M ammonium sulfate
↓
Agarose A-1.5m gel filtration using TGDEP containing 0.5 M ammonium sulfate
↓
Precipitate RNA polymerase fractions with 0.35 g/ml solid ammonium sulfate and suspend protein in TGDEP and 0.05 M ammonium sulfate
↓
DEAE cellulose chromatography
↓
CM Sephadex chromatography
↓
Phosphocellulose chromatography
↓
Sucrose density gradient centrifugation

(B)

Nuclei
↓
Solubilize RNA polymerase by sonication in TGDEP Buffer containing 0.5 M ammonium sulfate
↓
Centrifuge 50,000 RPM 1 hour and recover supernatant
↓
Precipitate supernatant protein with a 0.3 g/ml solid ammonium sulfate and suspend protein in a small volume of TGDEP containing 0.5 M ammonium sulfate
↓
Agarose A-5m gel filtration using TGDEP containing 0.5 M ammonium sulfate
↓
Precipitate RNA polymerase fractions with 0.3 g/ml solid ammonium sulfate and suspend protein in TGDEP and 0.125 M ammonium sulfate
↓
DEAE cellulose chromatography
↓
Phosphocellulose chromatography
↓
DEAE Sephadex chromatography
↓
Sucrose density gradient centrifugation

(C)

Embryonic Axes
↓
Homogenize embryos in TDEP buffer containing 0.075 M ammonium sulfate
↓
Centrifuge 10,000 × g for 30 minutes and recover supernatant
↓
Precipitate RNA polymerase activity with Polymin P
↓
Elute RNA polymerase activity from Polymin P with TDEP containing 0.2 M ammonium sulfate
↓
Ammonium sulfate precipitate RNA polymerase activity with 0.3 g/ml solid ammonium sulfate and suspend protein in TGDEP containing 0.125 M ammonium sulfate
↓
DEAE cellulose chromatography
↓
Phosphocellulose chromatography
↓
DEAE Sephadex chromatography
↓
Sucrose density gradient centrifugation

T = 50 mM Tris (pH 8.0); G = 30% glycerol; D = 1 mM dithiothreitol; E = 0.1 mM EDTA; P = 0.5 mM PMSF

Fig.4. Procedures for the purification of soybean RNA polymerases (A) I, (B) IIb, and (C) IIa.

bands are small amounts of type III RNA polymerase subunits since DEAE cellulose rather than DEAE Sephadex was used in the purification (24, 25). This possibility is suggested by the presence of two closely migrating bands between a and b (Fig. 6) which are similar in molecular weight to the published molecular weights of the two large subunits of RNA polymerase III in animals and yeast (6, 9, 59).

b. <u>RNA polymerase II</u>. RNA polymerase II from soybean has been purified to homogeneity by two different methods which yield two distinct forms of the enzyme, IIa and IIb (45, 60). The first method

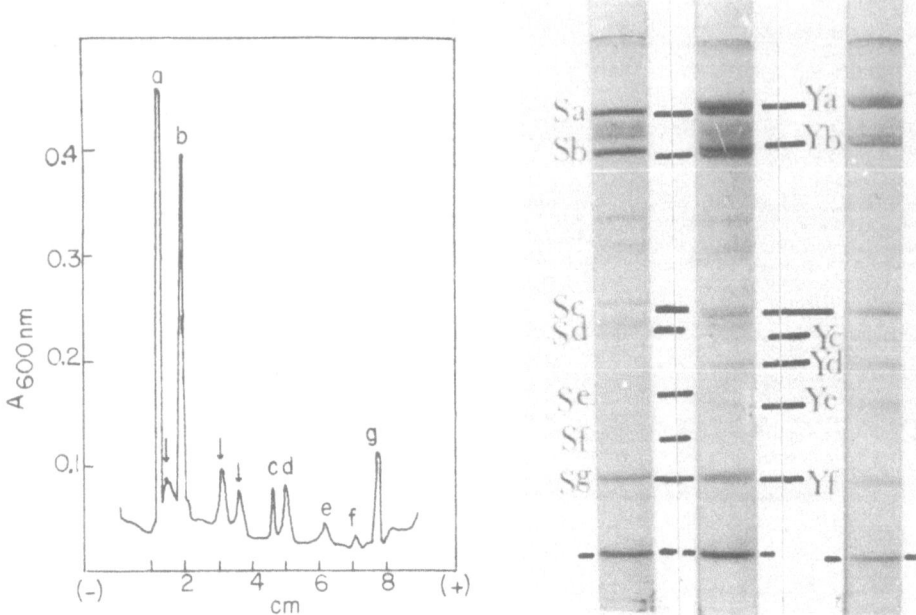

Fig.5.(left) SDS polyacrylamide gel scan of purified soybean RNA polymerase I. Soybean RNA polymerase I was purified through the sucrose density gradient step and samples were denatured, electrophoresed, and stained with Coomassie blue according to the methods of Schwartz and Roeder (7). SDS gels were 8.75% acrylamide, and polypeptides below the molecular weight of 25,000 daltons were not resolved on these gels. The gels were scanned at 600 nm with a Gilford Spectrophotometer equipped with a linear transport attachment. Putative subunits are labeled a to g and arrows indicate possible contaminating polypeptides.

Fig.6.(right) SDS polyacrylamide gels of soybean and yeast RNA polymerase I enzymes. Yeast RNA polymerase I was a generous gift of G. Hager and W. J. Rutter. Conditions for electrophoresis are described in Fig. 5. Soybean RNA polymerase I putative subunits are labeled Sa to Sg; yeast RNA polymerase I putative subunits are labeled Ya to Yf. The center gel is the co-electrophoreses of soybean and yeast RNA polymerase I enzymes. The dashed lines at the bottom of the gels indicate the position of the bromophenol blue tracking dye.

is similar to the procedure for the purification of soybean RNA
polymerase I, but isolated nuclei are used as the source of the
enzyme instead of pH 8 chromatin. A summary of this purification
procedure is given in Fig. 4B. RNA polymerases were solubilized
from nuclei by sonication in a TGDEP buffer containing 0.5 M ammonium sulfate and the insoluble protein was pelleted by centrifugation. The solubilized proteins were chromatographed on Agarose
A-5m at high ionic strength (0.5 M ammonium sulfate) resulting in
an approximately 10-fold purification of RNA polymerases I and II.
The RNA polymerase peak was then ammonium sulfate precipitated,
pelleted by centrifugation, and suspended in a TGDEP buffer at a
concentration of 0.125 M ammonium sulfate. At this salt concentration RNA polymerase I passes through DEAE cellulose while RNA
polymerase II is bound. RNA polymerase II was step eluted from
the DEAE cellulose column at 0.25 M ammonium sulfate. The enzyme
was then further purified by phosphocellulose chromatography. Using
a linear gradient from 0.05 to 0.5 M ammonium sulfate, RNA polymerase II activity elutes at approximately 0.135 M ammonium sulfate.
The enzyme was then loaded onto a small DEAE Sephadex column and
step eluted at 0.3 M ammonium sulfate. The concentrated enzyme
(1-2 mg/ml) was layered directly onto a 5-20% sucrose gradient containing TGDEP buffer and 0.3 M ammonium sulfate and centrifuged at
50,000 rpm in an SW 50.1 rotor for 36 hours. The sucrose purified
enzyme was dialyzed against TEDP buffer containing 50% glycerol and
stored at -80° C for six months without loss of activity. This enzyme is referred to as RNA polymerase IIb.

The second method of purification utilizes soybean embryonic
axes as the source of the enzyme. The soybean embryos are a very
rich source of RNA polymerase II; several mg of enzyme can be
purified from a kg of embryos. The soybean embryos were isolated
from dry seed by a modification of the method of Marcus et. al.
(61) for the preparation of wheat germ. The embryos were homogenized in a TEDP buffer containing 0.075 M ammonium sulfate, and
the RNA polymerase II was purified by a modification of the method
described by Jendrisak and Burgess (62). Fig. 4C outlines the purification procedure which is similar to the method outlined in Fig.
4B except that a Polymin P step is substituted for the Agarose gel
filtration step. Following sucrose gradient centrifugation, the enzyme was dialyzed and stored as described for RNA polymerase IIb.
The enzyme prepared by this method is referred to as RNA polymerase
IIa.

Utilizing these procedures both RNA polymerase IIa and IIb
have been purified to specific activities of about 200 nmoles UMP
incorporated/mg protein/30 min at 28 C using denatured DNA as template. These specific activities are similar to the specific activities reported for homogeneous RNA polymerase II from other
eukaryotes (8, 63, 64). Both soybean RNA polymerase IIa and IIb

migrate as single bands on 5% acrylamide gels electrophoresed under nondenaturing conditions. If enzymes IIa and IIb are electrophoresed together, they migrate as two distinct bands. Subsequent re-electrophoresis of the "native" bands under denaturing conditions and analysis of the subunits reveals that RNA polymerase IIb migrates faster than IIa on the "native" gels (Fig. 7). SDS slab gel electrophoresis of RNA polymerases IIa and IIb using an 8.75 to 15% acrylamide gradient is shown in Fig. 7 A and B. These SDS gels indicate that RNA polymerase IIa differs from IIb only with respect to its largest subunit (Fig. 7 A-E). Gel scans of RNA polymerases IIa and IIb are shown in Fig. 8. The putative subunits for soybean RNA polymerase IIa are a- 200,000 (1); c- 142,000 (1); d- 42,000 (1); e- 26,000 (2); f- 20,000 (1); g- 16,000 (1); h- 15,500 (1); and i- 14,000 (2) and for IIb are b- 170,000 (1); c- 142,000 (1); d- 42,000 (1); e- 26,000 (2); f- 20,000 (1); g- 16,000 (1); h- 15,500 (1); and i- 14,000 (2). Approximate molar ratios are given in parentheses. RNA polymerase IIa is similar to the major species of RNA polymerase II isolated from wheat germ (Fig. 7) while RNA polymerase IIb is similar in subunit structure to the RNA polymerase II purified from cauliflower nuclei (38). Other eukaryotes have similar subunit patterns for RNA polymerases IIa and IIb (8, 63, 64), but in other systems these enzymes are usually purified together and separated only by gel electrophoresis under nondenaturing conditions (8, 64). By choosing the appropriate isolation conditions and the appropriate developmental stage of the tissue, soybean RNA polymerases IIa and IIb can be purified and characterized independently

---

Fig.7. SDS gel electrophoresis of soybean RNA polymerase II enzymes. (A) 8.75-15% polyacrylamide linear gradient slab gel of soybean RNA polymerase IIa (a) and wheat germ RNA polymerase II (b). The putative subunit molecular weights of soybean RNA polymerase IIa (c) are given in kilodaltons. (B) 8.75-15% polyacrylamide linear gradient slab gel of soybean RNA polymerase IIb. The putative subunit molecular weights of the enzyme are given in kilodaltons. (C) Comparison of soybean RNA polymerase IIa and IIb on 10% SDS polyacrylamide disc gels. Molecular weights are given in kildaltons. (D) Comparison of the large molecular weight subunits of soybean RNA polymerase IIa and IIb on 7.5% SDS polyacrylamide disc gels. Molecular weights are given in kilodaltons. (E) Two mm sections from an overloaded 5% "native" gel containing both soybean RNA polymerase IIa and IIb were re-electrophoresed under denaturing SDS conditions using 7.5% acrylamide disc gels. The top of the "native" band (nearest the anode) when re-electrophoresed under denaturing conditions is shown in gel (a) and successive 2 mm sections migrating in the direction of the cathode on the "native" gel are shown in SDS gels (b) and (c). Polypeptide molecular weights are given in kilodaltons.

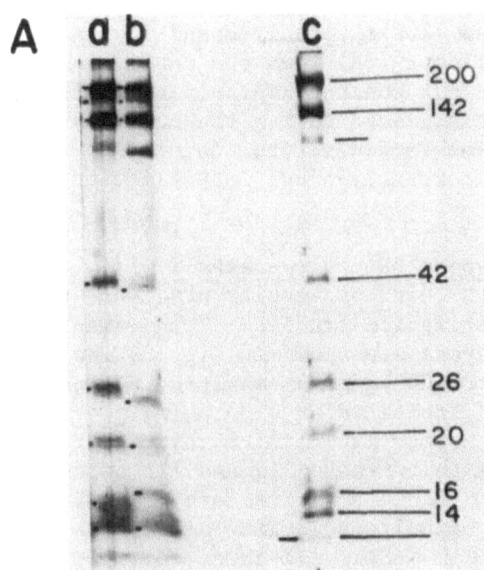

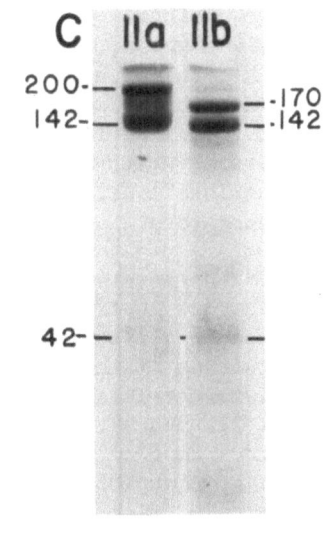

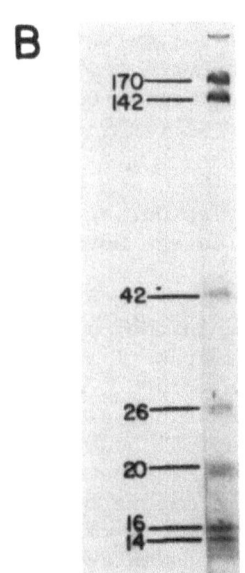

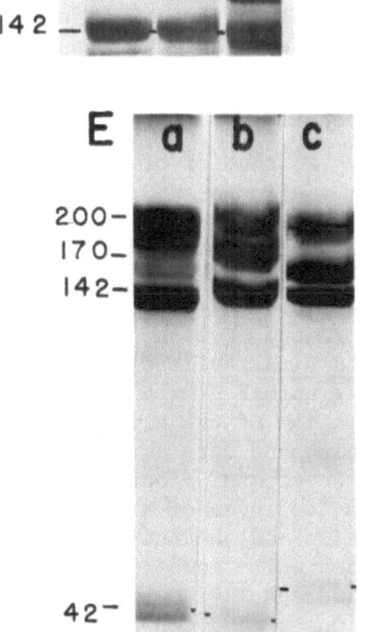

from one another. While RNA polymerase IIa appears to be a soluble
enzyme in soybean, RNA polymerase IIb is tightly bound to the chromatin in both soybean and cauliflower (38) when the chromatin or nuclear fraction is prepared at pH 6. Whether RNA polymerase IIb is
derived from IIa via proteolytic cleavage during the purification
procedure or whether the occurrence and distribution of the enzymes
are functionally significant is currently under investigation in
this laboratory.

    c. <u>Characterization of soybean RNA polymerases I, IIa, and
IIb</u>. Enzymes purified through the sucrose density gradient step
were used in all enzyme characterization studies. The α-amanitin
sensitivities of the soybean enzymes are shown in Fig. 9. While
RNA polymerase I is refractory to 100 μg/ml α-amanitin, RNA polymerases IIa and IIb are both 50% inhibited at 0.05 μg/ml α-
amanitin. None of the soybean enzymes is inhibited by either
rifamycin or cycloheximide. RNA polymerases IIa and IIb both have
a Km for ATP of approximately 30 μM and a Ki for corcycepin triphosphate of 2-3 uM (Guilfoyle, unpublished); RNA polymerase I has
a Km for ATP of approximately 20 μM (68). All three enzymes prefer
Mn^{++} over Mg^{++} with an optima of 1 mM for Mn^{++} and 5-10 mM for Mg^{++}.
With denatured DNA as template and Mn^{++} as the divalent cation, the
ammonium sulfate optima for RNA polymerase I, IIa, and IIb are 75mM,
50 mM, and 75 mM, respectively. All three enzymes prefer denatured
to native calf thymus DNA; however, the RNA polymerase II enzymes
show a much stronger preference for the denatured template. RNA
polymerase is able to transcribe poly (dC) at a 10-fold greater rate
than denatured DNA.

    3.   Modulation of RNA Polymerase Activities During Normal and
        Auxin-Induced Growth Transitions in the Soybean

    There are marked changes in RNA metabolism associated with
normal and auxin-induced growth transitions in the soybean seedling
(39-41, 65). We have assessed the RNA polymerase I and II activities in the soybean hypocotyl under various growth transitions where
the <u>in vivo</u> levels of RNA synthesis (accumulation) differ markedly.
During normal growth and development, the apical hook (zone A, Fig.
10) region of the etiolated soybean hypocotyl is the site of cell
division and very active RNA synthesis. The mature tissue in the
lower hypocotyl (zone B, Fig. 10) is not active in cell division and
has a low level of RNA synthesis. Following auxin treatment, cell
division and the accumulation of RNA in zone A are completely suppressed while cambial cell division and a marked accumulation of RNA
are induced in zone B. This change in the pattern of growth and
development induced by auxin is associated with an inhibition of
apical growth and a swelling of the mature hypocotyl (Fig. 10). In
the work which is summarized here, comparative studies were made of

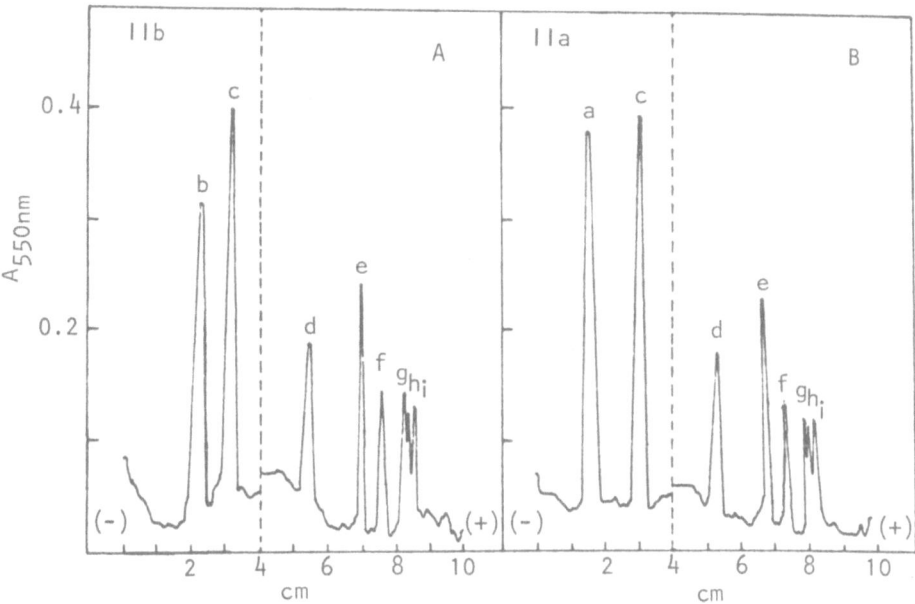

Fig.8. SDS polyacrylamide disc gel scans of purified soybean RNA
polymerases IIb (A) and IIa (B). Sucrose density gradient purified
enzymes were denatured, electrophoresed, stained with Coomassie blue,
and scanned at 550 nm as described in Fig. 5. The gels contained
two percentages of polyacrylamide (indicated by the dashed line)
with the upper half being 7.5% and the lower half 15% polyacrylamide. Putative subunits are labeled a to i.

the RNA synthetic activities of nuclei isolated from these tissues;
in addition, the levels of DEAE cellulose purified RNA polymerase I
and IIb activities which were solubilized from these nuclei were compared. A detailed analysis of auxin-induced modulation in chromatin-
bound, nuclear, and "soluble" RNA polymerase activities in zone B
has been published (42).

In the transition of the hypocotyl tissue from zone A to zone
B, the RNA polymerase I (α-amanitin-insensitive) activity expressed
in isolated nuclei drops precipitously; however, the RNA polymerase
II (α-amanitin-sensitive) activity expressed in the nuclei is relatively unchanged during this growth transition (Fig. 10). Thus the
4-fold higher level of RNA polymerase I activity expressed in zone
A nuclei relative to zone B nuclei is reflected by a change in the
ratio of RNA polymerase I to II from 1.3 to 0.35 in zone A and
zone B, respectively. When the RNA polymerases are solubilized from

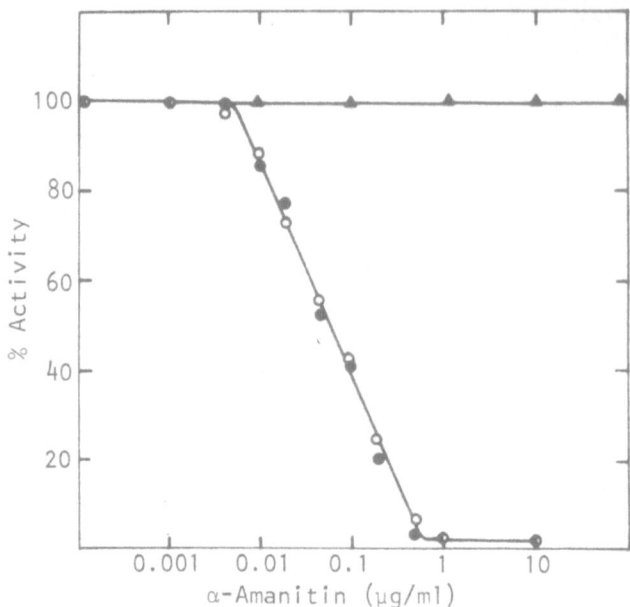

Fig.9. Sensitivities of soybean RNA polymerases I, IIa, and IIb to α-amanitin. (▲——▲), RNA polymerase I; (●——●), RNA polymerase IIa; and (○——○), RNA polymerase IIb.

Fig.10. Modulation of soybean RNA polymerase I activity under various growth transitions. (A) Diagram of (a) a three day old etiolated soybean seedling and four day old seedlings (b) after treatment with 2, 4-D for 24 hours and (c) untreated. (B) In vivo rate of RNA accumulation in (a) zone A of 2, 4-D treated, (b) zone B of 2, 4-D treated (c) zone A of untreated and (d) zone B of untreated 4 day old soybean seedlings. (C) In vitro levels of RNA polymerase activities assayed in nuclei from (a) zone A of 2, 4-D treated, (b) zone B of 2, 4-D treated, (c) zone A of untreated, and (d) zone B of untreated 4 day old soybean seedlings. The RNA polymerase II activities of the nuclei have been normalized since these activities are relatively constant during these growth transitions. (D) In vitro levels of RNA polymerase activities solubilized from nuclei and chromatographed on DEAE cellulose from (a) zone A of 2, 4-D treated, (b) zone B of 2, 4-D treated, (c) zone A of untreated, and (d) zone B of untreated 4 day old soybean seedlings. (●——●), minus α-amanitin; (○----○), plus 4 µg/ml α-amanitin.

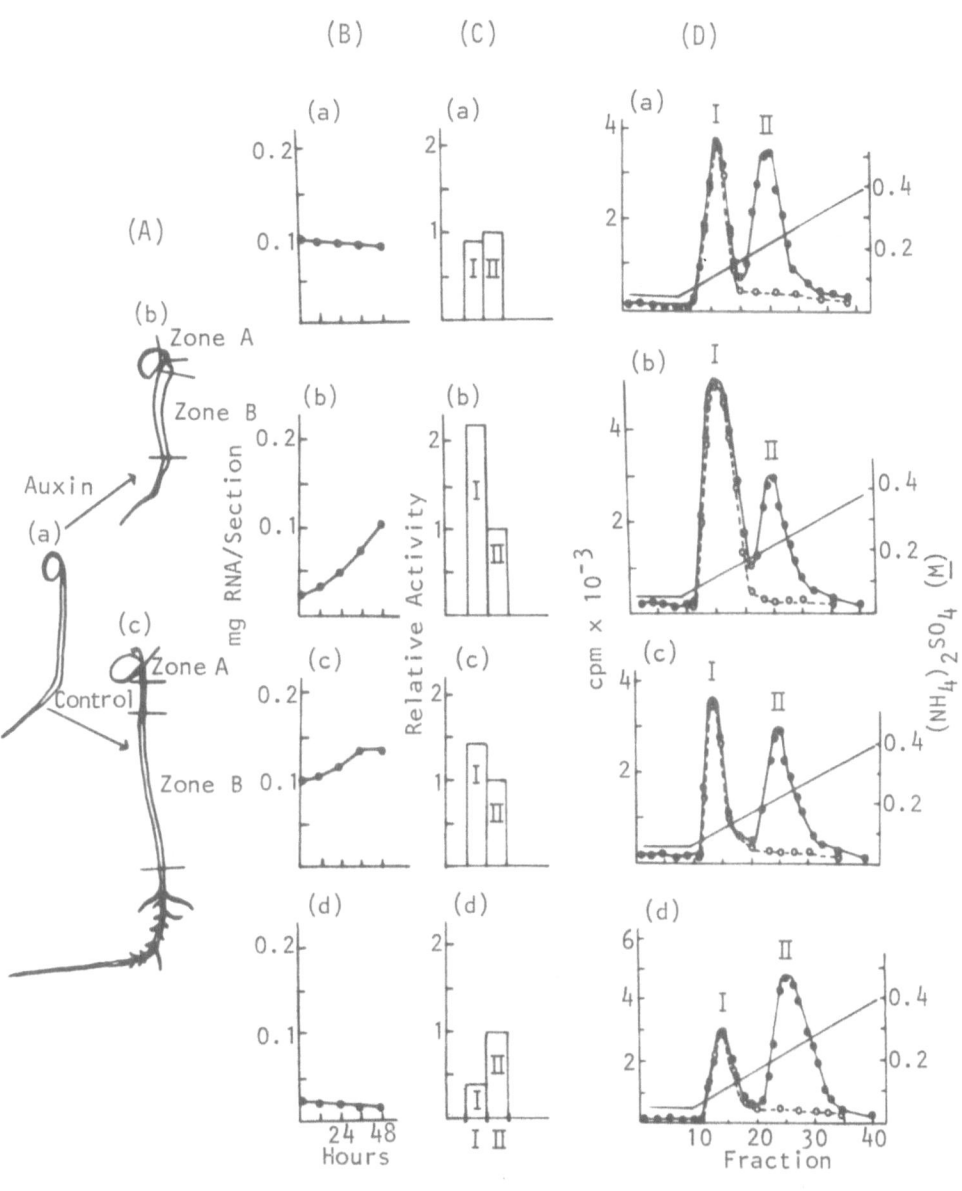

Fig. 10

the nuclei and fractionated on DEAE cellulose, the relative activities of RNA polymerase I and II are similar to those observed in the
nuclei. The level of solubilized RNA polymerase I is 2- to 3-fold
higher in zone A than in zone B while the level of solubilized RNA
polymerase II is similar between the two zones. The large decrease
in RNA polymerase I during the growth transition from the meristematic to the mature state shows a positive correlation with the
change in the relative level of in vivo rRNA synthesis which occurs
during this transition.

In contrast to normal growth and development in the soybean
hypocotyl, the application of auxin induces the normally quiescent
cells in zone B to swell and proliferate (65). This swelling and
proliferation in the mature tissue is associated with the massive
synthesis of rRNA and accumulation of ribosomes (65-67). Commen-
surate with the large accumulation of rRNA, there is a several-fold
increase in RNA polymerase I activity expressed in chromatin (42,
47) and nuclei (42, 43, 48) isolated from the auxin-treated tissue.
There is little or no change in the RNA polymerase II activity expressed in these chromatin or nuclear fractions. This results in an
increase in the RNA polymerase I to II ratio of nuclei from 0.35 in
normal zone B tissue to greater than 2 after 24 hours of auxin treatment. This increase in the ratio of RNA polymerase I to II is to a
large extent maintained following solubilization and DEAE cellulose
fractionation of the RNA polymerases from nuclei of normal and auxin-
treated mature tissue (42, 43). We have also observed that the level
of "soluble" (not pelleted in the chromatin or nuclear fractions) RNA
polymerase II does not change significantly in the mature tissue following auxin treatment (42).

Auxin treatment has an opposite effect on RNA synthesis in
zone A compared to zone B. In the apical hook tissue (zone A) which
normally has a high level of RNA polymerase I and a high ratio of
RNA polymerase I to II activity expressed in isolated nuclei, auxin
treatment causes a suppression of RNA polymerase I expressed in these
nuclei (Fig. 10). However, the activities of RNA polymerases I and
II after solubilization and fractionation on DEAE cellulose are
similar for control and auxin-treated zone A tissue. This suggests
a suppression of RNA polymerase I expressed in nuclei since similar
amounts of RNA polymerase I are solubilized from control and auxin-
treated apical, zone A tissue.

An even more dramatic example of this selective suppression of
RNA polymerase I activity is observed when the apical region is excised and incubated (43). There is a very dramatic and rapid decrease in the activity of RNA polymerase I expressed in isolated
nuclei associated with little or no change in the activity of RNA
polymerase II (Table 2). This results in a decrease in the RNA
polymerase I to II activity ratio from 1.3 to 0.25 after 12 hours

Table II. Modulation of RNA Polymerase Activities
during Excised Incubation of Zone A Soybean Hypocotyl

| | RNA Polymerase Activity* | | |
	I	II	I/II
Intact seedling (zero-time control)	80	62	1.3
Excised incubation (6 hours)	29	60	.5
Excised incubation (12 hours)	16	64	.25

* Nuclei were isolated from zone A tissue and assayed for RNA polymerase I and II activities as described (43). Data are expressed
as pmoles UMP incorporated/30 min/10^6 nuclei at 28° C. While the
activity of RNA polymerase I expressed in the nuclei decreased
dramatically during excised incubation, the relative levels of
RNA polymerase I and II did not change based on solubilization
of the activities and fractionation on DEAE cellulose (43).

of incubation. This decay in RNA polymerase I activity expressed in
nuclei has a half-life of about 2 hours during excised incubation of
the tissue (43, Guilfoyle, unpublished). In contrast to the rapid
decay of RNA polymerase I expressed in the nuclei, the amount of RNA
polymerase which is solubilized remains constant during the incubation period. Although the same level of RNA polymerase I is present
in the nucleus before and after incubation, the expression of RNA
polymerase I activity is strongly suppressed in incubated zone A
tissue. This suppression of RNA polymerase I activity within the
nucleus correlates with the very low level of rRNA synthesis in excised zone A tissue during incubation.

These studies show that RNA polymerase I activity undergoes
significant changes during normal and auxin-induced growth transitions while RNA polymerase II (IIb) activity associated with nuclei remains relatively constant. In preliminary experiments
(Guilfoyle, unpublished), RNA polymerase I activity expressed in
isolated nuclei was shown to increase many-fold in the embryonic
axis of soybean during 24 hours of germination; however, when the
RNA polymerase activity was solubilized from these nuclei and fractionated on DEAE cellulose, only a small increase in RNA polymerase
I was observed. Taken together these results show that RNA polymerase I activity is rapidly modulated up or down during growth
transitions in the soybean. At least two levels of control for
rDNA transcription are suggested: one involves the level of RNA
polymerase I enzyme and the other some nuclear control which modu-
lates the activity of those enzyme molecules. These studies indicate that while RNA polymerase I activity expressed in chromatin

or nuclei shows a strong correlation with the in vivo rate of rRNA
synthesis, this same correlation is not necessarily reflected in
the activity of solubilized RNA polymerase I.

DISCUSSION

RNA polymerase I and two forms of RNA polymerase II (IIa and
IIb) have been purified from soybean. Characteristic of RNA polymerase I enzymes from other eukaryotes, soybean RNA polymerase I
is localized in the nucleolus, is refractory to α-amanitin, transcribes genes which code for 18s and 25s rRNA, shows maximal activity at about 50 mM ammonium sulfate when assayed in the nucleus,
and elutes from ion-exchange resins at relatively low salt concentrations. Soybean RNA polymerase I has a subunit structure similar
to that of other eukaryotes (7, 55-57). The enzyme has a specific
activity of 200-300 nmoles UMP incorporated/30 min/mg protein at
28° C using denatured calf thymus DNA, comparable to other RNA
polymerase I enzymes. A 10- to 20-fold higher specific activity is
observed when poly (dC) is used as template.

While all of the RNA polymerase I activity is recovered with
nuclei or chromatin, only a small percentage of the RNA polymerase
II activity is recovered in association with nuclei or chromatin. RNA polymerase II activity was purified from hypocotyl nuclei and from embryonic axes where most of the enzyme is "soluble."
The subunit structure of the two enzymes is identical except for
the molecular weight of the largest subunit, similar to the form
IIa and IIb enzymes of animals (8, 15, 16, 58). The subunit structure of RNA polymerase IIa (embryonic axes enzyme) and IIb (nuclei-
associated hypocotyl enzyme) are: 200,000 (IIa) or 170,000 (IIb);
142,000; 42,000; 26,000; 20,000; 16,000; 15,500; 14,000. Soybean
RNA polymerase IIa is similar to the major RNA polymerase II purified from wheat germ (62 and Fig. 7), while the IIb enzyme has a
similar subunit structure to the RNA polymerase II purified from
cauliflower nuclei (38) and maize (34). RNA polymerase II from
parsley (36) tissue culture cells appears to have a similar subunit structure to soybean RNA polymerase IIa and IIb enzymes.

To date, we have not been successful in identifying a type
III RNA polymerase activity in soybean. Because of the methods
used in purification, a type III enzyme may be present as a minor
component (<10%) of our RNA polymerase I preparations. Guilfoyle
(38) was successful in demonstrating and partially characterizing
a type III RNA polymerase activity in cauliflower nuclei. An RNA
polymerase III from rye has also been reported (37), but not appreciably characterized. Low concentrations of the enzyme, lack
of stability, methodology, and/or insufficient properties to distinguish RNA polymerase III activity from RNA polymerase I may

relate to our lack of success in these studies. Subunit analysis
(6, 9, 16, 59) and identification of the reaction products
(16, 21) are required criteria for defining RNA polymerase III.

There is remarkably little evolutionary divergence among
RNA polymerase enzymes purified from a wide range of lower and
higher eukaryotes. The subunit structures of RNA polymerases
I (7, 16, 55-58), II (8, 16-17, 58), and III (6, 9, 59) from a
wide range of organisms are very similar although some variation
has been noted in their α-amanitin sensitivities (5, 13, 16, 59).
Higher plant RNA polymerases appear to have properties somewhat
intermediate between lower eukaryotes and higher animals. While
the subunit structures of RNA polymerase I from higher plants
(38, 44) are more similar to yeast (56, 57) than higher animals
(7, 16, 55), the plant enzymes are more similar to higher animals
with respect to α-amanitin insensitivity (5, 13, 16, 59). RNA
polymerase II from higher plants (34-36, 45) is similar to animal
(see 16) RNA polymerase II enzymes in subunit composition. Both
higher plants (36, 45) and animals (15-17) have two major forms
of RNA polymerase II which differ only in the molecular weight
of their largest subunits. The α-amanitin sensitivities of plant
class II enzymes (45) are similar to those of animals (16, 23)
but differ significantly from that of yeast (13). No work is
available on the subunit structure of higher plant RNA polymerase
III, but the α-amanitin sensitivity (38) is intermediate between
higher animals (16, 21, 23) and yeast (13, 59). The coconut
endosperm RNA polymerases are not characteristic of other plant
RNA polymerases. In addition to the unusual subunit structures,
the coconut RNA polymerase I and II enzymes appear unique in that
they are partially sensitive to rifamycin, elute in reverse order
from DEAE cellulose, and have intrinsic specific activities about
two orders of magnitude less than other eukaryotic RNA polymerases
(31, 32, 96).

The results presented here and those from earlier studies
(42, 43, 45) show that RNA polymerase I activity changes significantly during normal and auxin-induced growth transitions in the
soybean. RNA polymerase II activity changes much less if at all
during these same growth transitions. The level of RNA polymerase
I activity expressed in isolated nuclei generally correlates with
the in vivo level of rRNA synthesis of that tissue from which the
nuclei were isolated. However, the level of RNA polymerase I activity solubilized from these same nuclei does not necessarily
show this same correlation. Since the procedures employed solubilize greater than 90% of the RNA polymerase I from control or
treated nuclei, the differences observed between the level of
activity expressed within the nucleus and the level of the solubilized enzyme does not appear to be generated by differences in the
extractability of RNA polymerase I from control and treated nuclei.
Thus, in addition to changes in the apparent level or number

of RNA polymerase I molecules which occur during some of these
growth transitions, the data show that the activity of these RNA
polymerase I molecules may be significantly regulated within the
nucleus. The activity of RNA polymerase I has also been shown to
be selectively modulated during growth transitions or hormone
treatments in other eukaryotic systems (69-74). The selective modulation of RNA polymerase I relative to RNA polymerase II may be related to the genes transcribed by the two enzymes. Since RNA polymerase I selectively transcribes a single repetitive set of genes,
the rDNA cistrons, a change in initiation frequency and/or propa-
gation rate by RNA polymerase I molecules would result in dramatic
changes in the rate of synthesis (accumulation) of a single product, rRNA. In contrast, RNA polymerase II transcribes a vast set
of unique genes which code for unique mRNAs, and modulation in the
transcription of a single gene or even many different genes would
hardly result in dramatic changes in the overall level of RNA polymerase II actively involved in transcription. It would appear
that modulation in the transcription of unique genes could only
be detected by RNA: DNA hybridization using purified mRNAs or cDNA
probes. In addition, it appears likely that this modulation in the
transcription of unique genes would be regulated at the chromatin
level rather than at the level of the enzyme.

It has been suggested that the activity of RNA polymerase
I may be modulated by changes in the amount and/or activity
of a short half-life subunit or regulatory factor (75, 76). A
protein factor which may alter both chain initiation and transcription rate by RNA polymerase I has been studied in rat liver
(77). While we have no information on such factors in soybean,
the activation of rRNA synthesis by auxin seems to relate to a
change in both the number and activity of RNA polymerase I molecules. There is suggestive evidence that the rate of chain propa-
gation by RNA polymerase I on soybean chromatin may be increased
by auxin (78) similar to observations made with hormone-induced
systems in animals (79, 80). In contrast, an altered rate in the
initiation frequency of RNA polymerase I in vivo appears to account for the change in rate of rRNA synthesis associated with
amino acid deprivation in animal cells (81). There is in the soybean (see 43) as well as in animal (e.g. 82) and bacterial (e.g.
83) systems a very close coupling between rRNA synthesis, ribosome
function, and cell proliferation. In bacterial systems this
tight coupling appears to be accomplished by nucleotides (83, 84
and inclusive references) which are synthesized by ribosomes that
are inactive in protein synthesis. While the studies with bacterial systems provide an intriguing model for relating rRNA synthesis to ribosome function, no definitive data are available for
the operation of such a mechanism in eukaryotes.

A different approach to the study of the auxin regulation
of RNA synthesis relates to so-called "factor" experiments.

Some in vitro studies have been interpreted to show that the enhancement of RNA synthesis by auxin is accomplished by "factors" which in some way interact with auxin (e.g. 85-92). The mechanism of the enhancement of the RNA synthetic activity by the "factor" was not detailed in any of these studies. Furthermore, none of these studies utilized pure preparations of the putative factor or of RNA polymerase, and these studies generally used heterologous DNA rather than chromatin or homologous DNA. Recent studies with animal systems have shown that fidelity in the transcription of specific genes occurs only in the presence of homologous RNA polymerase and chromatin and generally not on heterologous or even homologous DNAs (93-95). While such "factor" studies may be suggestive of avenues for defining how auxin enhances RNA synthesis, the available data are at best circumstantial and must be viewed with the utmost caution.

ACKNOWLEDGMENTS

The authors wish to acknowledge the participation of C. Y. Lin, Y. M. Chen, W. B. Gurley, H. Chang, and K. Ester in portions of this work. This research was supported by Public Health Service Research Grant number CA-11624 from the National Cancer Institute to J. L. Key and a postdoctoral fellowship to T. J. Guilfoyle.

REFERENCES

1. Roeder, R. G., and Rutter, W. J. (1969) Nature (London) 224, 234-237.
2. Roeder, R. G., and Rutter, W. J. (1970) Proc. Natl. Acad. Sci. U. S. 65, 675-682.
3. Kedinger, Cl, Gniazdowski, M., Mandel, J. L., Gissinger, F., and Chambon, P. (1970) Biochem. Biophy. Res. Commun. 38, 165-171.
4. Kedinger, C., Nuret, P., and Chambon, P. (1971) FEBS Lett. 15, 169-174.
5. Chambon, P. (1974) In: The Enzymes, Vol. 10, pp. 333-374 (Boyer, P. D., ed.) Academic Press, New York.
6. Sklar, V. E. F., Schwartz, L. B., and Roeder, R. G. (1975) Proc. Natl. Acad. Sci. U. S. 72, 348-352.
7. Schwartz, L. B. and Roeder, R. G. (1974) J. Biol. Chem. 249, 5898-5906.
8. Schwartz, L. B. and Roeder, R. G. (1975) J. Biol. Chem. 250, 3221-3228.
9. Sklar, V. E. F. and Roeder, R. G. (1976) J. Biol. Chem. 251, 1064-1073.
10. Jacob, S. T., Sajdel, E. M., Munro, H. N. (1970) Biochem. Biophys. Res. Commun. 38, 765-770.

11. Reeder, R. H. and Roeder, R. G. (1972) J. Mol. Biol. 67, 433–441.
12. Lindell, T. J., Weinberg, F., Morris, P. W., Roeder, R. G., and Rutter, W. J. (1970) Science 170, 447–449.
13. Schultz, L. D. and Hall, B. D. (1976) Proc. Natl. Acad. Sci. U. S. 73, 1029–1033.
14. Zylber, E. A., and Penman, S. (1971) Proc. Natl. Acad. Sci. U. S. 68, 2861–2865.
15. Greenleaf, A. L., and Bautz, E. K. F. (1975) Eur. J. Biochem. 60, 169–179.
16. Roeder, R. G., Chou, S., Jaehning, J. A., Schwartz, L. B., Sklar, V. E. F., and Weinmann, R. (1975) In: The Isozymes Vol. III – Developmental Biology pp. 27–44 (Markert, C. L., ed.) Academic Press, New York.
17. Kedinger, C., Gissinger, F., Gniazdowski, M., Mandel, J. L., and Chambon, P. (1972) Eur. J. Biochem. 28, 269–276.
18. Dezelee, S., Sentenac, A., and Fromageot, P. (1972) FEBS Lett. 21, 1–6.
19. Seifart, K. H. and Benecke, B. J. (1974) Eur. J. Biochem. 53, 293–300.
20. Wilhelm, J., Dina, D., and M. Crippa (1974) Biochemistry 13, 1200–1208.
21. Weinmann, R. and Roeder, R. G. (1974) Proc. Natl. Acad. Sci. U. S. 71, 1970–1974.
22. Price, R. and Penman, S. (1972) J. Mol. Biol. 70, 435–450.
23. Schwartz, L. B., Sklar, V. E. F., Jaehning, J. A., Weinmann, R., and Roeder, R. G. (1974) J. Biol. Chem. 249, 5889–5897.
24. Austoker, J. L., Beebee, T. J. C., Chesterton, C. J., and Butterworth, P. H. W. (1974) Cell 3, 227–234.
25. Sergeant, A. and Krsmanovic, V. (1973) FEBS Lett. 35, 331–335.
26. Strain, G. C., Mullinix, K. P., and Bogorad, L. (1971) Proc. Natl. Acad. Sci., U. S. 68, 2647–2651.
27. Glicklick, D., Jendrisak, J. J., Becker, W. M. (1974) Plant Physiol. 54, 356–359.
28. Horgen, P. A. and Key, J. L. (1973) Biochim. Biophys. Acta. 294, 227–235.
29. Jendrisak, J. J. and Becker, N. M. (1973) Biochim. Biophys. Acta. 319, 48–54.
30. Gore, J. R. and Ingle, J. (1974) Biochem. J. 143, 107–113.
31. Biswas, B. B., Mondal, H., Ganguly, A., Das, A., and Mandal, R. K. (1974) In: Control of Transcription pp. 279–293. (Biswas, B. B., Mandal, R. K., Stevens, A., and Cohn, W. E., eds.) Plenum, New York.
32. Mondal, H., Mandal, R. K., and Biswas, B. B. (1972) Eur. J. Biochem. 25, 463–470.
33. Sasaki, Y., Sasaki, R., Hashizume, T., and Yamada (1973) Biochem. Biophys. Res. Commun. 50, 785–792.
34. Hardin, J. W., Apel, K., Smith, J., and Bogorad, L. (1975) In: The Isozymes Vol. I – Molecular Structure pp. 55–67. (Markert, C. L., ed.) Academic Press, New York.

35. Jendrisak, J. J. and Becker, W. M. (1974) Biochem. J. 139, 771–777.
36. Link, G. and Richter, G. (1975) Biochim. Biophys. Acta. 395, 337–346.
37. Fabisz-Kijowska, A., Dullin, P., and Walerych, W. (1975) Biochim. Biophys. Acta. 390, 105–116.
38. Guilfoyle, T. J. (1976) Plant Physiol., In Press.
39. Key, J. L. (1969) Ann. Rev. Plant Physiol. 20, 449–474.
40. Key, J. L. and J. Ingle (1968) In: Biochemistry and Physiology of Plant Growth Substances pp. 711–722 (Wightman, F. and Setterfield, G., eds.) The Runge Press Lts., Ottawa, Canada.
41. Key, J. L. and Vanderhoef, L. N. (1973) In: Developmental Regulation - Aspects of Cell Differentiation pp. 49–83 (Coward, S. J., ed.) Academic Press, New York.
42. Guilfoyle, T. J., Lin, C. Y., Chen, Y. M., Nagao, R. T., and Key, J. L. (1975) Proc. Natl. Acad. Sci. U. S. 72, 69–72.
43. Lin, C. Y., Chen, Y. M., Guilfoyle, T. J., and Key, J. L. (1976) Plant Physiol., In Press.
44. Guilfoyle, T. J., Lin, C. Y., Chen, Y. M., and Key, J. L. (1976) Biochim. Biophys. Acta. 418, 344–357.
45. Guilfoyle, T. J. and J. L. Key (1976) In: Molecular Mechanisms in the Control of Gene Expression (Nierlich, D. P. and Rutter, W. J., eds.,) In Press.
46. Lin, C. Y., Guilfoyle, T. J., Chen, Y. M., Nagao, R. T., and Key, J. L. (1974) Biochem. Biophys. Res. Commun. 60, 493–508.
47. Guilfoyle, T. J. and Hanson, J. B. (1973) Plant Physiol., 51, 1022–1025.
48. Chen, Y. M., Lin, C. Y., Chang, H., Guilfoyle, T. J., and Key, J. L. (1975) Plant Physiol. 56, 78–82.
49. Chang, H. (1976) M. S. Thesis, University of Georgia, Athens, Ga.
50. Widnell, C. C. and Tata, J. R. (1964) Biochim. Biophys. Acta. 87, 531–533.
51. Jacob, S. T., Sajdel, E. M., and Munro, H. N. (1970) Biochem. Biophys. Res. Commun. 38, 765–770.
52. Johnson, J. D., Jant, B. A., Kaufman, S. and Sekoloff, L. (1971) Arch. Biochem. Biophys. 142, 489–500.
53. Lin, C. Y., Guilfoyle, T. J., Chen, Y. M., and Key, J. L. (1975) Plant Physiol., 56, 850–852.
54. Gurley, W., Lin, C. Y., Guilfoyle, T. J., Nagao, R. T., and Key, J. L. (1976) Biochim. Biophys. Acta. 425, 168–174.
55. Gissinger, F. and Chambon, P. (1972) Eur. J. Biochem. 28, 277–282.
56. Valenzuela, P., Weinberg, F., Bell, G., and Rutter, W. J. (1976) J. Biol. Chem. 251, 1464–1470.
57. Buhler, J. M., Sentenac, A., and Fromageot, P. (1974) J. Biol. Chem. 249, 5963–5970.
58. Kedinger, C., Gissinger, F., and P. Chambon (1974) Eur. J. Biochem. 44, 421–436.

59. Valenzuela, P., Hager, G., Weinberg, F. and Rutter, W. J.
     (1976) Proc. Natl. Acad. Sci. U. S. 73, 1024-1028.
60. Guilfoyle, T. J. and Key, J. L. (1976) Plant Physiol. Suppl.
     57, 13.
61. Marcus, A., Efron, D., and Weeks, D. P. (1973) In: Methods
     in Enzymology Vol. 20. pp. 749-754. (Colowick, S. P. and
     Kaplan, N. O., eds.) Academic Press, New York.
62. Jendrisak, J. J. and Burgess, R. R. (1975) Biochemistry 14,
     4639-4645.
63. Weaver, R. F., Blatti, S. P., and Rutter, W. J. (1971) Proc.
     Natl. Acad. Sci. U. S. 68, 2994-2999.
64. Kedinger, C. and Chambon, P. (1972) Eur. J. Biochem. 28,
     283-290.
65. Key, J. L., Lin, C. Y., Gifford, E. M., Jr., Dengler, R.
     (1966) Bot. Gaz. 127, 87-94.
66. Chrispeels, M. J. and Hanson, J. B. (1962) Weeds 10, 123-125.
67. Key, J. L. and Hanson, J. B. (1961) Plant Physiol. 36, 145-
     152.
68. Guilfoyle, T. J. (1974) Ph. D. Thesis, University of Illinois,
     Urbana, Illinois.
69. Yu, F-L. and Feigelson, P. (1973) Biochem. Biophys. Res. Commun.
     53, 754-760.
70. Sebastian, J., Mian, F. and Halvorson, H. O. (1973) FEBS Lett.
     30, 65-70.
71. Teissere, M., Penon, P., and Ricard, J. (1973) FEBS Lett. 30,
     65-70.
72. Fuhrman, S. and Gill, G. N. (1974) Endocrinology. 94, 691-700.
73. Jaehning, J. A., Stewart, C. C., and Roeder, R. G. (1975)
     Cell. 4, 51-57.
74. Yu, F-L. (1975) Biochem. Biophys. Res. Commun. 64, 1107-1115.
75. Yu, F-L. and Feigelson, P. (1971) Proc. Natl. Acad. Sci. U. S.
     68, 2177-2180.
76. Seifert, K. H., Ferencz, A., and Benecke, B. (1973) In:
     Regulation of Transcription and Translation in Eukaryotes
     pp. 134-162. (Bautz, E. K. F., Karlson, P., and Hersten, H.)
     Springer Verlag, New York.
77. Goldberg, M. I., Perriard, J. C., Hager, G., Hallick, R. B.,
     and Rutter, W. J. (1973) In: Control of Transcription, pp.
     241-256 (Biswas, B. B., Mandal, R. K., Stevens, A., Cohn,
     W. E., eds.) Plenum, New York.
78. Guilfoyle, T. J. and Hanson, J. B. (1974) Plant Physiol. 53,
     110-113.
79. Barry, J. and Gorski, J. (1971) Biochemistry 10, 2384-2390.
80. Fuhrman, S. and Gill, G. N. (1975) Biochemistry 14, 2925-2933.
81. Grummt, I., Smith, V. A., and Grummt, F. (1976) Cell. 7,
     439-445.
82. Green, H. (1974) In: Control of Proliferation of Animal Cells.
     (Clarkson, E. and Baserga, R., eds.) pp. 743- 755. CSH Labora-
     tory, New York.

83. Cashel, M. (1975) Ann. Rev. Microb. Vol. 29, 301–318.
84. Gallant, J., Shell, L., and Bittner, R. (1976) Cell. 7, 75–84.
85. Hardin, J. W., Cherry, J. H., Morre, D. J., and Lembi, C. A. (1972) Proc. Natl. Acad. Sci. U. S. 68, 3146–3150.
86. Hardin, J. W., O'Brien, T. J., and Cherry, J. H. (1970) Biochim. Biophys. Acta. 224, 667–670.
87. Matthysse, A. G. and Phillips, C. (1969) Proc. Natl. Acad. Sci. U. S. 63, 897–903.
88. Venis, M. A. (1971) Proc. Natl. Acad. Sci. U. S. 68, 1824–1827.
89. Mondal, H., Mandal, R. K., and Biswas, B. B. (1972) Biochem. Biophys. Res. Commun. 29, 306–311.
90. Mondal, H., Mandal, R. K., and Biswas, B. B. (1972) Nature New Biol. 240, 111–113.
91. Teissere, M., Penon, P., Van Huystee, R. B., Azou, Y., and Ricard, J. (1975) Biochim. Biophys. Acta. 402, 391–402.
92. Penon, P., Teissere, M., Azou, Y., and Ricard, J. (1975) Physiol. Veg. 13, 813–829.
93. Roeder, R. G., Reeder, R. H., and Brown, D. D. (1970) CSHSQB 35, 727–735.
94. Parker, C. S., Sklar, V. E. F., Ng, S., and Roeder, R. G. (1976) In: Molecular Mechanisms in the Control of Gene Expression (Nierlich, D. P. and Rutter, W. J., eds.) In Press.
95. Beebee, T. J. C. and Butterworth, P. H. W. (1974) Eur. J. Biochem. 45, 395–406.
96. Ganguly, A., Das, A., Mondal, H., Mandal, R. K., and Biswas, B. B. (1973) FEBS Lett. 34, 27–30.

TRANSCRIPTION OF THE NUCLEAR GENOME OF

ACETABULARIA

Hans-Georg Schweiger

Max-Planck-Institut für Zellbiologie

2940 Wilhelmshaven, Germany

INTRODUCTION

As early as 50 years ago it was suggested that there are chemical transmitters between the genetic information of the cell nucleus and morphogenesis. Hämmerling was one of the very first who developed such a hypothesis and called these transmitters "morphogenetische Substanzen" (Hämmerling, 1932). The existence of such "morphogenetische Substanzen" was concluded from experiments which were performed on the unicellular and uninuclear green alga Acetabularia. The most important among these experiments lead to the observation that even in the absence of the nucleus, the green alga Acetabularia is still capable of performing morphogenesis. The morphogenetic agent was localized within the cytoplasm and was shown to exhibit a pronounced polarity with a high activity being found in the apical part and a low activity in the basal part of the cell. Since Hämmerling's early experiments, a number of investigations yielded indirect evidence that "morphogenetische Substanzen" are identical with mRNA; however, final proof for this hypothesis is still lacking.

At the present stage in our understanding of morphogenesis in Acetabularia, the following problems would seem of importance:
1. Identification of the regulatory mechanisms at the different steps of gene expression and their role in morphogenesis.
2. Regulation of transcription of the nuclear genome.
3. Chemical identification of the "morphogenetische Substanzen".

This contribution is particularly concerned with the transcription of the nuclear genome.

MORPHOLOGICAL BASIS

Prior to the discussion of the problems mentioned above, it is necessary to recall the morphological situation in the cell under discussion (Schweiger, 1969; Puiseux-Dao, 1970; Schweiger et al., 1975; Schweiger, 1976). At the end of the vegetative phase of its life cycle, the Acetabularia cell has formed a stalk which is up to 5 cm long in some species and even up to 20 cm long in others. At the apical end of the stalk is a cap-like structure, and at the basal end, a system of branches called the rhizoid. One of the branches of the rhizoid contains the cell's single nucleus. At this point it is useful to remember that the major part of the cell volume, particularly of the stalk, is occupied by the thick cell wall and a big central vacuole. The cytoplasm, which is localized adjacent to the plasma membrane covering the inner surface of the cell wall, forms strands which are interconnected in a net-like pattern. If regulatory mechanisms of gene expression in this cell are to be discussed, one has to keep in mind that transport over the long intracellular distances may play a major role (Kloppstech and Schweiger, 1975 a and b).

The cell nucleus of Acetabularia exhibits an extraordinary size. The diameter may be as great as 150 to 200 μm (Schweiger et al., 1974). A detailed investigation of the morphology of the nuclear membrane and adjacent regions of the perinuclear cytoplasm shows that the membrane contains numerous pores linking the two compartments (Berger and Schweiger, 1975 b; Franke et al., 1974). This raises the question of whether the nuclear membrane ever functions as a well defined boundary between the two compartments and therefore whether the nuclear membrane as well as the adjacent regions could be sites for regulation of gene expression (Berger and Schweiger, 1975 b).

RIBOSOMAL RNA

Any study of the regulation of gene expression must include a study of the products of transcription. Such products are either released into the cytoplasm directly, they are temporarily stored and partially degraded within the cell nucleus before being released, or they are totally degraded and never released. Consider first rRNA as a product of transcription.

Electron microscopy (Schweiger et al., 1974) as well as sedimentation in the ultracentrifuge (Kloppstech and Schweiger, 1973 a) reveal 80S cytosol ribosomes in Acetabularia. They are more or less evenly distributed within the cell. No significant differences in concentration can be detected between apical and basal fragments (Kloppstech and Schweiger, 1975 b) (Fig. 1). There is hardly any other organism in which it is possible by so simple a means to prove that the rRNA of 80S ribosomes is synthesized in the nucleus. Only nucleate but not anucleate cells are capable of incorporating precursors of RNA into the RNA of 80S ribosomes (Kloppstech and Schweiger, 1973 a) (Fig. 2).

It is of some interest to know what the fate of rRNA is after being exported from the nucleus. Although the cytoplasm exhibits a strong cytoplasmic streaming which can easily be detected under the microscope (Kamiya, 1959), the newly synthesized RNA is transported through the cell only very slowly (Fig. 3). There is good evidence that after being released from the nucleus, the rRNA is incorporated into 80S ribosomes which then migrate

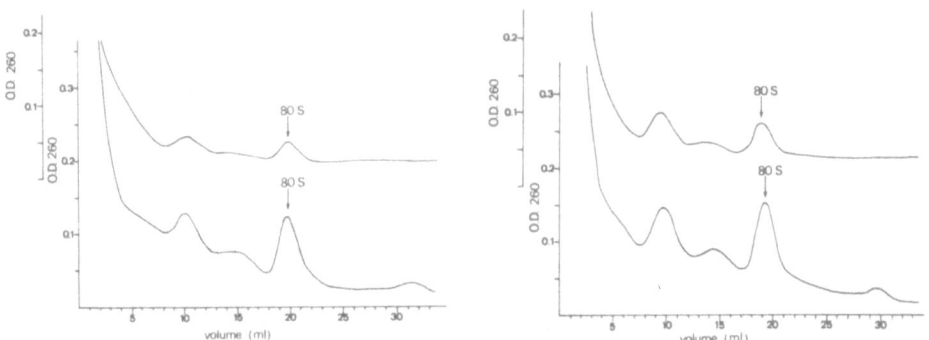

Fig. 1. Distribution of ribosomes in A. major (200 cells each) (Kloppstech and Schweiger, 1975 b). Left side: Apical halves, right side: Basal halves. Upper curves: Isolation in the absence of detergent. Lower curves: Isolation in the presence of 1 % detergent (Nonidet P40).

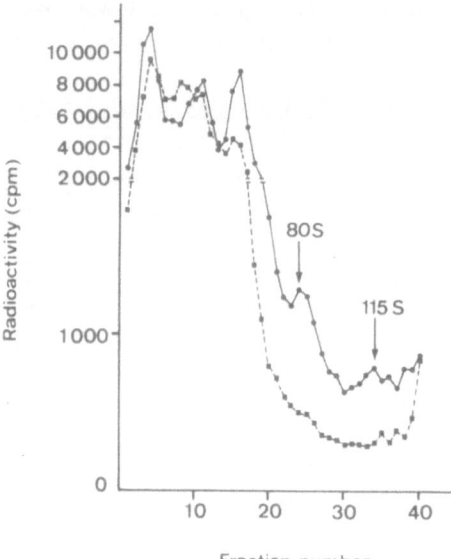

Fig. 2. Incorporation of labelled uridine into cytosol ribosomes from A. major (Kloppstech and Schweiger, 1973 a). The positions of the monomeric 80S and the dimeric 115S particles are marked. Time of labelling: 22 hours. ●—●—● nucleate cells, □·····□·····□ anucleate cells.

from the basal part to the apex. The speed of transport is about two to four mm per day and as such is roughly twice that of the cellular growth rate. From fractionation experiments it can be concluded that the ribosomes which are transported are membrane bound (Kloppstech and Schweiger, 1975 b).

Although the rRNA of 80S ribosomes is subjected to some turnover in the cytoplasm the half life of the rRNA is 80 days or more (Kloppstech and Schweiger, 1975 b). This extremely low turnover means that the cell is capable of retaining almost the total amount of rRNA which is synthesized. Correspondingly, the rate of degradation is extremely low. On the basis of this low turnover the rate of synthesis can be estimated. The rate of polymerization of rRNA is about 4×10^7 nucleotides per second per nucleus (Kloppstech and Schweiger, 1975 b). Supposing that the number of polymerases which are active on one cistron at the same time is roughly 100 (for references see Hamkalo and Miller,

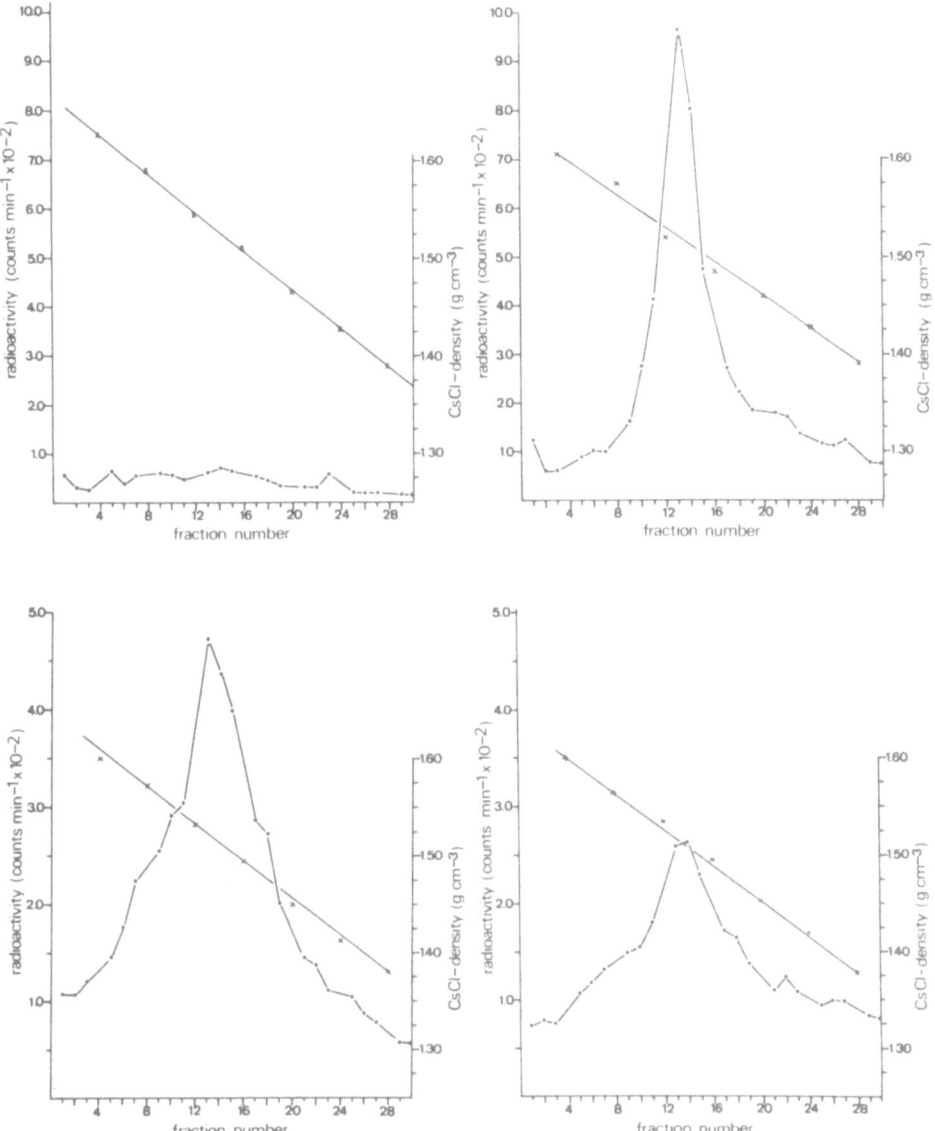

Fig. 3. Intracellular transportation of 80S ribosomes in A. major (Kloppstech and Schweiger, 1975 b). Cells were labelled for 22 hours with [³H] uridine and incubated for 3 days (left side) and 24 days (right side) respectively in "cold" medium. Cells were homogenized and subjected to sucrose density gradient centrifugation, the 80S peak was isolated and after fixation recentrifuged in a CsCl gradient. Upper curves: Apical fragments; lower curves: Basal fragments.

(1973) one may conclude that one polymerase should polymerize at least 10^5 nucleotides per second. This number is larger by several orders of magnitude than that which is known from E. coli and other organisms. Therefore one may assume that one nucleus contains more than one cistron for rRNA. A rough calculation indicates that the redundancy of rDNA should be in the order of 10^4 or more (Kloppstech and Schweiger, 1975 b).

The redundancy of the rRNA cistrons can be visualized in the electron microscope in preparations which have been spread by the technique developed by Miller (Hamkalo and Miller, 1973). For this purpose the nuclei are separated from Acetabularia cells and mechanically disintegrated, the nucleoli are isolated and are spread in a solution of extremely low ionic strength. By this technique one obtains characteristic pictures in which the cistrons, with radiating fibrils of newly synthesized RNA, are clearly seen on the DNA strands (Berger and Schweiger, 1975 d; Trendelenburg et al., 1974; Kloppstech et al., 1974; Woodcock et al., 1975) (Fig. 4). On the basis of the number of nucleoli per cell nucleus and the maximum number of cistrons on one DNA strand, one can estimate that the redundance of rDNA in Acetabularia is larger than 10^4 (Berger and Schweiger, 1975 d). This estimation is in good agreement with that obtained from the rate of synthesis of rRNA.

Every cistron contains RNA fibrils whose true length is identical with the distance between the initiation site and the position of the corresponding polymerase on the DNA strand. The total length of the cistron is correlated then with the length of the primary transcription product. Within a given species, the length of the rRNA cistrons is constant; however, the length of the rRNA cistrons is different for different species (Berger and Schweiger, 1975 c) (Fig. 5). Similarly, the non-transcribed spacers between cistrons are of a constant length for individuals of the same

Fig. 4. Repetitive rRNA cistrons from B. oerstedii (with spacer).
x 7700

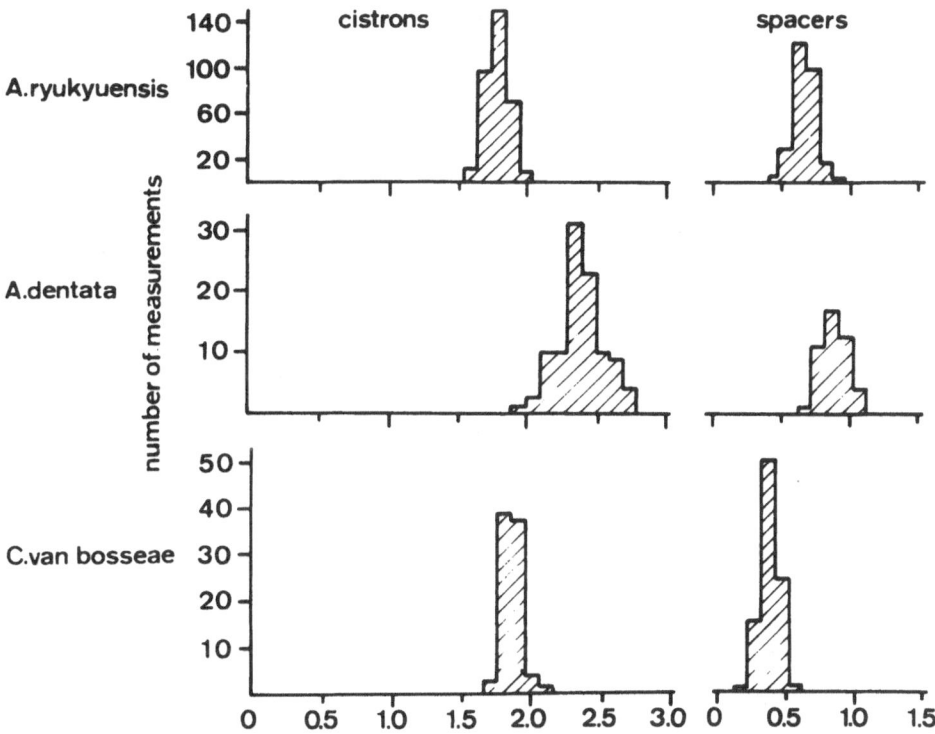

Fig. 5. Distribution pattern of length of actively transcribed
rRNA cistrons and their non-transcribed spacers in 3 species
of Dasycladaceae (from Berger and Schweiger, 1975 c).

species, but exhibit large differences between different species
(Berger and Schweiger, 1975 c). An extreme case is Batophora
oerstedii in which the cistrons are so close together that no non-
transcribed spacers are detectable by the electron microscopic
method (Berger and Schweiger, 1975 a). Besides this "spacer-free"
another type of rDNA occurs in the same species which has
spacers.

One might ask whether under special conditions the cell is
capable of increasing the number of nucleotides which are polymerized per unit time. That this is possible, can be concluded from
the substantially increased rate of rRNA synthesis observed when
regeneration is induced by surgically separating the rhizoid from

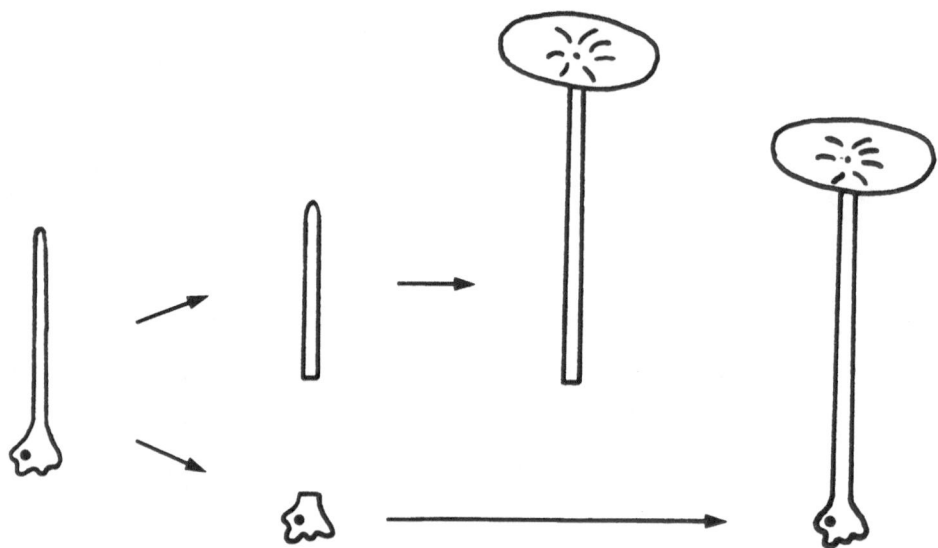

Fig. 6. Schematic presentation of regeneration of a rhizoid and of morphogenesis of an anucleate cell of <u>Acetabularia</u>.

the greater part of the cytoplasm (Kloppstech and Schweiger, 1973 b) (Fig. 6). Under these conditions the nucleus starts to compensate for the loss of cytoplasm and increases the rate of incorporation of nucleotides into the RNA of the 80S ribosomes (Fig. 7). This increase reaches its maximum on the third day after surgery, at which time the rate of incorporation goes up twentyfold relative to the control. Under the same conditions, the rate of synthesis of chloroplast rRNA is also increased; however, this increase is delayed and reaches its maximum on day 7 (Kloppstech and Schweiger, 1973 b).

The next question is how the cell increases the rate of synthesis. The following is a discussion of the various possibilities.

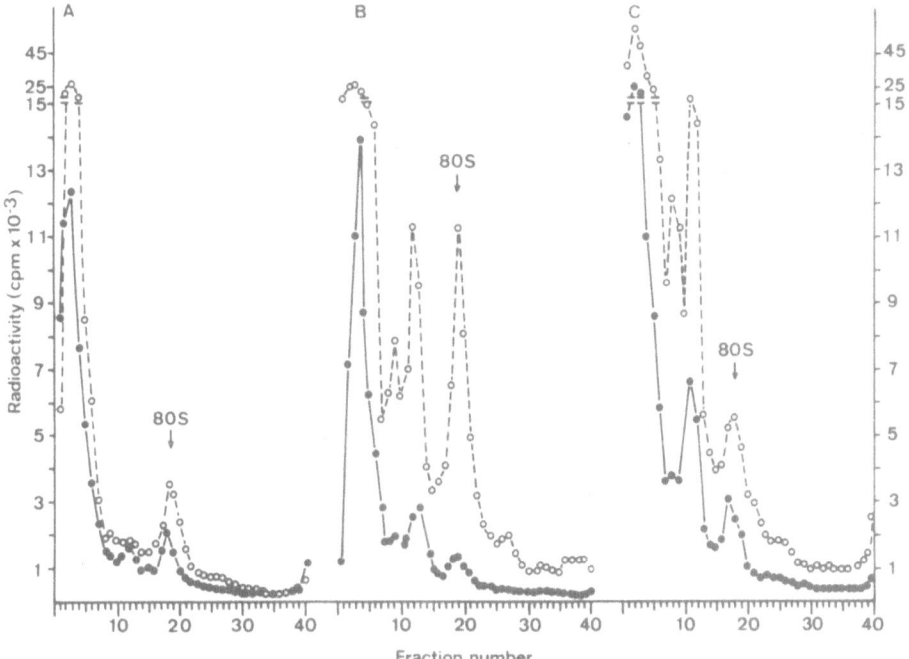

Fig. 7. Incorporation of [³H] uridine into ribosomes during regeneration of A. mediterranea cells (Kloppstech and Schweiger, 1973 b). Cells were labelled for 24 hours with [³H] uridine 0 (left), 3 (middle) and 6 (right) days after cell surgery in "cold" medium. The position of the 80S ribosomes is marked. The two subunits of the 70S organelle ribosomes show maximum activity on day 6.

1. The number of polymerases which are active on one cistron is increased. This would result in the distance between two polymerase molecules being decreased. So far there is no evidence for such a change under the conditions of an increased rate of synthesis. The least one can say is that this possibility is extremely improbable.
2. The rate of polymerization per polymerase is increased. This type of regulation cannot be ruled out completely. If such an increase in the rate of polymerization occurs, then one would have to ask subsequently whether this happens by an increased concentration of critical components or by activation of rate limiting partial steps of transcription; however, again there is

no evidence for this type of regulation.

3. Additional cistrons which under normal conditions are inactive, become activated. Such a regulation would include the possibility that either additional individual cistrons or whole groups of cistrons are activated. The activation of individual cistrons is an attractive hypothesis (Hackett and Sauerbier, 1975; McKnight and Miller, 1976). If the increase in the rate of transcription were due to the activation of individual cistrons one should see such inactive non-transcribed cistrons in the spreading preparations.

Indeed, from non-regenerating cells of Acetabularia a certain number of "dead" cistrons can be seen which lack RNA fibrils (Berger and Schweiger, 1975 c; Berger and Schweiger, 1975 d); however, the number of these "dead" cistrons would be absolutely insufficient to explain a twenty-fold increase in transcriptional rate.

If one thinks more thoroughly about the possibility that whole groups of cistrons are activated, this alternative again seems to be improbable for such a hypothesis would mean that there are activator sites on the rDNA by which a whole group of rRNA cistrons say a whole rDNA thread can be regulated (Perry and Kelley, 1970). Although such a possibility is imaginable it is questionable that there would be only one regulatory site for so many cistrons (one DNA thread: > 100 cistrons) or that transcription would be regulated at sites other than the individual cistron sites.

4. The apparent rate of transcription is regulated by degradation of rRNA. In the cytoplasm such a possibility can be ruled out, since the half life of rRNA is extremely long (Kloppstech and Schweiger, 1975 b). Based on pulse experiments, it is also highly improbable that within the cell nucleus degradation processes of rRNA or of their precursors play a significant role (Kloppstech, Richter and Schweiger, in preparation).

5. The number of cistrons available is increased during regeneration by amplification. Under conditions of regeneration the magnitude as well as kinetics for the increase might indicate that amplification indeed plays an important role.

Two possibilities of how such an amplification could occur are discussed here. One possibility is that amplification could occur at only a single site per cell. The other possibility is that there might be a multiplicity of amplification sites in the cell and in this case it would be appealing to assume that each nucleolus contains its own amplification site. A semiquantitative estimation shows that the second possibility is more probable.

If one assumes that the increase in transcriptional rate under the conditions of regeneration is due to amplification of the rDNA, the amount of rDNA has to be increased by an amount which corresponds to 4×10^9 nucleotides (Berger and Schweiger, 1975 d; Kloppstech and Schweiger, 1973 b). If one assumes that there is only one amplification site per cell and the rate of polymerization at 20^0 to be 100 nucleotides per second (Kornberg, 1974), then a total of 4×10^7 seconds or 1×10^4 hours are needed, which is much too long. With a period of only 70 hours during which amplification occurs, then one may conclude that there must be available a total of 200 sites of amplification. This number corresponds quite well to the number of nucleoli which are found in a cell at the developmental stage under consideration. It is tempting to assume that if regulation occurs via amplification, such an amplification occurs in more or less each nucleolus. Further experiments are necessary to show whether this working hypothesis is correct.

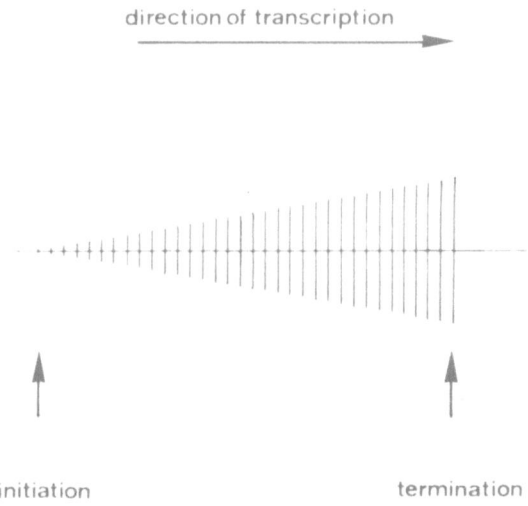

Fig. 8. Schematic presentation of initiation and termination sites as possible sites of regulation.

Another aspect which follows from the electron microscopic pictures is the following. In the spread preparations, RNA fibrils of different lengths are observed bound to each cistron with polymerase apparently mediating the binding. The lengths of the fibrils exhibit a gradient with increasing length from the initiation to the termination site. It might well be that the situation as presented in the spread preparation reflects a static rather than a dynamic condition. A static situation could mean that at least some of the apparently active cistrons are not being actively transcribed at all but are inhibited at the level of termination (McKnight and Miller, 1976) (Fig. 8). The special conditions of regeneration would then be required to suspend the inhibition at the level of termination.

Although the situation is far from clear there is some evidence that such an inhibition does not occur at the level of termination. This evidence comes from experiments with actinomycin D. If regulation at the termination level were to play a major role, one might think that in the presence of actinomycin D, some cistrons would show regular christmas-tree-like structures which could be distinguished from controls. These unchanged patterns could represent cistrons which indeed are in a static situation. In these cistrons the actinomycin might be intercalated into the DNA without changing the total pattern. Other cistrons should show an influence of the inhibitor. Irregular patterns would represent cistrons which are being actively transcribed. Caution should be exercised, however, since we do not know whether actinomycin D could interfere with the normal pattern under static conditions. In the presence of actinomycin there is no significant occurrence of such cistrons with regular fibril patterns. This might indicate that more or less all the cistrons were in a dynamic state and were indeed being actively transcribed (Berger, in preparation).

The transcription product of the rRNA cistron is subjected to some changes before it is released from the nucleus and is integrated into the 80S ribosomes. From a number of organisms it is known that there is a discrepancy between the lengths of the primary and the final transcription product (Tartof, 1975). This conversion of the primary into the final transcription product is called processing or maturation and in the case of rRNA represents a sequential splitting of various lengths of RNA from the primary transcription product. The length of the primary transcription product can be estimated in two ways. The first is to measure the length of the rRNA cistron. Such values are obtained from electron photomicrographs of spread preparations. The second way is to determine the molecular weight by electrophoretic methods of the first transcription product in pulse experiments. Both methods have been used in different species of Acetabularia (Berger and Schweiger, 1975 c; Kloppstech, Richter and Schweiger, in preparation). These experiments revealed a good agreement between the length of a cistron estimated from electron photomicrographs and the molecular weight obtained from pulse experiments. It was shown however, that at least in one of the species of Acetabularia which were investigated, the RNA fragment which is split off during maturation must be short.

At this point a brief comment might be made on the electron microscopic estimation of the length of a cistron. Apparently the DNA thread is shortened at the site of active transcription. This observation has been made already by Miller (Hamkalo and Miller, 1973) and we were able to confirm his observation. We

	Length of cistrons (μm)	Molecular weight of primary transcriptional product (daltons x 10^{-6})	
		predicted	estimated
A. mediterranea	1.9	2.2	2.3
A. dentata	2.4	2.8	2.8

Tab. 1. Molecular weight of the primary transcription product of rRNA cistrons in two species of Acetabularia as predicted from electron microscopy and estimated in pulse experiments (Kloppstech, Richter and Schweiger, in preparation; Richter, 1976).

found that this shortening is 20 % of the total length of a cistron even under optimal spreading conditions.

The molecular weight of the primary transcription product was predicted by the electron microscopic method as 2.2×10^6 daltons in Acetabularia mediterranea and 2.8×10^6 daltons in Acetabularia dentata, and was estimated by pulse labelling experiments to be 2.3×10^6 for Acetabularia mediterranea and 2.8×10^6 daltons for Acetabularia dentata (Tab. 1). In both species the predicted and estimated values were in good agreement. The difference in the molecular weight of the primary transcription product in the two species under consideration is a striking observation, since it means that even in species of the same family the primary transcription product shows some variability in length. The question is raised whether the processing still exhibits common features.

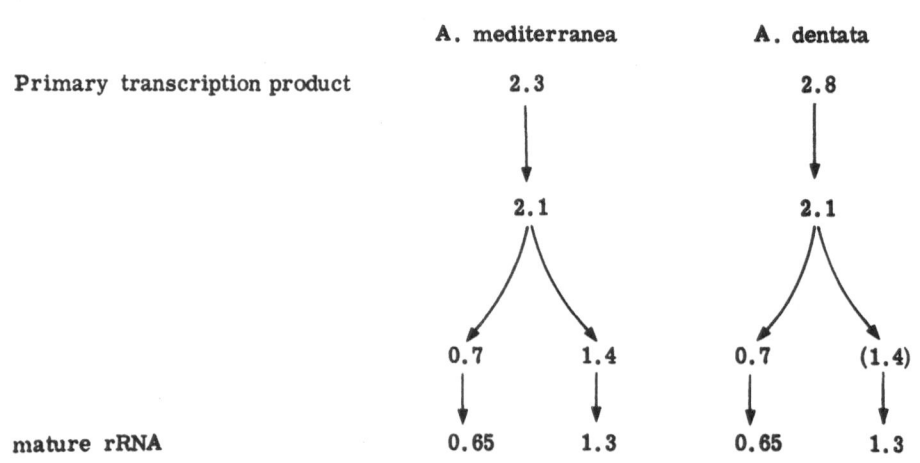

Fig. 9. Schematic presentation of processing of rRNA in A. mediterranea and A. dentata (Kloppstech, Richter and Schweiger, in preparation; Richter, 1976).

In pulse experiments with Acetabularia mediterranea a number of labelled components were found with molecular weights
of 2.4, 2.1, 1.4, and 0.7 x 10^6 daltons, and finally the mature
ribosomal ribonucleic acids with molecular weights of 1.3 and
0.65 x 10^6 daltons (Fig. 9). In Acetabularia dentata similar
pulse experiments revealed essentially the same spectrum of
labelled products with the exception that the primary transcription product had a molecular weight of 2.8 x 10^6 daltons (Kloppstech, Richter and Schweiger, in preparation; Richter, 1976).
The fact that the 0.7 x 10^6 dalton compound was not found in
Acetabularia dentata might be explained by a high intranuclear
activity of the enzyme which converts the 0.7 x 10^6 into the
0.65 x 10^6 dalton compound. The similarity of the processing
scheme between the two species (Fig. 9) might indicate that
there are identical DNA stretches in the two species which are
split by family specific enzymes during processing. Furthermore, because of the enormous interspecies differences in length
of the products which are split off during the processing of the
primary transcription product, one may assume that these products are of no essential importance for the organism.

Another interesting question is whether the species with the
larger or the one with the smaller primary transcription product is the evolutionarily older species, i. e., whether during
evolution there is a lengthening or shortening of the RNA moiety
which is split off during processing. If one compares the length
of splitting products in the lower organism Acetabularia with the
higher organisms of the vertebrates one finds that during evolution a lengthening might occur.

"MORPHOGENETISCHE SUBSTANZEN" AND mRNA

Fifty years ago when Hämmerling postulated the existence of
"morphogenetische Substanzen" (Hämmerling, 1932) he attributed
the following properties to them: They originate in the nucleus;
they are arranged in the cytoplasm in an apical-basal gradient;
they exhibit an enormous stability in the cytoplasm and they are
species-specific. It is hypothesized that the "morphogenetische
Substanzen" are identical with mRNA. Several results are in
agreement with this; however, only indirect evidence is available so far. In order to clarify the identity of the "morphogenetische Substanzen" poly(A) RNA was isolated from Acetabularia. The properties of this poly(A) RNA, as tested so far, are

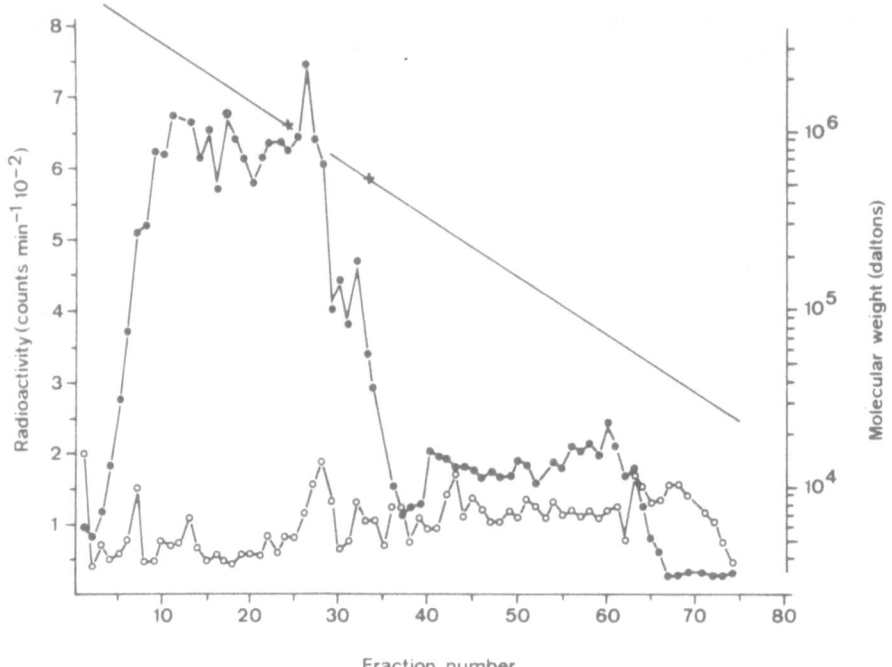

Fig. 10. Synthesis of poly(A) RNA in cells of A. mediterranea (Kloppstech and Schweiger, 1975 a). ●—●—● nucleate cells, o—o—o anucleate cells.

in accordance with the properties which Hämmerling had postulated (Kloppstech and Schweiger, 1975 a; Kloppstech, in preparation). The poly(A) RNA is synthesized in the cell nucleus.
After removal of the nucleus, RNA precursors are no longer
incorporated into the poly(A) RNA (Fig. 10). The poly(A) RNA
is stable in the cytoplasm with a half life of about 10 days or
more depending on the developmental stage. Preliminary experiments indicate that the poly(A) RNA is species-specific. In
a cell-free system from germinating wheat poly(A) RNA from
Acetabularia is capable of inducing protein synthesis. The electrophoretic behavior of the newly synthesized polypeptides is
different from those whose synthesis is induced by globin messenger and tobacco mosaic virus RNA. The poly(A) RNA is
transported through the cell by a mechanism which is different
from that for rRNA. The speed of the transport is quite high.
Preliminary experiments have also shown that there is an apical-

basal gradient for poly(A) RNA as was shown by Hämmerling for the morphogenetic capacity of his "morphogenetische Substanzen".

On the basis of the specific activity of the intracellular nucleoside triphosphate it was possible to roughly estimate the number of nucleotides which are polymerized into poly(A) RNA per second per cell. This rate of polymerization is 10^7 to 10^6 nucleotides per second depending on the stage of development (Kloppstech and Schweiger, in preparation). This number is slightly smaller than that which has been estimated for ribosomal RNA. One has, however, to keep in mind that the rate of polymerization was estimated on the basis of poly(A) RNA which was released into the cytoplasm. So far, it is not possible to estimate the real synthetic capacity of the cell nucleus, for one might expect that at least part of the newly synthesized RNA is subjected to degradation within the cell nucleus (Kloppstech and Schweiger, 1973 a; for a detailed discussion see Harris, 1970).

Under conditions of regeneration, that is after removal of the greater part of the cytoplasm, the synthesis of poly(A) RNA is increased (Kloppstech and Schweiger, in preparation). While the maximum rate of synthesis is found on day 3 for rRNA, the maximum rate for poly(A) RNA is on day 5 to 7. In contrast to the twenty-fold rate increase for rRNA synthesis, the increase in the rate during regeneration is only three- to five-fold for poly(A) RNA. This result indicates that the normal cell produces poly(A) RNA at a high rate with a limited capacity for increasing the rate of synthesis, while the production of rRNA occurs at a lower rate but with a greater capacity for increasing of the synthesis rate during regeneration.

CONCLUDING REMARKS

The experiments which were described give some information about the synthetic capability of the cell nucleus under normal conditions. This information is not only qualitative but to a certain extend also quantitative. The results are a good foundation for further experiments by which a number of important questions may be answered. One of the most important questions which should be considered is whether transcription plays a role in the regulation of gene expression. This question can be explored directly since in Acetabularia we are able to study the behavior of a cell in the absence of the cell nucleus. Under these conditions the cell is capable of performing morphogenesis for 4 to 6 weeks after enucleation without any visible disturbances. This result indicates that the role of transcription in the regulation of gene expression can be only very limited. The following conclusions may be drawn:

1. Morphogenesis can be performed without simultaneous transcription of the nuclear genome.

2. All the ribonucleic acids, including mRNA's, which are necessary for post enucleation morphogenesis can be synthesized and released into the cytoplasm before the nucleus is removed.

3. All the different types of mRNA's which might be involved in morphogenesis can be transcribed at about the same time prior to morphogenesis and the removal of the cell nucleus, instead of in a sequential manner as morphogenesis proceeds.

4. The transcription products must contain in a coded form all the information necessary for the temporal sequence of gene expression.

It would, however, be premature to assume that this limited role of transcription in morphogenesis is unique to Acetabularia.

REFERENCES

Berger, S. and H. G. Schweiger (1975 a) Molec. gen. Genet. 139: 269.

Berger, S. and H. G. Schweiger (1975 b) in "Molecular Biology of Nucleocytoplasmic Relationships", S. Puiseux-Dao, Ed., p. 243. Elsevier Scientific Publishing Company, Amsterdam.

Berger, S. and H. G. Schweiger (1975 c) Planta 127: 49.

Berger, S. and H. G. Schweiger (1975 d) Protoplasma 83: 41.

Brachet, J. and P. Malpoix (1971) in "Adv. Morphog.", Vol. 9, p. 263.

Franke, W. W., S. Berger, H. Falk, H. Spring, U. Scheer, W. Herth, M. F. Trendelenburg and H. G. Schweiger (1974) Protoplasma 82: 249.

Hackett, P. B. and W. Sauerbier (1975) J. Mol. Biol. 91: 235.

Hämmerling, J. (1932) Biol. Z. 52: 42.

Hämmerling, J. (1953) Int. Rev. Cytol. 2: 475.

Hamkalo, B. A. and O. J. Miller jr. (1973) Ann. Rev. Biochem. 42: 379.

Harris, H. (1970) "Nucleus and Cytoplasm", Clarendon Press, Oxford.

Kamiya, N. (1959) in "Protoplasmatologia, Handbuch der Protoplastenforschung", Band VIII, L. V. Heilbrunn and F. Weber, Eds., Springer-Verlag, Wien.

Kloppstech, K., S. Berger and H. G. Schweiger (1974) J. Cell Biol. 63: 173 a.

Kloppstech, K. and H. G. Schweiger (1973 a) Biochim. Biophys. Acta 324: 365.

Kloppstech, K. and H. G. Schweiger (1973 b) Differentiation 1: 331.

Kloppstech, K. and H. G. Schweiger (1975 a) Differentiation 4: 115.

Kloppstech, K. and H. G. Schweiger (1975 b) Protoplasma 83: 27.

Kornberg, A. (1974) "DNA Synthesis", W. H. Freeman and Company, San Francisco.

McKnight, S. L. and O. L. Miller jr. (1976) Cell 8: 305.

Perry, R. P. and D. E. Kelley (1970) J. Cell Physiol. 76: 127.

Puiseux-Dao, S. (1970) "Acetabularia and Cell Biology", Logos Press Ltd., London.

Richter, G., this volume.

Sandakhchiev, L., R. Niemann and H. G. Schweiger (1973) Protoplasma 76: 403.

Schweiger, H. G. (1969) Curr. Top. Microbiol. Immunol. 50: 1.

Schweiger, H. G. (1976) in "Handbook of Genetics", R. C. King, Ed., Vol. 5, p. 451. Plenum Publishing Corporation, New York.

Schweiger, H. G., H. Bannwarth, S. Berger and K. Kloppstech (1975) in "Molecular Biology of Nucleocytoplasmic Relationships", S. Puiseux-Dao, Ed., p. 203. Elsevier Scientific Publishing Company, Amsterdam.

Schweiger, H. G., S. Berger, K. Kloppstech, K. Apel and M. Schweiger (1974) Phycologia 13: 11.

Stich, H. and W. Plaut (1958) J. Biophys. Biochem. Cytology 4: 119.

Tartof, K. D. (1975) Ann. Rev. Genet. 9: 355.

Trendelenburg, M. F., H. Spring, U. Scheer and W. W. Franke (1974) Proc. Nat. Acad. Sci. USA 71: 3626.

Woodcock, C. L. F., J. E. Stanchfield and R. R. Gould (1975) Plant Science Letters 4: 17.

MODE OF SYNTHESIS OF RIBOSOMAL RNA IN VARIOUS PLANT ORGANISMS

G. Richter, R. Misske, R. Stempka, W. Dirks

Institut für Botanik, Technische Universität

D-3000 Hannover 21 (G.F.R.), Herrenhäuserstr. 2

INTRODUCTION

In a number of eukaryotes the pathway of synthesis of ribosomal RNA has been extensively studied and well documented: The two species are transcribed in the nucleolus as components of a single large precursor molecule from which they are released through a succession of intermediates. This holds true for certain animal cells and yeast. In other eukaryotes, especially in higher and lower plants, details of rRNA synthesis are less well understood. A major difficulty consists in the diversity of high molecular RNA molecules mainly observed in those plant cells which dispose of additional systems of rRNA synthesis, i.e. in chloroplasts as well as in mitochondria. One approach to this problem is provided by the selection of a plant organism with features suitable for the preferential investigation of one of these various pathways of rRNA synthesis. The present communication considers details of rRNA synthesis in the nucleolus, in chloroplasts and in mitochondria of eukaryotic plant cells.

CYTOPLASMIC RIBOSOMAL RNA

Gel electrophoresis of total RNA from cell cultures of Petroselinum lacking chloroplasts shows that during a 20 min incubation with $[^3H]$ uridine added to the culture medium six high molecular weight rapidly labelled RNA species are formed (Fig. 1). Judged by their mobility in agarose-polyacrylamide gels the following apparent molecular weights were calculated for these species: $3.0 - 2.9 \times 10^6$, 2.4×10^6, 1.9×10^6, 1.4×10^6, 1.0×10^6 and 0.75×10^6 daltons (for methods, see RICHTER, 1973). Thus they can be clearly distinguished from the two ribosomal RNA (rRNA) species (1.3×10^6 and 0.7×10^6 daltons) which contained only traces of radioactivity. With [3H] adenosine and [3H] orotic acid, respectively, as precursors the radioactivity scan of total RNA after a 30 min labelling period was very similar. Distinct minor peaks with apparent molecular weights from $3.2 - 4.0 \times 10^6$ daltons appear additionally in the radioactivity profile. Label from [3H] methylmethionine as a methyl donor was rapidly incorporated into all six main components when the cells had been previously cultured in a medium lacking sucrose, amino acids and SO_4^{2-} for 12 h. Incubation for $45 - 120$ min with [3H] uridine of the cells lead to an accumulation of radioactivity predominantly in the 1.4×10^6 and 0.75×10^6 daltons RNA as well as in the mature rRNA species.

Further evidence for the functioning of these rapidly labelled RNA species as precursors and intermediates in the synthesis of rRNA came from cell fractionation experiments. Fig. 2a illustrates an electrophoretic analysis of RNA extracted from isolated and partially purified nuclei after labelling the cells with [3H] uridine for 30 min. It yielded six labelled components the molecular weights of which compare well with those found in the total RNA fraction from pulse-labelled cells. However, the pattern differs insofar as practically no radioactivity appeared in the mature rRNA species. After 10 min exposure to [3H] uridine only the $3.0 - 2.9 \times 10^6$ and the 2.4×10^6 daltons RNAs were labelled. Prolongation of the labelling

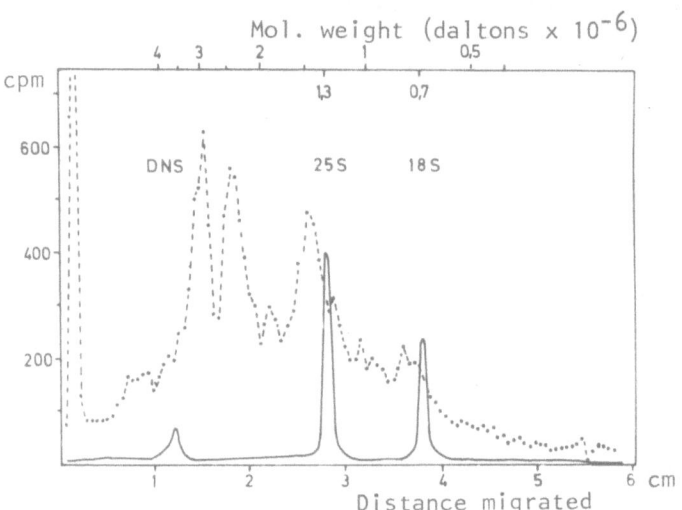

Fig. 1. Electrophoresis of total nucleic acids from Petroselinum cells in 2% agarose-polyacrylamide gel (2.5 h at 5 mA and 4°C; TPE buffer + 0.2% SDS). Labelling with [3H] uridine (2 μCi/ml) for 20 min. ——, Absorbance at 265 nm; ●--●, radioactivity (cpm).

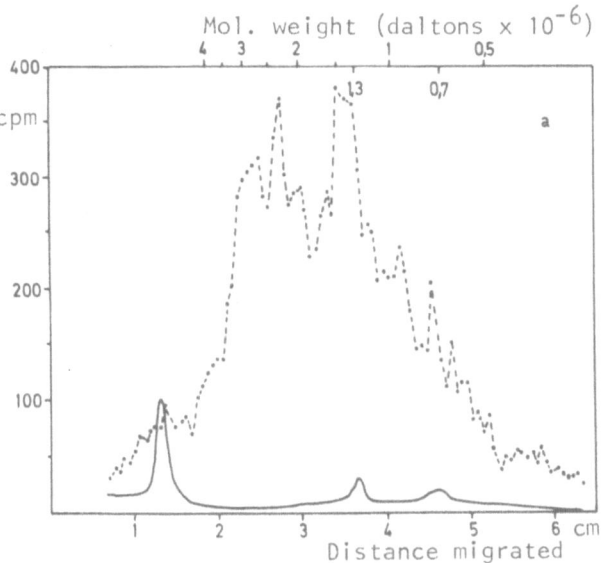

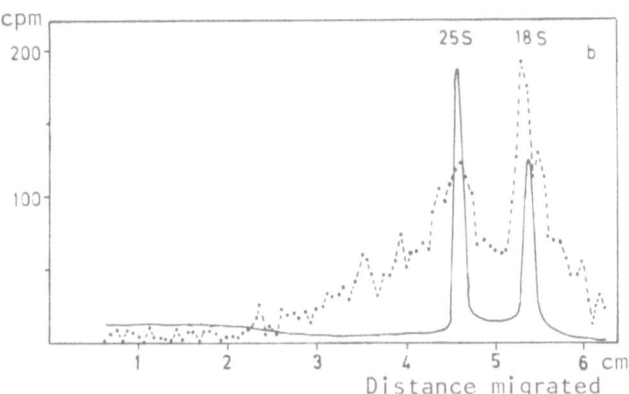

Fig. 2. Gel electrophoresis of the nucleic acids from nuclear fraction (a) and cytoplasmic ribosomes (b) of <u>Petroselinum</u> cells. Labelling with [^{3}H] uridine (2 µCi/ml).
For details of separation see Fig. 1.

time to 60 min did not significantly change the radioactivity profile of the nuclear RNA. On the other hand, after a 30 min pulse
with [3H] uridine the cytoplasmic ribosomes accumulated radioactivity in their 1.3 x 10^6 and 0.7 x 10^6 daltons RNA species (Fig. 2b).
These findings together with the kinetic analyses of the flow of
radioactivity through and into various RNA species are strong evidence that the rapidly synthesized nuclear RNA species are precursors (pre-rRNA) to cytoplasmic rRNA. Those with molecular weights
of 2.4 x 10^6 and 1.0 x 10^6 daltons have counterparts in cultured
sycomore cells (COX a. TURNOCK, 1973) and in several plant tissues
(GRIERSON a. LOENING, 1974). Together with the 0.75 x 10^6 daltons
RNA they evidently serve as precursors of the mature rRNA species.

As far as the larger pre-rRNA species are concerned a true
difference seems to exist in various plant tissues, i.e. in the leaf
and the root (GRIERSON a. LOENING, 1972) since the molecular weights
found range from 2.9 to 2.5 x 10^6 daltons. This apparent heterogenity
in size of the primary transcript of genes for cytoplasmic rRNA poses
a problem as to the pathway of precursor conversion to the 1.3 x 10^6
and 0.7 x 10^6 daltons rRNAs. One possible explanation is that the
precursor molecule transcribed from the gene is cleaved in alternate
ways thus giving rise to intermediates of different molecular sizes.
Support for this type of argument comes from studies on rRNA synthesis in the giant nucleus of the green alga Acetabularia (see below)
and from inhibitor experiments with cycloheximide in Petroselinum
plant cell cultures. [3H] uridine labelling of the latter in the
presence of this drug (5 µg/ml) for 30 to 120 min not only reduces
the incorporation of [3H] into total RNA to about 30% but also brings
about significant changes in the labelling pattern of the nuclear
RNA species (Fig. 3): The predominant radioactivity peak in the high
molecular weight region corresponds to the 2.4 x 10^6 daltons species;
little or no radioactivity is detectable in the region of the 3.0 -
2.9 x 10^6 and 1.9 x 10^6 daltons pre-rRNAs. On the other hand, the
synthesis of the 1.4 x 10^6, 1.0 x 10^6 and 0.75 x 10^6 daltons RNA
species apparently remained largely unimpaired by the cycloheximide
treatment. Although the mode of action of this inhibitor on nucleolar rRNA synthesis is unknown the results obtained indicate that
under these conditions the maturation of rRNA still proceeds with
the 2.4 x 10^6 daltons pre-rRNA as the predominant precursor. Since
the 3.0 - 2.9 x 10^6 and the 2.4 x 10^6 daltons components do not exhibit the same kinetics of accumulation during short labelling
periods the larger component may serve as a direct precursor of the
2.4 x 10^6 daltons pre-rRNA. The absence of the 1.9 x 10^6 daltons RNA
in the radioactivity profile may be easily explained by the fact
that the alternate cleavage of the 3.0 - 2.9 x 10^6 daltons precursor
did not occur which in the control cells normally led to the formation of the rapidly labelled 1.9 x 10^6 daltons intermediate. Although the data do not allow a definitive conclusion as to the pathway of precursor conversion the best fit would be the following
hypothetical scheme.

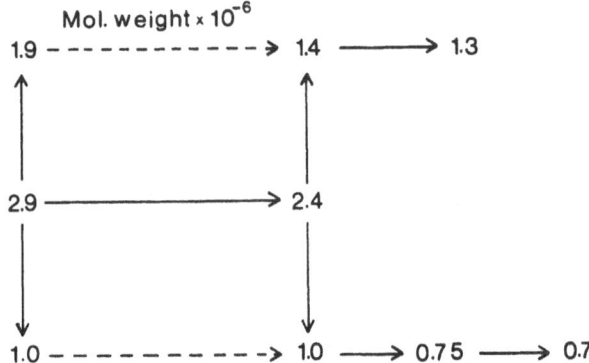

Scheme 1. Proposed steps of processing of the pre-rRNA in _Petro-
selinum_ cells.

    An indication that the two large pre-rRNA species found in
Petroselinum cell cultures are probably not alternative transcription products of the genes for cytoplasmic rRNA comes from studies
on the rRNA synthesis in the primary nucleus of the giant unicellular green alga _Acetabularia mediterranea_. As revealed by previous
electron microscopic studies of spread nucleolar material from this
organism the active rRNA cistrons have an average length of 1.9 μm
which apparently represents the majority of the active regions on
the rDNA (BERGER a. SCHWEIGER, 1975; SPRING et al., 1974). Accor-
dingly, the transcription of pre-rRNA molecules with molecular
weights up to 2.5×10^6 daltons would be compatible with this cistron length. Indeed, gel-electrophoretic separation of rapidly labelled RNA from manually isolated nuclei of young plants (20 - 30 mm)
pulse-labelled for 1 h with $[^3H]$ uridine (5 μCi/ml) like-wise revealed the existence of two major components with molecular weights of
about 2.4×10^6 and 2.2×10^6 daltons (Fig. 4). They most likely
represent the largest stable pre-rRNA species. With the labelling
time increasing (3 - 7 h) the majority of radioactivity appeared in
the 2.2×10^6 daltons pre-rRNA as well as in a 1.45×10^6 daltons
component; in addition other RNA species with molecular weights of
about 1.65×10^6 and 0.75×10^6 daltons form minor radioactivity
peaks. While the latter and the 1.45×10^6 daltons component function as the immediate precursors of the mature rRNA species, the
1.65×10^6 daltons intermediate presumably represents a cleavage
product of the heavy pre-rRNA species. These data together with the
evidence cited above lend support to a preliminary processing scheme
similar to that proposed for _Petroselinum_ insofar as an alternate
cleavage of the larger pre-rRNA transcript is assumed.

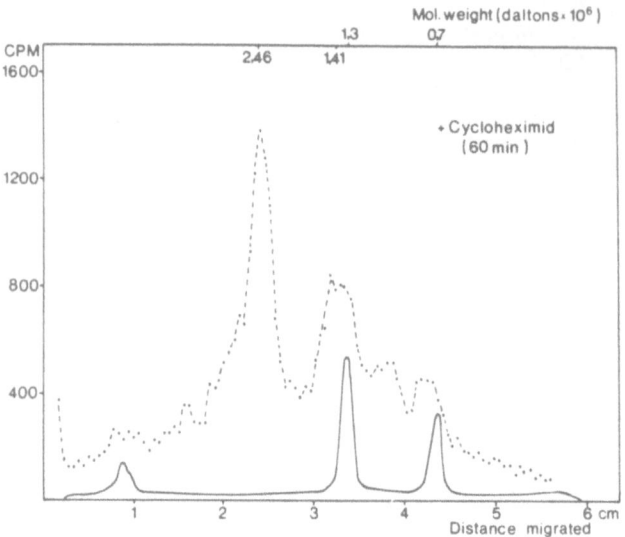

Fig. 3. Effect of cycloheximide (5 µg/ml) on the labelling pattern of total nucleic acids of <u>Petroselinum</u> cells. Labelling for 60 min with [^{3}H] uridine (2 µCi/ml) in the presence of the inhibitor. For details of separation see Fig. 1.

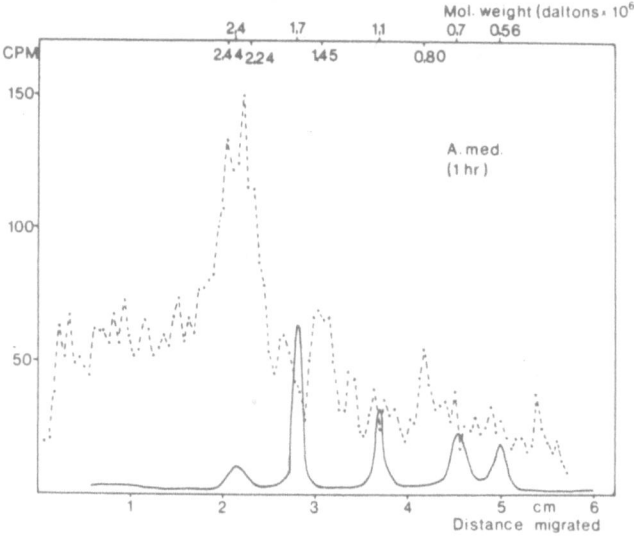

Fig. 4. Gel-electrophoretic separation of total nuclear RNA of <u>Acetabularia mediterranea</u>. 52 nuclei were manually isolated from young plants (20 - 30 mm length) previously incubated with [^{3}H] uridine (5 µCi/ml Erdschreiber medium) for 1 h, and their nucleic acids immediately extracted. For details of separation see Fig. 1.

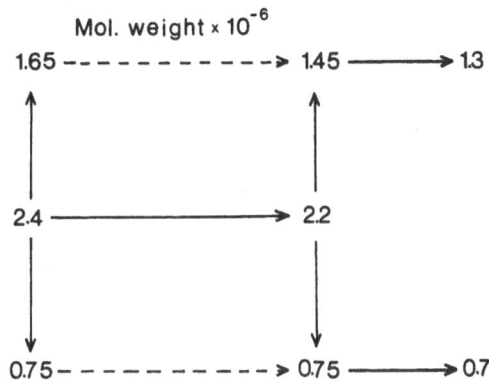

Scheme 2. Proposed steps of processing of the pre-rRNA in _Acetabu-laria_ _mediterranea_.

CHLOROPLAST RIBOSOMAL RNA

        Studies on leaf chloroplasts (HARTLEY a. ELLIS, 1973; MUNSCHE
a. WOLLGIEHN, 1973; GRIERSON a. LOENING, 1974) and on algal chloroplasts of _Chlamydomonas_ (MILLER a. McMAHON, 1974) and _Euglena_
(HEIZMANN, 1974) have revealed that the chloroplast rRNAs (chl-rRNA)
are apparently synthesized from discrete immediate precursors. The
involvement of a high molecular RNA species serving as a common
precursor for both, however, is still a controversial question.

        In an attempt to obtain further information about the mecha-
nism of chl-rRNA synthesis cells of the unicellular thermophilic
alga _Cyanidium_ _caldarium_, a member of the _Rhodophyceae_, as well as
isolated seedling roots exhibiting chloroplast development in blue
light were examined. Pulse-labelling with $[^3H]$ uridine or $[^3H]$ uracil of the _Cyanidium_ cells for 1 h under autotrophic conditions resulted in a number of labelled RNA components when separated by gel
electrophoresis (Fig. 5a). At least four major peaks of radioactivity were present corresponding to molecular weights of about
2.8×10^6, 1.8×10^6, 1.6×10^6 and 0.75×10^6 daltons (STEMPKA,
unpubl. results). In contrast to several minor peaks they had no
counterparts in the optical density scan of cytoplasmic and chloroplast rRNAs; the latters contained relatively little radioactivity.
The incorporation of label into the various RNA components as a
function of the labelling time indicate that the 2.8×10^6 daltons
component is the primary precursor of the cytoplasmic rRNA species
(molecular weights: 1.4×10^6 and 0.69×10^6 daltons) which is

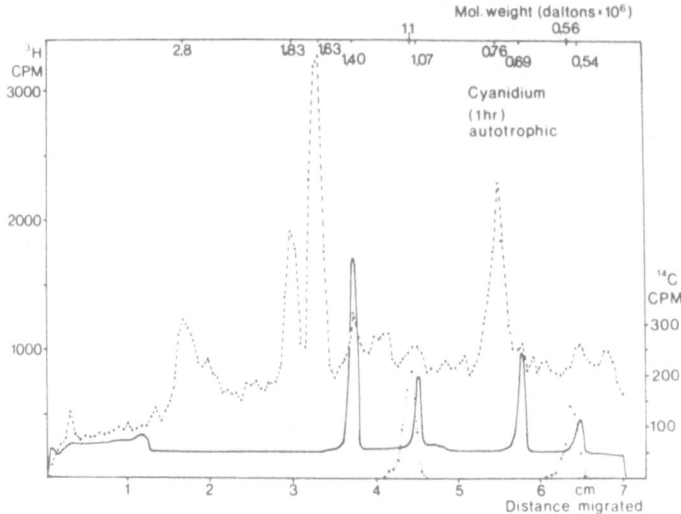

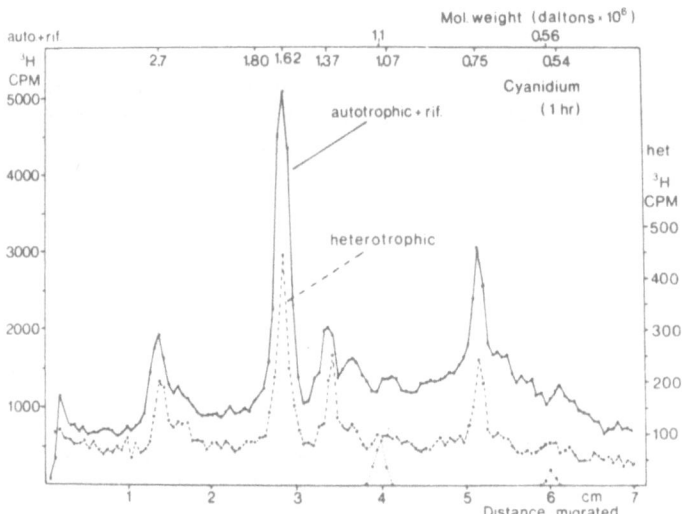

Fig. 5. Electrophoresis of total nucleic acids from cells of Cyanidium caldarium after previously labelling with [3H] uracil (2 µCi/
ml) for 1 h under autotrophic conditions (a) as well as in the presence of rifampicin and under heterotrophic conditions, respectively (b). 2.4% agarose-polyacrylamide gel; for other details see
legend to Fig. 1.

processed to form their immediate precursors with molecular weights
of about 1.6 x 10^6 and 0.8 x 10^6 daltons. The possible role of the
1.8 x 10^6 daltons component is indicated by the results of [^{3}H]label-
ling of autotrophic cells in the presence of rifampicin (20 ug/ml),
and of heterotrophic cells of Cyanidium (Fig. 5b). In both cases no
label appeared in the region of the 1.8 x 10^6 daltons RNA but in all
other typical precursor species. Definite proof of the precursor
function of this RNA component will require other lines of experimentation, e.g. hybridization competition experiments. Preliminary
data from similar studies with the green alga Chlorella pyrenoidosa
appear to support the findings obtained with Cyanidium.

Excised roots of pea seedlings (Pisum sativum va. "Alaska")
which exhibit differentiation of functional chloroplasts from leucoplasts when irradiated with blue light (DIRKS a. RICHTER, 1973)
have the great advantage that they can easily be cultured under
well-defined and sterile conditions. Moreover, the transition from
leucoplasts to chloroplasts in blue light is a relatively slow process which therefore presents a favourable system tò study the
plastid RNA metabolism involved. After short labelling periods
(1 - 5 h) the majority of [^{3}H]radioactivity incorporated into plastid
RNA was present in two components with molecular weights of 1.2 x
10^6 and 0.65 x 10^6 daltons (Fig. 6). Considering also the results of

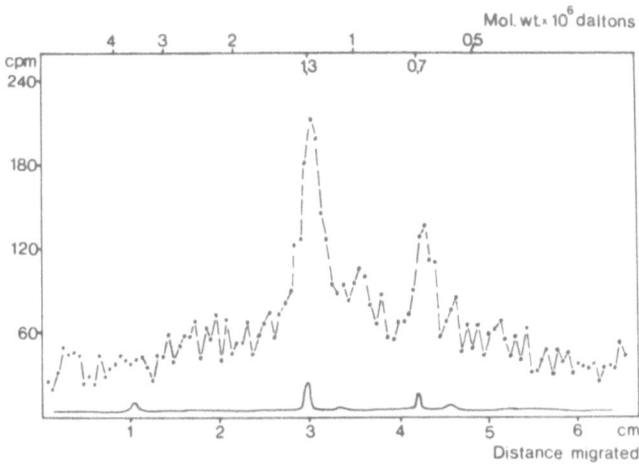

Fig. 6. Electrophoresis of plastid rRNA from isolated seedling roots
of Pisum sativum which were cultured for 11 days in blue light, then
incubated with [^{3}H] uridine (2 μCi/ml) for 5 h. Details of separation as described in the legend to Fig. 1, except: 2.4% agarose-
polyacrylamide gel; 3.75 h running time.

kinetic analyses we are confident that we are observing here the
true immediate precursors of the two chloroplast rRNA species. Under
the experimental conditions applied rapidly labelled high molecular
weight RNA corresponding to a 2.9 - 2.5 x 10^6 daltons chl-rRNA precursor could not be detected in the greening root leucoplasts.

MITOCHONDRIAL RIBOSOMAL RNA

Little is known about rRNA synthesis in mitochondria of higher
plants. By a more refined experimentation we succeeded in isolating
and purifying of mitochondria from Petroselinum plant cell cultures
(MISSKE, 1976). Their typical fine structure as revealed by electron-
microscopy as well as their capacity to perform still specific biochemical reactions (oxidative phosphorylation, cytochrome c oxidore-
ductase activity) indicated their intactness. By incubation with
[^{3}H] uridine of the intact cells for 3 h two rapidly labelled high
molecular weight RNA species appeared among the nucleic acids of the
isolated mitochondria when separated in 2.4% agarose-polyacrylamide

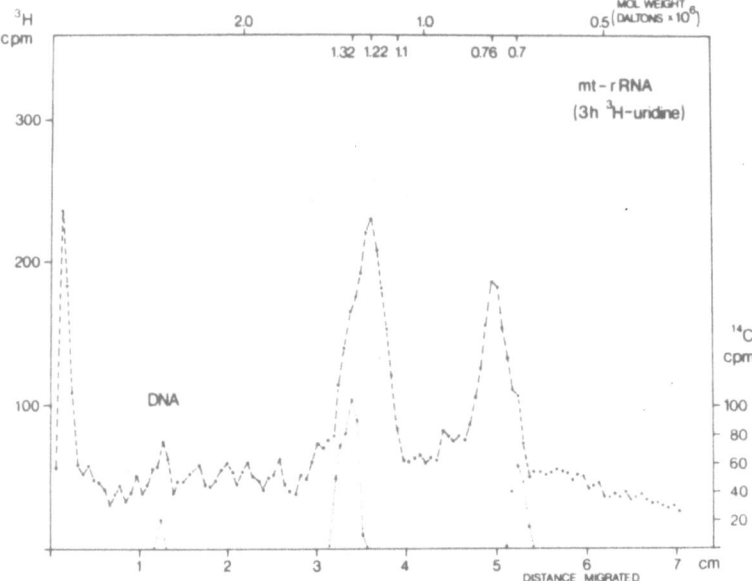

Fig. 7. Electrophoresis of rRNA from isolated mitochondria of Petroselinum cells previously labelled with [^{3}H] uridine (2 μCi/ml) for
3 h. 2.4% agarose-polyacrylamide gel. [^{14}C] labelled cytoplasmic
rRNA of Petroselinum added as marker prior to separation. For other
details see Fig. 1.

gels (Fig. 7). According to their molecular weights of about 1.22×10^6 and 0.76×10^6 daltons they did not correspond to cytoplasmic rRNA species since their radioactivity peaks did not coincide with those of the $[^{14}C]$ labelled cytoplasmic rRNA added as marker prior to separation. They were not identical either with the mature mt-rRNA species because the molecular weights of the latters are different: 1.1×10^6 and 0.73×10^6 daltons for the <u>Petroselinum</u> cells. About 8 h after the onset of labelling the larger 1.22×10^6 daltons component had accumulated more radioactivity (Fig. 8), but concomitantly a shoulder on the light side of its radioactivity peak appeared which coincided with the absorbance peak of the heavy mature mt-rRNA species of 1.1×10^6 daltons. In contrast, relatively less radioactivity was found in the 0.76×10^6 daltons component giving rise to a shoulder on the heavy side of a peak now coinciding with the absorbance peak of the mature light mt-rRNA species. After longer periods of labelling (10 - 12 h) radioactivity became prominently associated with the mature mt-rRNA species. It seems reasonable to conclude that the two rapidly labelled 1.22×10^6 and 0.76×10^6 daltons RNA components represent the immediate precursors of the mature mt-rRNA species. A discrepancy seems to exist between both insofar as the 1.1×10^6 daltons mt-rRNA is obviously processed more slowly than the 0.73×10^6 daltons one; this would account for the converse labelling of precursor and mature product

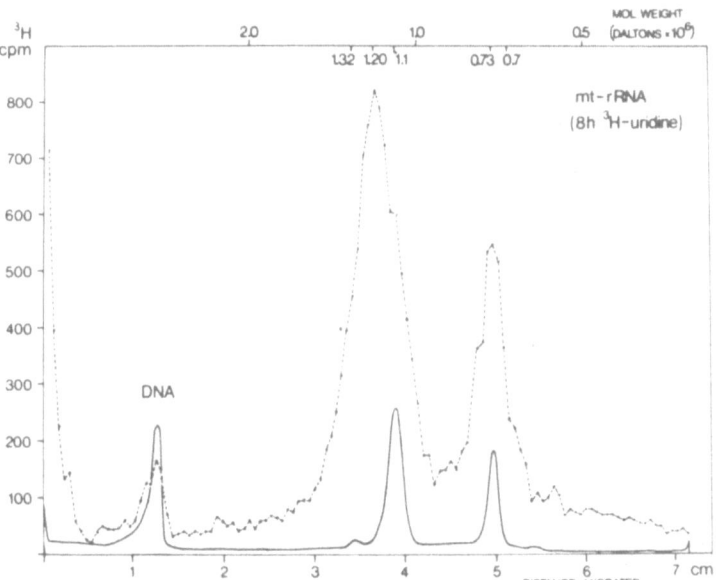

Fig. 8. As described in the legend to Fig. 7, except: incubation of the cells with $[^3H]$ uridine for 8 h.

in each (see Fig. 8). This preferential synthesis of the smaller
rRNA species was also observed in the processing of the cytoplasmic
species in <u>Petroselinum</u> cells (RICHTER, 1973) as well as in <u>Aceta-</u>
<u>bularia</u> (see Fig. 4). No indication for the existence of a common
precursor of both the mt-rRNA species has been found in the mitochondria of the <u>Petroselinum</u> cells up to now.

These studies were supported by the Deutsche Forschungsgemein-
schàft and the Stiftung Volkswagenwerk; to both we wish to express
our gratitude.

REFERENCES

Berger, S., H.G. Schweiger, Protoplasma 83, 41-50 (1975)
Cox, B.J., G.Turnock, Eur. J. Biochem. 37, 367-376 (1973)
Dirks, W., G. Richter, Biochem. Physiol. Pflanzen 168, 157-166 (1975)
Grierson, D., U. Loening, Nat. New Biol. 235, 8o-82 (1972)
Grierson, D., U. Loening, Eur. J. Biochem. 44, 5o1-5o7 (1974)
Hartley, M.R., R.J. Ellis, Biochem. J. 134, 249-262 (1973)
Heizmann, P., Biochem. Biophys. Res. Commun. 56, 112-118 (1974)
Miller, M.J., D. McMahon, Biochim. Biophys. Acta 366, 35-44 (1974)
Misske, R., Dissertation Hannover 1976
Munsche, D., R. Wollgiehn, Biochim. Biophys. Acta 294, 1o6-117 (1973)
Richter, G., Planta (Berl.) 113, 79-96 (1973)
Spring, H., M.F.Trendelenburg, U. Scheer, W.W. Franke, W. Herth,
        Cytobiol. 1o, 1-65 (1974).

tRNAs AND AMINOACYL-tRNA SYNTHETASES IN PLANT CYTOPLASM, CHLORO-

PLASTS AND MITOCHONDRIA

J.H. WEIL, G. BURKARD, P. GUILLEMAUT, G. JEANNIN,
R. MARTIN and A. STEINMETZ
Institut de Biologie Moléculaire et Cellulaire
Université Louis Pasteur
15, rue Descartes, 67000 Strasbourg, France

In a plant cell, protein synthesis takes place in three compartments, namely in the cytoplasm, in the chloroplasts and in the
mitochondria. In many respects, protein synthesis in the organelles
(mitochondria and chloroplasts) resembles protein synthesis in prokaryotic organisms (especially bacteria) rather than that taking
place in the surrounding cytoplasm, as illustrated, for instance,
by the small size of the organellar ribosomes, the formylation of
organellar initiator methionyl-tRNA, or the sensitivity of organellar protein synthesis to certain antibiotics such as chloramphenicol
(1). These observations are in agreement with the theory that the
organelles have an endosymbiotic origin and have evolved from prokaryotic ancestors. As you will see, more arguments in favor of
this theory can be obtained from studies on organellar tRNAs and
aminoacyl-tRNA synthetases.

The first two steps of protein synthesis, namely i) aminoacid
activation and ii) aminoacid attachment to the cognate tRNA (or
tRNA aminoacylation) are summarized in fig. 1. These two steps are
very important for the fidelity of translation because in the later
steps of protein synthesis, the aminoacyl-tRNA is recognized through
its polynucleotide moiety (and especially its anticodon), so that
any mistake in the nature of the aminoacid attached to a tRNA would
result in an error in the protein(s) synthesized, as shown by the
experiments of Chapeville et al.(2)

The correct aminoacylation of a tRNA is catalyzed by an enzyme
appropriately named "activating enzyme" or "aminoacyl-tRNA synthetase". In a bacterial cell, there are 20 different aminoacyl-tRNA
synthetases (one specific for each aminoacid). But there are about

97

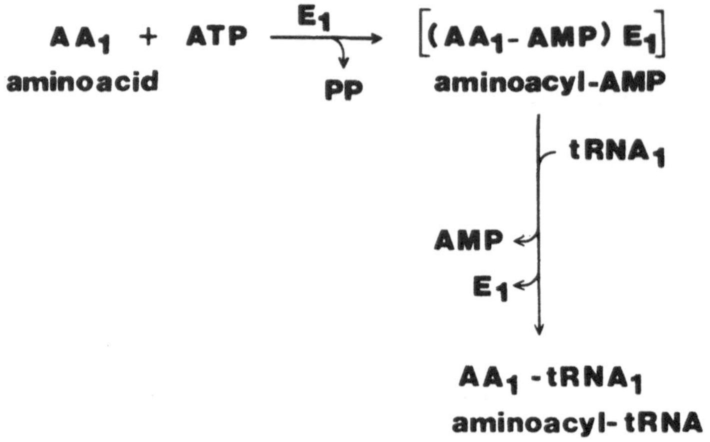

$$\text{AA}_1 + \text{ATP} \xrightarrow[\text{PP}]{\text{E}_1} \left[(\text{AA}_1 - \text{AMP})\, \text{E}_1 \right]$$

aminoacid   aminoacyl-AMP

tRNA$_1$

AMP

E$_1$

AA$_1$ - tRNA$_1$
aminoacyl- tRNA

Fig. 1. The first two steps of protein synthesis : aminoacid activation and aminoacid attachment to the cognate tRNA.

60 different tRNAs (obviously more than one for each aminoacid). The
various tRNAs (in general, 2 or 3) which accept the same aminoacid
are called "isoacceptors"; although they are recognized by the same
aminoacyl-tRNA synthetase, their sequences differ : sometimes only
one or two nucleotides are different (out of a total of about 75-80),
sometimes the structural differences are more important; sometimes
two isoacceptors have the same anticodon, sometimes their anticodons
are different so that they recognize different triplets (coding for
the same aminoacid).

      The situation becomes somewhat more complicated when one con-
siders a eukaryotic cell, because, as shown by the pioneering studies of Barnett et al.(3,4) mitochondria contain tRNAs and aminoacyl-tRNA synthetases which are different from the cytoplasmic counterparts, and the situation becomes even more complicated in a plant
cell, because it contains, in addition, chloroplast tRNAs and aminoacyl-tRNA synthetases, which, as we shall see, can be different
from the cytoplasmic and from the mitochondrial counterparts.

I. SIMILARITIES BETWEEN BACTERIAL AND PLANT ORGANELLAR tRNAs AND AMINOACYL-tRNA SYNTHETASES

1) Cross-Aminoacylation Reactions

The fractionation of cytoplasmic, mitochondrial and chloroplast tRNAs and aminoacyl-tRNA synthetases specific for a number of aminoacids have opened the possibility to perform homologous and heterologous aminoacylation reactions and therefore to compare the recognition of the tRNAs from the three plant cell compartments by homologous and heterologous (including bacterial) aminoacyl-tRNA synthetases.

Such studies were performed for instance on the $tRNA^{Pro}$ and prolyl-tRNA synthetases found in the three compartments of bean Phaseolus vulgaris. Chloroplasts were obtained from lyophilized leaves by centrifugation in a non-aqueous discontinuous gradient(5). Mitochondria were prepared from dark-grown hypocotyls (6). The tRNAs and aminoacyl-tRNA synthetases were obtained from chloroplasts, from mitochondria, or from hypocotyl cytoplasm by methods previously described (5).

A) Fractionation of the aminoacyl-tRNA synthetases from the three cell compartments. The aminoacyl-tRNA synthetases specific for the same aminoacid but located in the three cell-compartments can be separated on the basis of their chromatographic mobilities, as illustrated in the case of bean prolyl-tRNA synthetases.

When a cytoplasmic enzymatic preparation is chromatographed on hydroxyapatite and when cytoplasmic tRNA is used as a substrate to determine the prolyl-tRNA synthetase activity of the fractions, only one peak of activity is observed (peak II, Fig.2a). When chloroplast tRNA is used as a substrate two peaks of activity are observed : one is superimposable on that obtained with cytoplasmic tRNA as a substrate (peak II); the presence of the other one, which is eluted first and is smaller (peak I), suggests the existence in the crude cytoplasmic enzymatic preparation of another prolyl-tRNA synthetase, presumably of organellar origin, because it can aminoacylate the chloroplast tRNA or the E.coli tRNA, but not the cytoplasmic tRNA (Fig. 2a).

Upon hydroxyapatite chromatography of chloroplast or mitochondrial enzymatic preparations two peaks of activity are also obtained when chloroplastic or mitochondrial tRNA is used as substrate (Fig. 2b). But here the peak which elutes first (peak I) is much more important, confirming the organellar origin of this enzyme. The second peak (peak II) which aminoacylates the cytoplasmic tRNA (and

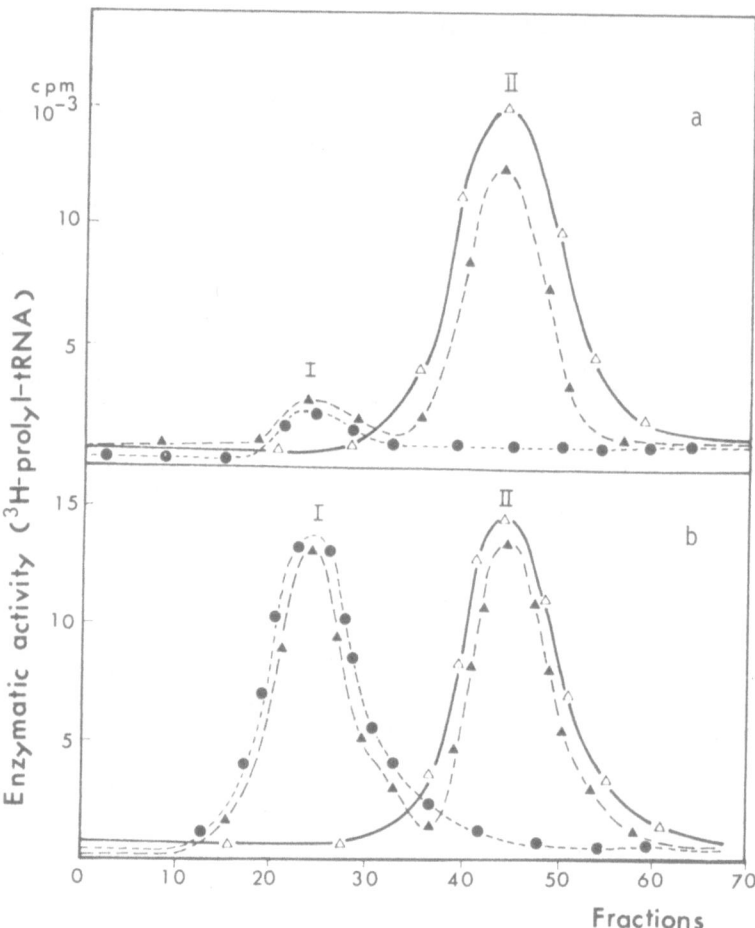

Fig. 2. Fractionation of cytoplasmic, chloroplastic and mitochondrial prolyl-tRNA synthetases on hydroxyapatite. About 100 mg total protein (crude enzymatic preparation) were put on a column (8 x 1 cm) of hydroxyapatite. Elution was performed using a total volume of 2 x 40 ml phosphate gradient (0.005-0.4 M) pH 7.5, 1 ml fractions were collected and assayed for enzymatic activity. (a) Fractionation of a cytoplasmic enzymatic preparation; (b) Fractionation of a chloroplastic or a mitochondrial preparation.△——△,enzyme activity revealed by using total cytoplasmic tRNA;▲---▲,enzyme activity revealed by using total chloroplastic tRNA;●---●,enzyme activity revealed by using E.coli tRNA

also the chloroplast or mitochondrial tRNAs because they contain the cytoplasmic isoaccepting species (see Fig 3b and c)) is also present in the chloroplastic or mitochondrial enzymatic preparations.

It therefore appears that there are at least two prolyl-tRNA synthetases in bean. The first eluting species (peak I) aminoacylates chloroplastic, mitochondrial and bacterial tRNAs and is a major peak in the extracts from both types of organelles (Fig. 2b); small amounts of this enzyme are also present in cytoplasmic extracts (Fig. 2a), probably because it is almost impossible to prevent the lysis of some organelles during the preparation of these extracts.

We have tried to separate chloroplastic peak I from mitochondrial peak I by polyacrylamide gel electrophoresis, but these attempts have not been successful, so it is not possible to decide whether we have the same prolyl-tRNA synthetase in both organelles or if there are two different organellar enzymes.

Identical results were obtained in our laboratory in the case of the methionyl-tRNA synthetases(7), and of the lysyl-tRNA synthetases (8), where organellar (chloroplast and mitochondrial) enzymes could be separated from the cytoplasmic enzyme, and in the case of the leucyl-tRNA synthetases where a chloroplast-specific enzyme could be separated from the cytoplasmic counterpart (9). Other authors, working on other plants, have also been able to characterize chloroplast-specific or mitochondrial-specific aminoacyl-tRNA synthetases (10-18), and in the case of Euglena, it has been shown that the chloroplast-specific isoleucyl-tRNA synthetase is induced by light (10). These organelle-specific enzymes differ from their cytoplasmic counterparts in their intracellular localization, in their chromatographic behavior and in their substrate (tRNA) specificity. But it is not known whether the chloroplast enzyme is coded by the chloroplast DNA and synthesized in the organelle, or if it is coded by a nuclear gene, made on cytoplasmic ribosomes and imported into the chloroplasts (19-21). There is even a possibility that the two enzymes are products of the same (nuclear) gene and that the chloroplast enzyme is modified (for instance by limited proteolysis) upon entering the organelle. Complete purification of a cytoplasmic aminoacyl-tRNA synthetase and of its chloroplast counterpart, followed by comparative studies of the structural and kinetic properties of the two enzymes, should provide an answer to this problem.

B) Aminoacylation of the tRNAs from the three compartments by homologous and heterologous enzymes. Here again we shall concentrate on one example, and study the aminoacylation properties of the tRNAs^{Pro}. The tRNAs from each compartment were aminoacylated with {³H}or{¹⁴C}-proline at pH 7.2, which is the optimal pH for the organellar (peak I) and the cytoplasmic (peak II) enzymes, using a

Mg^{++}/ATP ratio of I.5, in the presence of either the plant cytoplasmic, or the plant organellar, or the bacterial enzymes. The resulting {3H}or {^{14}C}-prolyl-tRNAs were than co-chromatographed on a reverse-phase chromatography (RCP-5) column (22,23), a method particularly suitable for the fractionation of aminoacyl-tRNAs and especially of isoacceptor tRNAs, as its allows, in some instances the separation of two tRNAs which differ only by one nucleotide, or only in the level of post-transcriptional modification (methylation, etc..) undergone during the process of tRNA maturation.

When charged by cytoplasmic peak II prolyl-tRNA synthetase, cytoplasmic tRNAs yield two peaks upon RPC-5 chromatography (Fig. 3a). No charging of cytoplasmic tRNAs occurs when chloroplastic (or mitochondrial) peak I enzyme or E. coli enzyme is used.

When chloroplastic tRNAs are aminoacylated by the chloroplast (or mitochondrial) peak I enzyme, or by E. coli enzyme, one chloroplast-specific prolyl-tRNA peak is observed. When the cytoplasmic peak II enzyme is used, this chloroplast-specific tRNA is not charged, but the two isoaccepting species found in the cytoplasm (which are also present in the chloroplasts) are aminoacylated (Fig. 3b).

In the mitochondria there are two specific prolyl-tRNAs different from the two cytoplasmic and from the chloroplast-specific species. These two mitochondria-specific tRNAs^{Pro} are charged only by the chloroplast (or mitochondrial) peak I enzyme, or by E. coli enzyme, but not by cytoplasmic peak II enzyme; this cytoplasmic enzyme only recognizes the two tRNAs^{Pro} found in the cytoplasm and which are also present in the mitochondria (Fig. 3c).

These aminoacylation experiments show that chloroplast-specific and mitochondria-specific tRNAs^{Pro} can be charged by chloroplastic, mitochondrial or E. coli prolyl-tRNA synthetase, but not by the cytoplasmic enzyme. On the contrary the two cytoplasmic tRNAs^{Pro} can be charged by the cytoplasmic enzyme, but not by the enzymes from the organelles or E. coli.

Cross-aminoacylation reactions between chloroplastic and prokaryotic tRNAs and aminoacyl-tRNA synthetases have also been shown by other studies performed on bean (7-9), cotton (14) or algae (24,25). Mitochondrial aminoacyl-tRNA synthetases in some cases resemble the chloroplastic and bacterial enzymes, as illustrated in the case of bean mitochondrial prolyl-tRNA synthetase (see above), lysyl-tRNA synthetase (8) or methionyl-tRNA synthetase (7). But in other cases, the mitochondrial enzyme is closer to the cytoplasmic one, as shown for instance for bean leucyl-tRNA synthetases (9), tobacco leucyl-tRNA synthetases (18) and Euglena isoleucyl-tRNA synthetases (16).

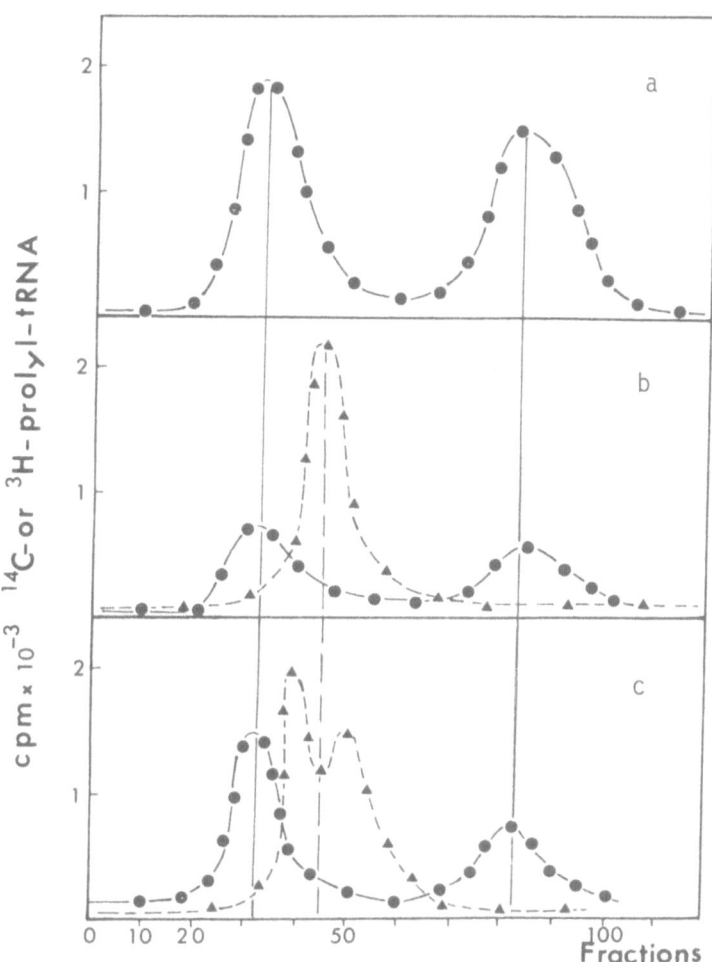

Fig. 3. RPC-5 chromatography of cytoplasmic (a), chloroplastic (b) and mitochondrial (c) prolyl-tRNAs after aminoacylation with homologous or heterologous enzymes.●———●, aminoacylation performed with cytoplasmic peak II prolyl-tRNA synthetase;▲---▲, aminoacylation performed with organellar peak I or E. coli prolyl-tRNA synthetase Na Cl gradient from 0.45 to 0.65 M (2 x 75 ml) pH 4.7 Column size 55 x 0.45 cm. Fractions of 1.25 ml were collected.

In general, these cross-aminoacylation reactions reveal a cross-recognition pattern between plant organellar and bacterial tRNAs and aminoacyl-tRNA synthetases, suggesting a similarity between organellar and bacterial tRNAs. This is only indirect evidence, but that such a similarity exists was confirmed by the discovery of structural analogies between chloroplastic and bacterial tRNAs.

2) Comparison of the Base Composition of Bacterial and Bean Chloroplast tRNA^{Phe}

Until 1976 there was no information available on the structure of chloroplast-specific tRNAs, but recently,two base-composition analyses were published, that of Euglena chloroplast tRNA^{Phe} (26) and that of bean chloroplast tRNA$_1^{Phe}$ (27). Such an analysis can only be performed on a pure tRNA species, and we shall first see how chloroplast-specific tRNA$_1^{Phe}$ from bean was purified.

Upon BD-cellulose chromatography of bean leaf total tRNA, determination of phenylalanine accepting activity in the fractions reveals 3 peaks (Fig. 4). The first two peaks correspond to chloroplast-specific tRNAs^{Phe}, as shown by the following features : i) the two peaks are only revealed using an enzyme from E. coli or from bean chloroplasts, and we know that in many instances (see above) chloroplast-specific tRNAs can be aminoacylated by chloroplast or bacterial enzymes, but not by cytoplasmic enzyme ; ii) peak 1 is predominant in preparations obtained from purified bean chloroplasts ; iii) both peak 1 and peak 2 hybridize with chloroplast DNA, whereas peak 3 does not (see below). Peak 3 is the cytoplasmic tRNA^{Phe} species.

Chloroplastic tRNA^{Phe} peak 1 could be purified by two further chromatographic steps. In the first step, the fractions corresponding to tRNA$_1^{Phe}$ from the previous column (Fig. 4) were chromatographed (without prior aminoacylation) on a BD-cellulose column, using a NaCl-ethanol gradient. Chloroplast tRNA$_1^{Phe}$ was eluted at 0.77 M NaCl, 9% ethanol (Fig. 5a). The fractions showing phenylalanine accepting activity (shown by the arrow on fig. 5a) were pooled; at this stage this chloroplastic tRNA$_1^{Phe}$ can accept about 300 pmol/OD unit. This tRNA$_1^{Phe}$ was aminoacylated as previously described (5),re-extracted and re-chromatographed on BD-cellulose using the same gradient as in the first step. In this second step, the chloroplastic {^{14}C}-phenylalanyl-tRNA$_1^{Phe}$ is eluted at 0.85 M NaCl, 12% ethanol (Fig.5b). The fractions containing more than 1650 pmol of phenylalanyl-tRNA/OD unit were collected, re-extracted and used for analysis. In this way 1 mg of chloroplastic tRNA$_1^{Phe}$ was obtained from 250 mg of bean leaf total tRNA. It was shown to be pure and homogenous by two-dimensional polyacrylamide gel electrophoresis (28).

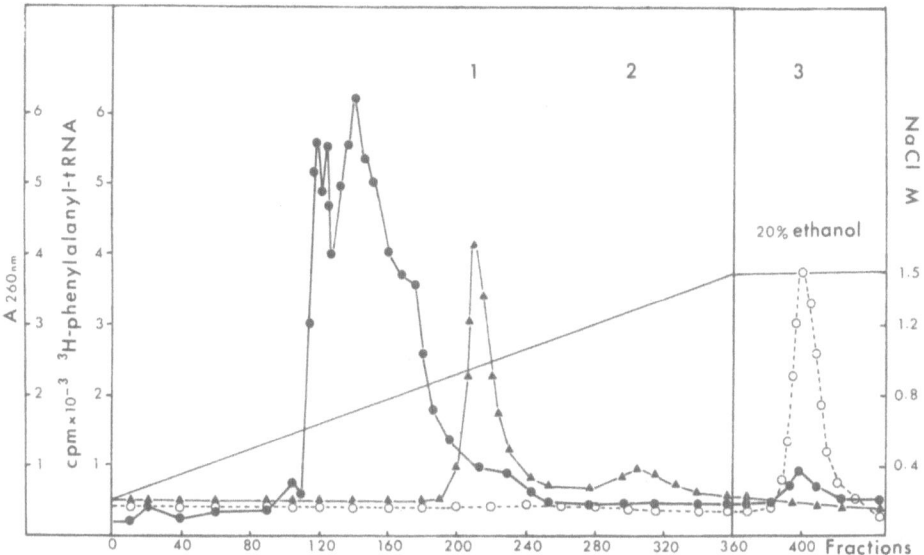

Fig. 4. Fractionation of bean leaf total tRNA on BD-cellulose. Column size : 140 x 1.6 cm. 250 mg tRNA were loaded on the column. Elution with a 2 x 2 l gradient of NaCl 0.2→1.5 M,Tris-HCl 0.01 M pH 7.5, Mg++ 0.01 M, then by 1 l of 20% ethanol in 1.5 M NaCl. Fractions of 11 ml were collected. (●———●)A260nm; (▲———▲) {3H}phenylalanine accepting activity revealed using a chloroplast or E. coli enzyme; (O--O) {3H}phenylalanine accepting activity revealed using a cytoplasmic enzyme.

        After enzymatic hydrolysis of 5 OD units of this purified chloroplastic tRNA$^{Phe}_1$, the resulting nucleosides were fractionated by two-dimensional thin-layer chromatography as shown on fig. 6. The nucleosides were characterized by their position on the chromatogram by comparison with the map of Rogg et al. (29); they were eluted with water and identified by their spectra in 0.1 N HCl and in 0.1 N KOH, which allowed quantitative determination. Results are given in table I, where the nucleoside composition of E. coli and of wheat germ (cytoplasmic) tRNAs^{Phe} have been listed for comparison.

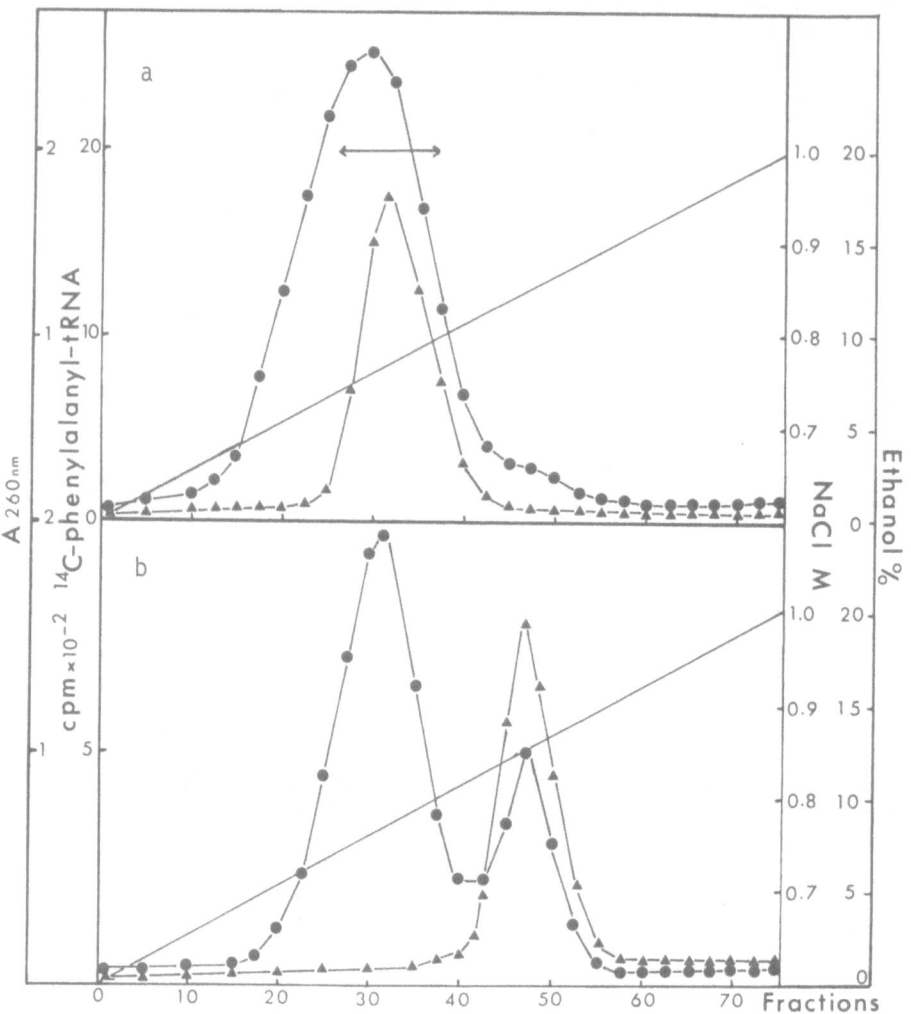

Fig. 5. Purification of chloroplastic tRNA$_1^{Phe}$ on BD-cellulose. (a)
Fractions corresponding to peak 1 on fig. 4 were pooled, precipitated
with ethanol, re-extracted with phenol and loaded on a 40 x 1 cm column without prior aminoacylation. Elution with 2 x 150 ml gradient
of NaCl 0.6→1.0 M and ethanol 0→20% in sodium acetate 0.01 M pH 4.5,
Mg^{++} 0.01 M. Fractions of 4 ml were collected. (●—●)A$_{260nm}$;(▲—▲)
{^{14}C} phenylalanine accepting activity revealed using a E. coli
enzyme. (b) Fractions corresponding to the peak (marked with an arrow)
on fig. 2a were pooled, precipitated, re-extracted, aminoacylated with
{^{14}C} phenylalanine (12,000 cpm/nmole) using a E. coli enzyme, re-
extracted and loaded on a 40 x 1 cm column. Elution with the same
gradient as in fig. 2a. Fractions of 4 ml were collected. (●—●)A$_{260nm}$;
(▲—▲) Radioactivity due to {^{14}C}phenylalanyl-tRNA measured on a 25 μl
aliquot.

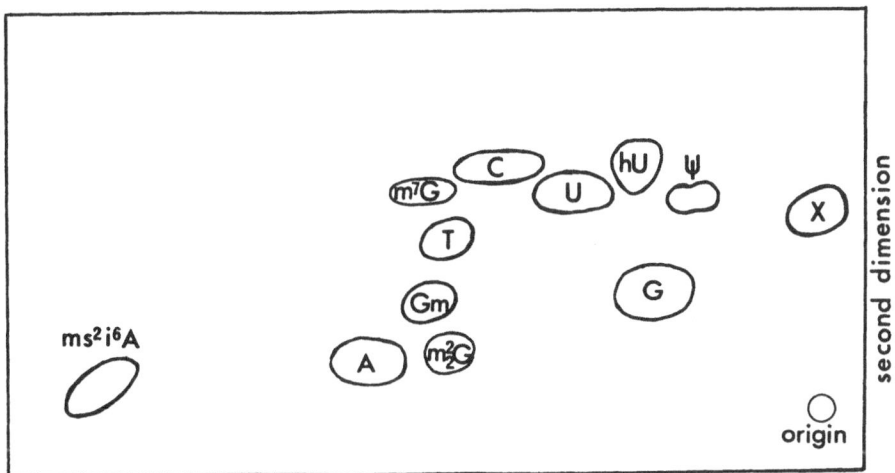

first dimension

Fig. 6. Thin-layer chromtography of the enzymatic hydrolysate of 5 OD units pure chloroplastic tRNA$_1^{Phe}$ on a 50 x 20 cm cellulose plate (DC-Alurolle Cellulose, Merck, Darmstadt). First dimension : n-butanol-isobutyric acid-25% ammonia-water (150 : 75 : 5 : 50, by vol). Second dimension : satured ammonium sulfate-0.1 M sodium acetate pH 6, -isopropyl alcohol (79 : 19 : 2, by vol). X = acp^{3}U = 3-(3-amino-3-carboxypropyl) uridine.

Chloroplast tRNA$_1^{Phe}$ resembles <u>E. coli</u> tRNA^{Phe} in that they have the same two hypermodified nucleosides, namely ms^{2}i^{6}A and acp^{3}U (sometimes still called X), and in that they both lack the nucleoside Y which is present in cytoplasmic tRNAs^{Phe}(the structures of these three hypermodified nucleosides are shown on fig.7). But chloroplast tRNA$_1^{Phe}$ differs from <u>E. coli</u> tRNA^{Phe} in that it contains no s^{4}U (present in <u>E. coli</u> tRNA^{Phe}) and does contain m$_2^2$G and Gm (absent in <u>E. coli</u> tRNA^{Phe}). On the other hand, chloroplast tRNA^{Phe} resembles wheat germ cytoplasmic tRNA^{Phe} which also contains one m^{7}G, one m$_2^2$G and one Gm, but differs in that it has no Y, no m^{1}A and no m2G. It could be argued that chloroplast tRNA$_1^{Phe}$ resembles prokaryotic tRNA^{Phe} by some features and plant cytoplasmic tRNA^{Phe} by other features, but if the hypermodified nucleosides (which are highly characteristic features) are taken as criteria, bean chloroplast tRNA$_1^{Phe}$ and <u>E. coli</u> tRNA^{Phe} appear quite similar. A better comparison of chloroplast and bacterial tRNAs^{Phe} will be possible, when the nucleotide sequence of <u>Euglena</u> chloroplast tRNA^{Phe}, which has been determined by Barnett, Rajbhandary and their co-workers, will be published.

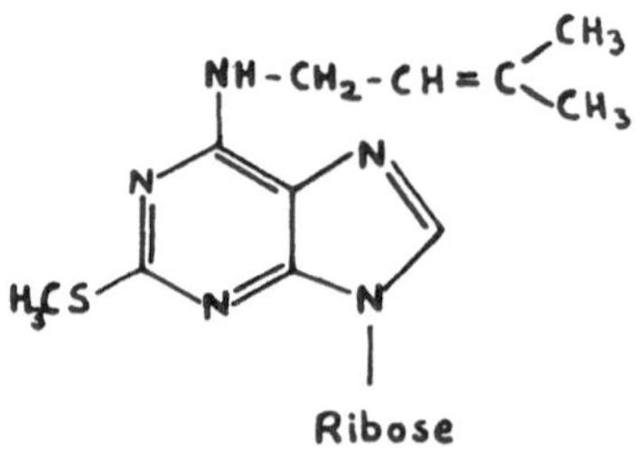

N^6-(Δ^2-Isopentenyl)-2-methylthioadenosine. (ms^{2}i^{6}A)

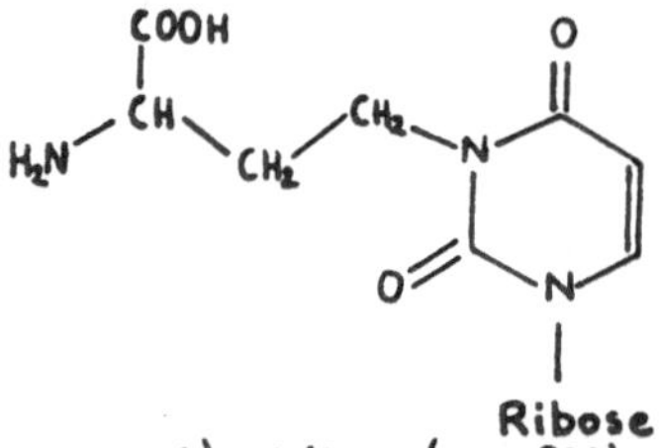

3-(3-Amino-3-carboxypropyl)uridine.(acp^{3}U)

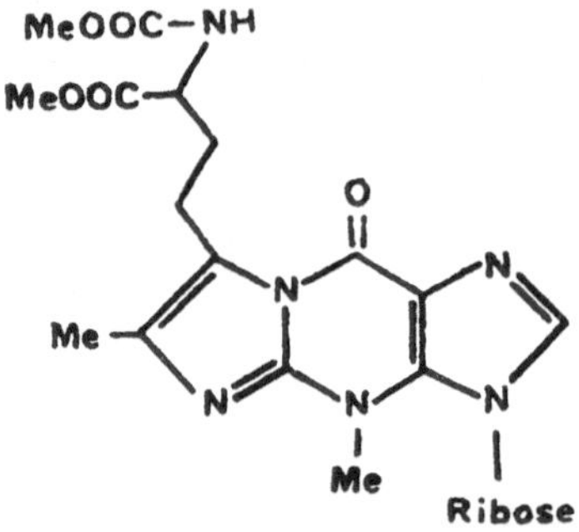

Nucleoside "Y"

Fig. 7. Structures of the three hypermodified nucleosides found in plant cytoplasmic, plant chloroplast or E. coli tRNA^{Phe} (see table I).

Table 1

Nucleoside composition of chloroplastic $tRNA^{Phe}_1$ as compared to that of E. coli and of wheat germ (cytoplasmic) $tRNAs^{Phe}$

	E. coli $tRNA^{Phe}$(30)	Bean chloroplastic $tRNA^{Phe}_1$	Wheat germ (cyto-plasmic)$tRNA^{Phe}$ (31)
A	14	17	16
U	8	10 or 11	7
C	21	19	19
G	23	20	20
ms^2i^6A	1	1	-
m^1A	-	-	1
T	1	1	1
ψ	3	2	3
hU	2	2	3
acp^3U	1	1	-
s^4U	1	-	-
Cm	-	-	1
m^2G	-	-	1
m^2_2G	-	1	1
m^7G	1	1	1
Gm	-	1 or 2	1
Y	-	-	1
	76	76 (?)	76

The results are expressed as moles of nucleosides/mole tRNA. In the case of chloroplast $tRNA^{Phe}$ the calculations were made with the assumption that this tRNA is also 76 nucleosides long and the values were rounded off to the closest unit.

II. ORIGIN OT THE CHLOROPLAST-SPECIFIC tRNAs

    The existence of chloroplast-specific tRNA species, different
from their cytoplasmic counterparts, raises the problem of their
origin : are these chloroplastic tRNAs coded for by chloroplast DNA
or by nuclear genes ? This question can be answered by DNA-tRNA hybridization studies, as performed by several authors, either using
total chloroplast tRNA aminoacylated with one (32,33) or with a
mixture of aminoacids (34), or using fractionated aminoacyl-tRNA
isoacceptors (35,36). We have studied the hybridization of fractionated bean chloroplast leucyl-and phenylalanyl-tRNAs, in order to
determine which DNA(nuclear or chloroplastic) is coding for these
tRNAs and to see whether chloroplast isoacceptor tRNAs are coded
for by the same or by different genes.

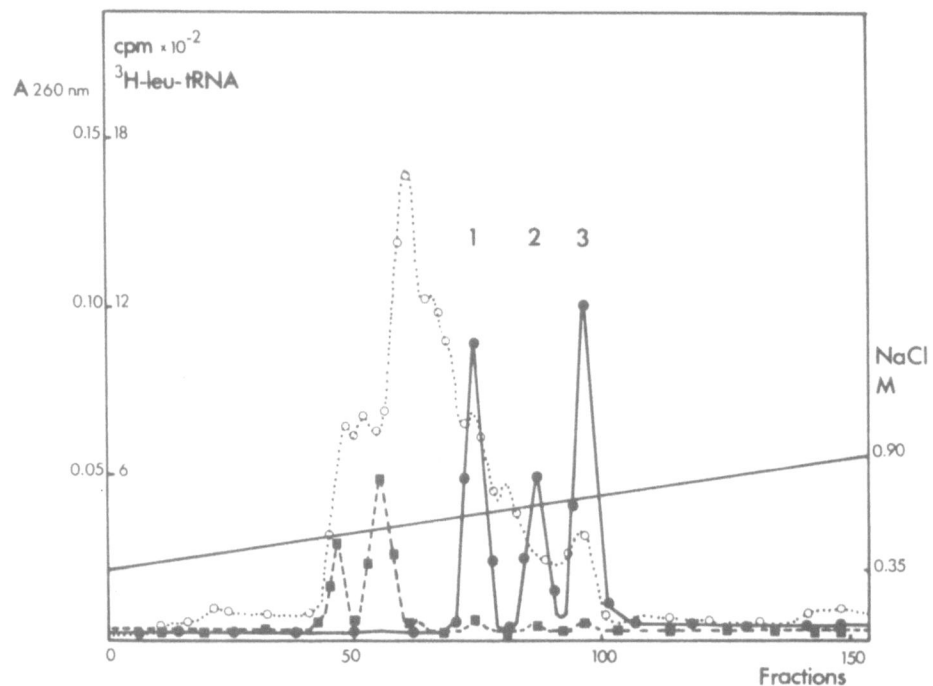

Fig. 8. Fractionation of bean leaves tRNAs^{Leu} by RPC-5 chromatography. Column size : 60 x 0.6 cm. NaCl gradient from 0.35 to 0.9 M
(2 x 85 ml) in Tris-HCl 0.05 M pH 7.4, Mg Cl$_2$ 0.01 M. Fractions of
1 ml were collected. o········o A$_{260nm}$; ■---■ {^{3}H}leucyl-tRNAs charged
by the cytoplasmic enzyme ; ●——● {^{14}C}leucyl-tRNAs charged by the
E. coli enzyme (these are the 3 chloroplast-specific tRNAs^{Leu}).

The hybridization of a population of tRNAs can be studied after labelling them in-vivo with ^{32}P or in vitro with ^{125}I, but the hybridization of a specific tRNA can only be studied after complete purification of that tRNA species or after labelling it specifically with the corresponding (radioactive) aminoacid. We have studied the hybridization of {^{3}H}-aminoacyl-tRNA to DNA under conditions selected to prevent deacylation and loss of the label (see below).

Bean chloroplast $tRNA^{Phe}_1$ and $tRNA^{Phe}_2$ were obtained as described above, while chloroplast $tRNA^{Leu}_1$, $tRNA^{Leu}_2$ and $tRNA^{Leu}_3$ were fractionated by reverse phase chromatography using the RPC-5 system at pH 7.4. As shown on fig. 8, the three chloroplast-specific $tRNAs^{Leu}$ are eluted later than the cytoplasmic $tRNAs^{Leu}$ and are separated from each other.

These fractionated $tRNAs^{Phe}$ and $tRNAs^{Leu}_3$ were then aminoacylated using high specific radioactivity {^{3}H}-phenylalanine or leucine (about 30 Ci/mmole).

Chloroplast DNA was prepared according to Kolodner and Tewari (37) except that after phenol extraction (at pH 8) the aqueous phase was dialyzed, brought to 0.18 M phosphate buffer pH 6.8 and passed through a hydroxyapatite column (38). Elution was performed with 0.4 M phosphate buffer pH 6.8 and the solution was dialyzed prior to denaturation of DNA (39).

The denatured DNA was immobilized (40) on Sartorius nitrocellulose filters (0.45 µ) and the hybridization with{^{3}H}-aminoacyl-tRNA was performed at 40° for 2 h.(unless otherwise stated) in the presence of 8 M urea (41) at pH 5 in order to prevent deacylation. The diameter of the filters was 6 mm (they contained up to 10-15 µg DNA) and the volume of the incubation was 50 µl. After incubation, the filters were washed twice with 10 ml 2 x SSC for 10 min., treated with 10 ml of a T_1 ribonuclease solution (15 units/ml) for 1 h, washed twice with 10 ml 2 x SSC, dried and counted in a solution of omnifluor in toluene (4g/l). Blanks(filters without DNA) were substracted.

Total tRNA from bean chloroplasts, or cytoplasm or from E. coli was aminoacylated with high specific radioactivity {^{3}H}-leucine or phenylalanine using homologous enzyme, reisolated by running the incubation mixture through a small RPC-5 column, and hybridization with chloroplast or E. coli DNA was measured as a function of time. Chloroplast leucyl- and phenylalanyl-tRNAs hybridize to chloroplast DNA and, to a much lesser extent to E. coli DNA (Fig. 9a, d). The reverse situation is true for E. coli leucyl- and phenylalanyl-tRNAs which hybridize preferentially to E. coli DNA (Fig. 9b, e). Cytoplasmic leucyl-and phenylalanyl-tRNAs do not hybridize to chloroplast or to E. coli DNA (Fig. 9c, f). It should pointed out that,

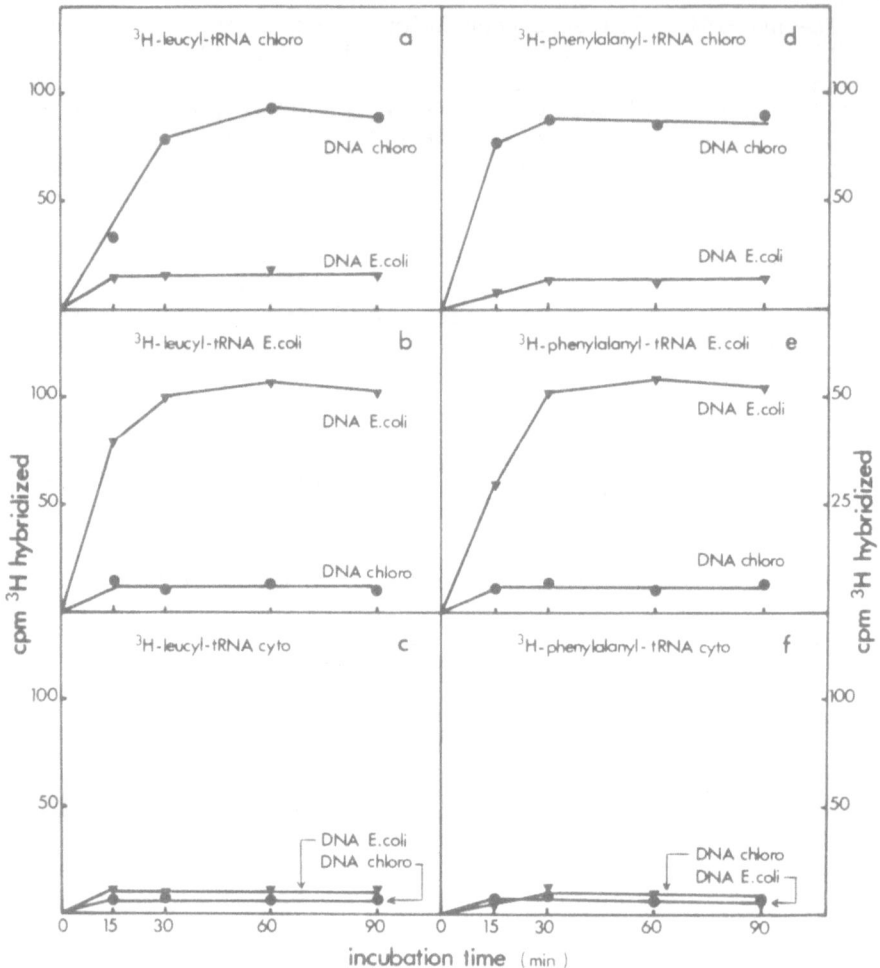

Fig. 9. Hybridization of {^{3}H}-leucyl- or {^{3}H}-phenylalanyl-tRNAs from bean chloroplasts or cytoplasm or from E. coli to chloroplast or E. coli DNA. 8 μg chloroplast or E. coli DNA was immobilized on each filter, and the input radioactivity was between 15,000 and 20,000 CPM.

when chloroplast tRNAs are aminoacylated with leucine or phenylala-nine using the homologous (chloroplast) aminoacyl-tRNA synthetase, only the chloroplast-specific isoacceptors are charged while the cytoplasmic tRNA species remain uncharged (9,27), so that the re-sults of fig. 9 suggest that chloroplast-specific tRNAs^{Leu} and tRNAs^{Phe}, in contrast to their cytoplasmic counterparts, are coded for by chloroplast DNA.

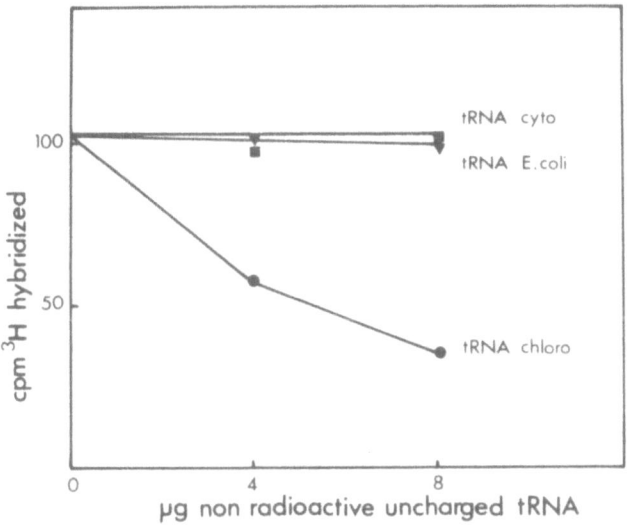

Fig. 10. Hybridization of chloroplast {^{3}H}-leucyl-tRNA to chloroplasts DNA : competition observed with increasing amounts of non-radioactive uncharged cytoplasmic, chloroplast or E. coli tRNA. 8 µg chloroplast DNA was immobilized on each filter and the input radioactivity was 20,000 CPM.

This is condirmed by the fact that uncharged cytoplasmic tRNA is not able to compete with chloroplast {^{3}H}-leucyl-tRNA in hybridization experiments with chloroplast DNA, whereas uncharged chloroplast tRNA does compete (Fig. 10).

Uncharged E. coli tRNA is not a competitor in the hybridization between chloroplast DNA and chloroplast {^{3}H}-leucyl-tRNA (Fig. 10). Reciprocally uncharged chloroplast tRNA does not compete with E. coli {^{3}H}-leucyl-tRNA in the hybridization to E. coli DNA (Fig. 11).

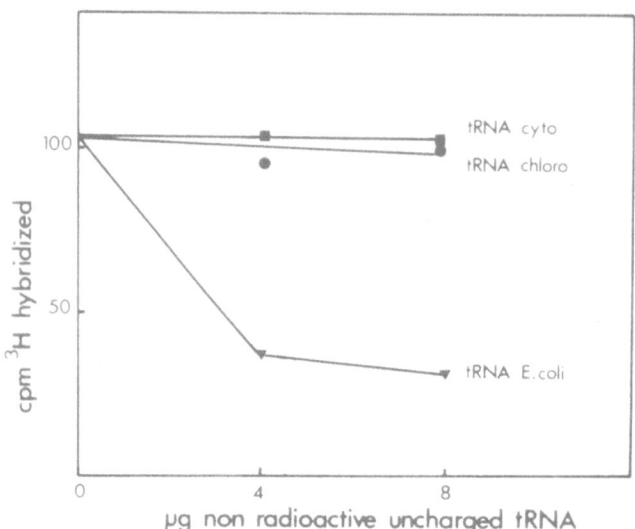

Fig. 11. Hybridization of E. coli {³H}-leucyl-tRNA to E. coli DNA :
Competition observed with increasing amounts of non-radioactive uncharged cytoplasmic, chloroplast or E. coli tRNA. 8 μg chloroplast
DNA was immobilized on each filter and the input radioactivity was
18,000 CPM.

    The hybridization experiments performed with total chloroplast
tRNAs, charged with leucine or phenylalanine (fig. 9), were confirmed using the fractionated chloroplast-specific tRNA species,
namely 3 isoacceptors in the case of tRNAs^{Leu} (9) and 2 in the case
of tRNAs^{Phe} (27). Chloroplast specific tRNA$_1^{Leu}$ (fig. 12a, b),
tRNA$_2^{Leu}$ (fig. 12a, c), tRNA$_3^{Leu}$ (fig. 12b, c), tRNA$_1^{Phe}$ (fig. 12d)
and tRNA$_2^{Phe}$ (fig. 12d) hybridize with chloroplast DNA.

    In order to determine whether the chloroplast-specific iso-
acceptors are coded for by the same or by different gene(s), experiments were performed to check if the hybridizations of the
isoacceptors are additive. As shown on fig. 12a, b, c, all three
combinations of two chloroplast-specific tRNAs^{Leu} show no additivity

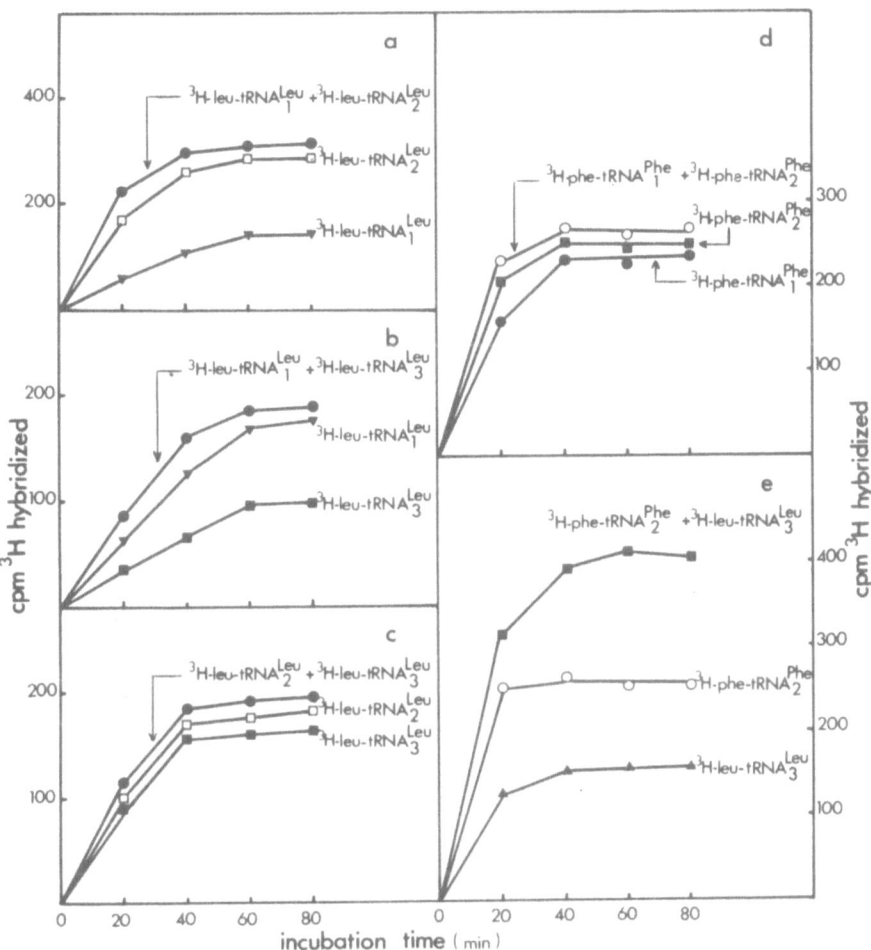

Fig. 12. Hybridization of chloroplast {^{3}H}-leucyl- and {^{3}H}-phe-nylalanyl-tRNA isoacceptors to chloroplast DNA : comparison of the hybridization values obtained with a single isoacceptor and with a mixture of two isoacceptors (either specific for the same aminoacid or for two different aminoacids). 8 μg chloroplast DNA was immobi-lized on each filter and the input radioactivity was between 15,000 and 20,000 CPM experiments with a single isoacceptor, and between 30,000 and 40,000 CPM in experiments where two isoacceptors were mixed.

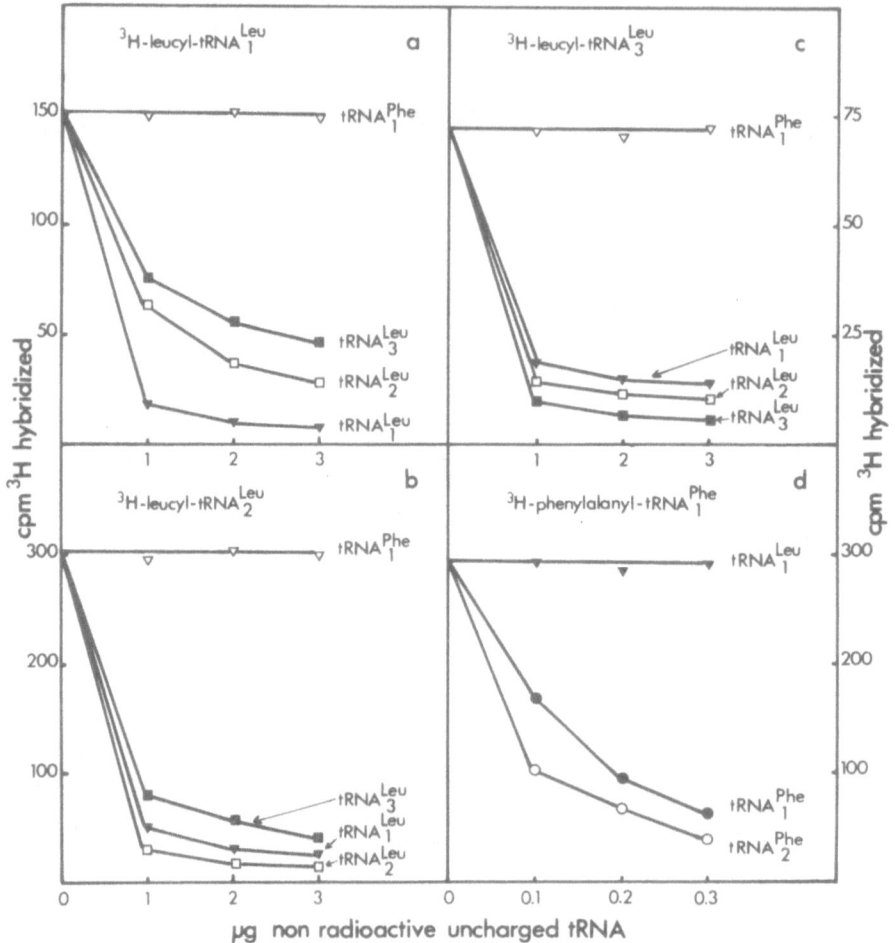

Fig. 13. Hybridization of chloroplast {³H}-leucyl- and {³H}-phe-
nylalanyl-tRNA isoacceptors to chloroplast DNA : Competition observed upon addition of increasing amounts of non-radioactive uncharged isoacceptors (either specific for the same aminoacid, or
specific for a different aminoacid). 8 µg chloroplast DNA was immobilized on each filter and the input radioactivity was between
15,000 and 20,000 CPM.

and the combination of the two chloroplast tRNAs^Phe does not show
any additivity either (fig. 12d), whereas control experiments
where one chloroplast tRNA^Leu and one chloroplast tRNA^Phe were mixed
showed the expected additivity (fig. 12e). These results suggest
that the 3 chloroplast tRNAs^Leu are coded by the same gene(s), and
that this is also the case for the 2 chloroplast tRNAs^Phe.

This is confirmed by experiments where the competition between
the chloroplast isoacceptors was determined in the hybridization
with chloroplast DNA. In these experiments increasing amounts of
non-radioactive uncharged tRNAs were added to each {^{3}H}-leucyl or
{^{3}H}-phenylalanyl-tRNA isoacceptor in the hybridization mixture.
The hybridization of each of the three chloroplast tRNAs^{Leu} is
decreased upon addition of the same isoacceptor or of any of the
other two isoacceptors, but not upon addition of a tRNA^{Phe} isoacceptor (Fig. 13a, b, c). Conversely, competition occurs when uncharged tRNA$^{Phe}_2$ is added to {^{3}H}-phenylalanyl-tRNA$^{Phe}_1$, but not upon
addition of a tRNA^{Leu} (fig. 13d).

The fact that the hybridization of the three chloroplast
tRNA^{Leu} isoacceptors or of the two chloroplast tRNA^{Phe} isoacceptors
shows no additivity, and that the chloroplast isoacceptors compete
with each other for the same sites on chloroplast DNA, suggests that
these isoacceptors are coded for either by the same gene(s) and
differ only in the extent of post-transcriptional modification,
or by very similar genes which may have evolved from one gene but
have undergone different mutations during evolution. It is known
that reverse-phase chromatography is able to separate tRNAs which
differ very little (even by one modified nucleotide only), so that
the isoacceptor chloroplast tRNAs which we have fractionated may
indeed be transcribed from the same gene(s). In a recent study of
the hybridization of Euglena chloroplast total tRNAs (labelled with
^{125}I) with Euglena chloroplast DNA, Barnett et al. were able to estimate that there are about 26 tRNA cistrons on Euglena chloroplast
DNA (42), and Tewari et al. found that there are between 30 and 40
tRNA cistrons on pea chloroplast DNA (43).

Our hybridization experiments (44) show that chloroplast leucyl-
and phenylalanyl-tRNAs do not hybridize appreciably with E. coli DNA
(Fig. 9a, d) and that E. coli leucyl- and phenylalanyl-tRNAs do not
hybridize with chloroplast DNA (Fig. 9b, e). Furthermore E. coli
tRNA is not a competitor in the hybridization of chloroplast leucyl-
tRNA to chloroplast DNA (fig.10), and chloroplast tRNA does not
compete with E. coli leucyl-tRNA in the hybridization to E. coli
DNA (Fig. 11). This is not in contradiction with the similarities
observed between chloroplast and E. coli tRNAs when studying cross-
aminoacylation reactions or base composition : some structural features could be identical, and allow cross-aminoacylation, but might
not be enough to allow appreciable cross-hybridization. The lack
of cross-hybridization does not rule out the possibility of an endosymbiotic origin of the chloroplasts, as the ancestor could have
been a photosynthetic prokaryote rather than a bacterium, and as
the tRNA structural genes may have been modified during evolution
so as to no longer allow appreciable cross-hybridization.

III. SIGNIFICANCE OF ORGANELLE-SPECIFIC tRNAs
AND AMINOACYL-tRNA SYNTHETASES

The existence of organelle-specific tRNAs, coded by organellar DNA, and which can only be aminaocylated by the cognate organellar (or bacterial) aminoacyl-tRNA synthetase, is now well documented. But why are organelle-specific tRNAs necessary ? Could not the cytoplasmic tRNAs species be imported into the organelles and allow organellar protein synthesis ? The presence of organelle-specific tRNAs is perhaps required to recognize certain codons which cannot be recognized by cytoplasmic tRNAs; this hypothesis is presently being tested in our laboratory : we are studying the binding of the two cytoplasmic and of the three chloroplast leucyl-tRNAs (see fig.8) to the six leucine codewords in the presence of ribosomes. But it is also possible that chloroplast-specific tRNAs are required for efficient protein synthesis within the organelle, because these tRNAs (and not the cytoplasmic ones) specifically interact with some constituents of the chloroplast protein synthesis machinery; to test this hypothesis we would need a cell-free protein synthesizing system from chloroplast which is tRNA-dependant, so as to allow a comparative study of the efficiency of chloroplast tRNAs and cytoplasmic tRNAs in the synthesis of a typical chloroplast protein made on chloroplast ribosomes.

But instead of asking the question why are there organelle-specific tRNAs and organelle-specific aminaocyl-tRNA synthetases, one could ask a more general question : why is there an organelle-specific protein synthesizing machinery ? And in view of the fact that some important organelle-specific proteins are coded by nuclear DNA and made on cytoplasmic ribosomes, why are some organellar proteins coded by organellar DNA and made on organellar ribosomes ?

Some of the polypeptides which are manufactured in the organelles are parts of organelle-specific enzymes (cytochrome oxydase or ATPase in the mitochondria, ribulose 1-5 diphosphate carboxylase in the chloroplasts) which must combine with other polypeptide chains, made in the cytoplasm, to form active enzymes. This of course raises the problem of the relationships between the nucleus, the cytoplasm and the organelles, a problem which will be discussed during the next few days.

REFERENCES

1) Boulter D., Ellis R. J. and Yarwood A. (1972) Biol. Rev. 47, 113-175.

2) Chapeville F., Lipmann F., Von Ehrenstein G., Weisblum B., Ray W. and Benzer S. (1962) Proc. Natl. Sci. U.S. 48, 1086-1092

3) Barnett W.E. and Brown D. (1967) Proc. Natl. Acad. Sci. U.S. 57, 452-458.

4) Barnett W. E., Brown D. and Epler J. (1967) Proc. Natl. Acad. Sci. U.S. 57, 1775-1781.

5) Burkard G., Guillemaut P. and Weil J. H. (1970) Biochim. Biophys. Acta 224, 184-198.

6) Guillemaut P., Burkard G. and Weil J. H. (1972) Phytochemistry 11, 2217-2219.

7) Guillemaut P. and Weil J. H. (1975) Biochim. Biophys. Acta 407, 240-248.

8) Jeannin G., Burkard G. and Weil J. H. (1976) Biochim. Biophys. Acta 442, 24-31.

9) Guillemaut P., Steinmetz A., Burkard G. and Weil J. H. (1975) Biochim. Biophys. Acta 378, 64-72.

10) Reger B., Fairfield S., Epler J. and Barnett W. E. (1970) Proc. Natl. Acad. Sci. U.S. 67, 1207-1213.

11) Aliyev K. and Philippovich I. (1968) Mol. Biol. 2, 364-373.

12) Wright R., Kanabus J. and Cherry J. (1974) Plant Sci. Lett. 2, 347-355.

13) Kanabus J. and Cherry J. (1971) Proc. Natl. Acad. Sci.U.S. 68, 873-876.

14) Merrick W. and Dure L. (1973) Biochemistry 12, 629-635.

15) Brantner J. and Dure L. (1975) Biochim. Biophys. Acta 414, 99-114.

16) Kislev U., Selsky M., Norton C. and Eisenstadt J. (1972) Biochim. Biophys. Acta 287, 256-269.

17) Krauspe R. and Parthier B. (1974) Biochem. Physiol. Pflanz.165, 18-36.

18) Guderian R., Pulliam R. and Gordon H. (1972) Biochim. Biophys. Acta 262, 50-65.

19) Parthier B. (1973) FEBS Lett. 38, 70-74.

20) Hecker L., Egan J., Reynolds R., Nix C., Schiff J. and Barnett W. E. (1974) Proc. Natl. Acad. Sci. U.S. 71, 1910-1914.

21) Lesiewicz J. and Herson D. (1975) Arch. Microbiol. 105,117-121.

22) Pearson R., Weiss J. and Kelmers A. (1971) Biochim. Biophys. Acta 228, 770-774.

23) Kelmers A. and Heatherly D. (1971) Anal. Biochem. 44, 486-495.

24) Beauchemin N., Larue B. and Cedergren R. (1973) Arch. Biochem. Biophys. 156, 17-25.

25) Parthier B. and Krauspe R. (1974) Biochem. Physiol. Pflanz. 165, 1-17.

26) Hecker L., Uziel M. and Barnett W. E. (1976) Nucl. Ac. Res. 3, 371-380.

27) Guillemaut P., Martin R. and Weil J. H. (1976) FEBS Lett. 63, 273-277.

28) Fradin A., Gruhl H. and Feldmann H. (1975) FEBS Lett. 50, 185-189.

29) Rogg H., Brambilla R., Keith G. and Staehelin M. (1976) Nucl. Ac. Res. 3, 285-295.

30) Barrell B. and Sanger F. (1969) FEBS Lett. 3, 275-278

31) Dudock B., Katz G., Taylor E. and Holley R. (1969) Proc. Natl. Acad. Sci. U.S. 62, 941-945.

32) Williams G. and Williams A. (1970) Biochem. Biophys. Res. Comm. 39, 858-863.

33) Guillemaut P., Steinmetz A., Burkard G. and Weil J.H. (1973) C.R. Soc. Biol. 167, 961-966.

34) Tewari K. and Wildman S. (1970) Soc. Experim. Biol. Sympos. 24, Cambridge Univ. Press 147-179.

35) Williams G., Williams A. and George S. (1973) Proc. Natl. Acad. Sci. U.S. 70, 3498-3501.

36) Schwartzbach S., Hecker L. and Barnett W. E. (1975) Plant Physiol. 56, S 384

37) Kolodner R. and Tewari K. (1972) J. Biol. Chem. 247, 6355-6364.

38) Bernardi G. (1969) Biochim. Biophys. Acta 174, 423-434.

39) Wetmur J. and Davidson J. (1968) J. Mol. Biol. 31, 349-370.

40) Gillespie D. and Spiegelman S. (1965) J. Mol. Biol. 12, 829-842.

41) Kourilsky P., Manteuil S., Zamansky M. and Gros F. (1970) Biochem. Biophys. Res. Comm. 41, 1080-1087.

42) Schwartzbach S., Hecker L. and Barnett W. E. (1976) Proc. Natl. Acad. Sci. U.S. 73, 1984-1988.

43) Tewari K., Kolodner R., Chu N. and Meeker R. (1976) these Proceedings

44) Steinmetz A. and Weil J. H. (1976) Biochim. Biophys. Acta (in press).

THE LOW-MOLECULAR-WEIGHT RNAs OF PLANT RIBOSOMES:

THEIR STRUCTURE, FUNCTION AND EVOLUTION

T.A. Dyer, C.M. Bowman and P.I. Payne

A.R.C. Unit of Developmental Botany

181A Huntingdon Road, Cambridge CB3 ODY, U.K.

INTRODUCTION

Why do the ribosomes of chloroplasts, mitochondria and the cytoplasm in plants differ from one another although they apparently have a common function in protein synthesis? Is this merely the result of their separate origins? If so, why have these differences been preserved? Does each type of ribosome have unique features which make it especially well fitted to synthesise particular proteins or does it reflect that, because of differences in factors such as pH or the concentration of particular ions, one type of ribosome could not work properly in the organelles and in the cytoplasm? A detailed comparison of the structures of the different types of plant ribosomes and the way in which they function forms the basis for trying to answer these questions.

Ribosomes which are not taking part in protein synthesis readily dissociate into two subunits [1, 2], the larger of which is about twice the size of the smaller. Each subunit contains one molecule of high-molecular-weight RNA which is its major single structural component and also numerous different proteins [3], most of which probably have specific catalytic functions [4]. In addition to high-molecular-weight RNA, the large subunits of plant ribosomes also contain low-molecular-weight RNAs. Recently increased attention has been paid to these low-molecular-weight RNAs as their small size and relatively easy purification makes it straightforward to study their nucleotide sequences.

In this paper we discuss, in fairly general terms, some of the characteristics of plant low-molecular-weight RNAs with particular reference to results obtained in our laboratory. By doing this it is hoped to highlight important features of these molecules and provide a comprehensive summary of what is known about them at present.

TYPES OF LOW-MOLECULAR-WEIGHT rRNA

5S rRNA. Of the low-molecular-weight ribosomal RNAs, 5S rRNA is the easiest to detect as it may be distinguished in total nucleic acid preparations fractionated either by chromatography on columns of methylated albumin kieselguhr [5, 6, 7], Sephadex [7, 8] or by gel electrophoresis [5, 9]. The cytoplasmic, chloroplast and mitochondrial ribosomes of flowering plants each contain one molecule of 5S rRNA. The plant mitochondrial ribosomes are perhaps different from mitochondrial ribosomes of fungi and mammals in which 5S rRNA was not detected [10, 11].

Although very similar in size, the 5S rRNAs from the chloroplasts, mitochondria and cytoplasm of flowering plants are obviously not identical. The cytoplasmic molecule contains 118 nucleotides and has a molecular weight of 37,500 while the chloroplast component is slightly larger and contains about 122 nucleotides [9]. In the mitochondrial 5S rRNA there are about 120 nucleotides [12, 13]. The flowering plant 5S rRNAs are similar in size to the 5S rRNA of other types of organism that have been studied. In most of these it contains 120 nucleotides [14]. However, 5S rRNA is much smaller than the high-molecular-weight rRNAs. The smallest of these, from the small subunit of chloroplast ribosomes, contains about 1,600 nucleotides and the largest, from the large subunit of cytoplasmic ribosomes, contains approximately 3,700 nucleotides. The tRNAs, in comparison, contain about 80 nucleotides on average.

After synthesis, 5S rRNA is integrated into the large ribosomal subunit. When these subunits from the cytoplasmic ribosomes of plants are disrupted by either a formamide or an EDTA treatment, the 5S rRNA is dissociated in the form of a ribonucleoprotein complex (Dyer and Zalik, unpublished results). The single protein which remains attached to 5S rRNA is very similar in size (mol. wt. 38,000) to the protein which remains attached to 5S rRNA when mammalian ribosomes are disrupted by similar treatments [15, 16]. This indicates that not only the 5S rRNA but also the protein with which it is associated has been conserved during evolution.

A protein - 5S rRNA complex has also been isolated from bacterial ribosomes [17]. The complex from both animal and bacterial ribosomes has GTPase activity and consequently, in intact

TABLE 1

Estimated molecular weights and nucleotide contents of low-molecular-weight RNA from the large subunits of plant ribosomes

RNA species	Cytoplasm		Chloroplast		Mitochondria	
	Mol. Wt.	No. Nucleotides	Mol. Wt	No. Nucleotides	Mol. Wt.	No. Nucleotides
5.8 S	50,400	157	-	-	-	-
5 S	37,500	118	39,400	122	38,750	120
4.5 S	-	-	25,000 or 33,000	80 or 103	-	-

ribosomes, the 5S rRNA and its associated protein may be involved
in the translocation step in protein synthesis [17, 18]. The 5S
rRNA itself probably interacts with the high-molecular-weight rRNA
as well as with protein in bacterial ribosomes [19]. In plants
there appears to be a sequence in the cytoplasmic 5S rRNA which is
complementary to a sequence in 18S rRNA as the two aggregate when
heated together to temperatures which only partially disrupt
hydrogen bonding [20].

5.8S rRNA. When the RNA of plant cytoplasmic ribosomes is
fractionated by polyacrylamide gel electrophoresis, the only low-
molecular-weight RNAs that can be distinguished are 5S rRNA and a
small amount of tRNA. The tRNA is probably that in transient
association with ribosomes participating in protein synthesis.
However, if the RNA has been subjected to a treatment which
dissociates hydrogen bonds (for example if the RNA is heated to a
high enough temperature) prior to electrophoresis, an additional
low-molecular-weight component is found [21]. This is now most
commonly referred to as 5.8S rRNA. It is hydrogen bonded with the
high-molecular-weight RNA of the large ribosomal subunit and is a
structural component of the plant cytoplasmic ribosome. The plant
5.8S rRNA is very similar in size to the homologous RNA in fungal
and mammalian cytoplasmic ribosomes but a comparable RNA was not
found in chloroplast, mitochondrial [13] or prokaryote ribosomes.

4.5S rRNA. We have recently found that the chloroplast
ribosomes of flowering plants also contain a low-molecular-weight
ribosomal RNA in addition to 5S rRNA. It was independently
discovered in Canberra by Whitfeld and his colleagues, and by
Herrmann and his associates in Duesseldorf. This RNA is present in
approximately equimolar amounts with the 5S rRNA and is therefore
probably an additional low-molecular-weight RNA component of the
chloroplast ribosome. In contrast to the other low-molecular-
weight RNAs, it does not appear to be of similar size in all
flowering plants. While in many it contains about 103 nucleotides,
and therefore fractionates to a position between the tRNA and 5S
rRNA during gel electrophoresis, in certain legumes there is a
smaller form of this molecule which fractionates to the same
position as tRNA. This RNA contains about 80 nucleotides and can
easily be mistaken for tRNA. It remains, in fact, to be shown that
the two forms of 4.5S rRNA are true homologues of one another.

SEQUENCE ANALYSIS OF LOW-MOLECULAR-WEIGHT RNAs

The sequence of nucleotides in RNAs determines how they are
folded after synthesis, and how they associate with other molecules.
Furthermore, as the homologous RNAs in living organisms appear to
have been derived by very slow change from a common ancestral type,

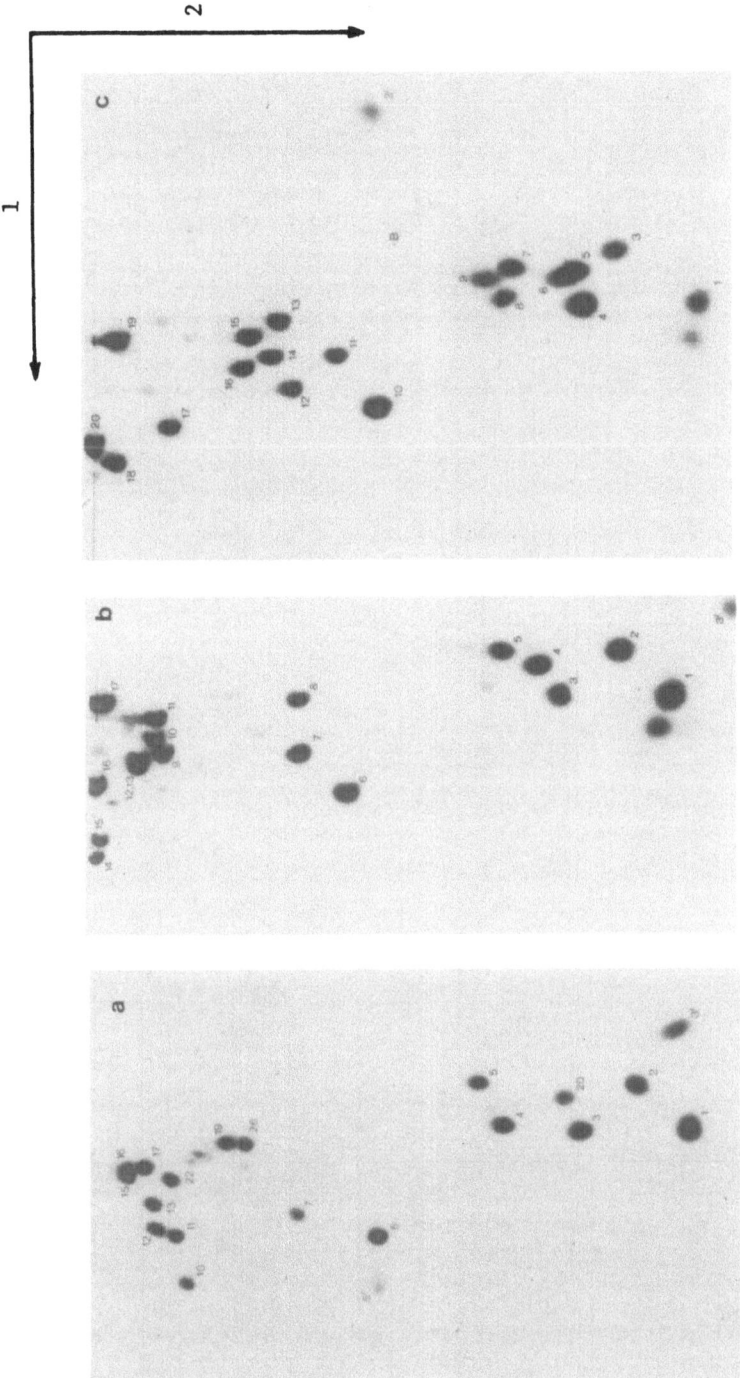

Fig. 1. Autoradiograph of the fractionated products of duckweed low-molecular-weight rRNA digested with T, ribonuclease: (a) cytoplasmic 5S rRNA, (b) chloroplast 5S rRNA, (c) chloroplast 4.5S rRNA. Fractionation was by electrophoresis, first in a strip of cellulose acetate (direction 1). The material was then transferred to a sheet of DEAE-cellulose paper which was wetted with 7% formic acid before electrophoresis in the second dimension (direction 2). The composition of each product is given elsewhere [27].

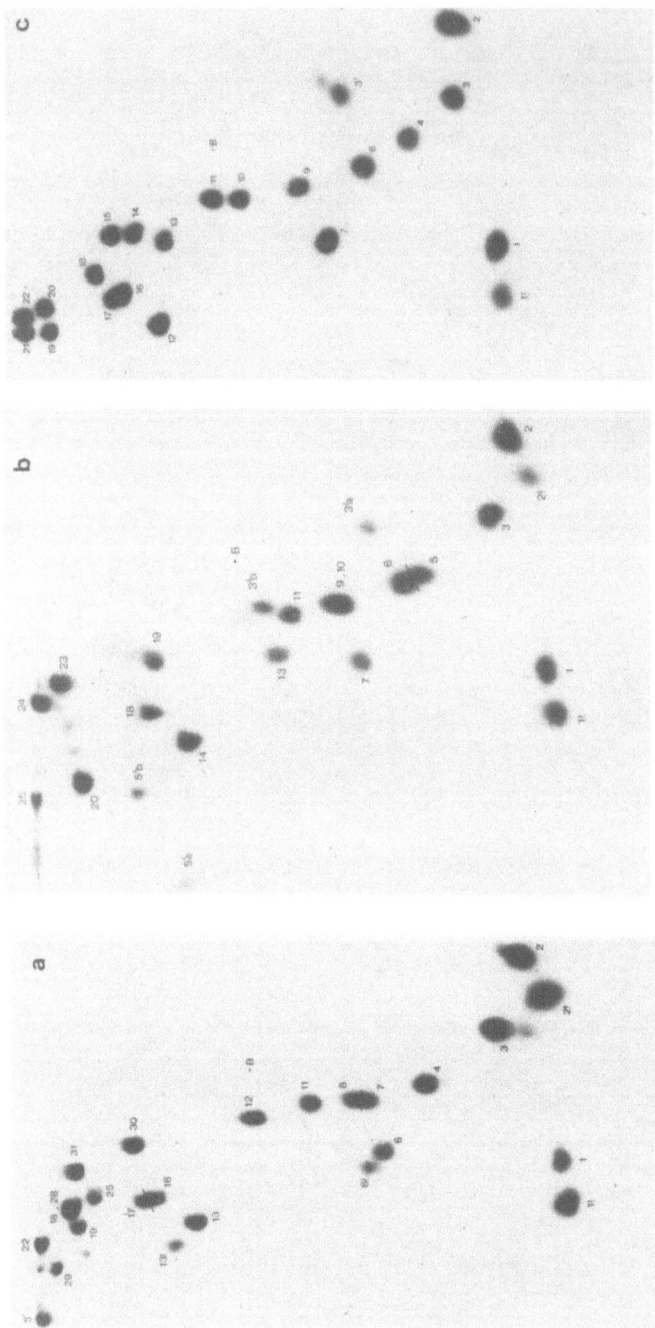

Fig. 2. Autoradiograph of the fractionated products of duckweed low-molecular-weight rRNA digested with pancreatic A ribonuclease: (a) cytoplasmic 5S rRNA, (b) chloroplast 5S rRNA, (c) chloroplast 4.5S rRNA.

sequence differences between homologous RNA molecules can provide information about the phylogenetic relationships between the organisms of which they are part [14, 22]. Although it would be advantageous to know the sequence of examples of all types of RNA from plant ribosomes, the vast effort that would be involved in determining the complete sequences of the high-molecular-weight rRNA does not seem to be warranted at present. Only the low-molecular-weight rRNAs can be sequenced comparatively easily.

Sequencing of RNA is usually accomplished in two distinct stages [23]. Firstly, the molecule is highly labelled with ^{32}P, isolated and then digested with an enzyme which cleaves it at specific residues. T ribonuclease, for instance, produces oligonucleotides which have -Gp at their 3' end and also mononucleotides (Gp) from sequences of G. Pancreatic A ribonuclease, on the other hand, produces oligonucleotides which have a pyrimidine at their 3' end and pyrimidine mononucleotides (Up and Cp) from sequences of pyrimidines. The resulting oligonucleotides are fractionated by electrophoresis in two dimensions and detected by autoradiography (Fig. 1 & 2). The nucleotide sequence of each oligonucleotide can then be determined by further enzymatic digestion. The oligonucleotides produced by digestion of different molecules can be compared at this stage (cataloguing) in order to get some idea as to how similar or different the molecules are [22, 24].

In the second and more difficult stage of the sequencing process the molecule is partially digested so that not every accessible bond is broken. By further analysis of the fragments produced it is possible to determine which of the primary oligonucleotides are adjacent to each other in the complete molecule. By using enzymes with different specificities, overlapping fragments are obtained and the complete sequence is established.

5S rRNA. So far we have determined the complete nucleotide sequence of 5S from the cytoplasmic ribosomes of rye [25]. It does not contain any modified bases and it is probably homogeneous as none of the olignucleotides produced by enzymic digestion of the molecule are present in sub-molar amounts.

We have also determined which oligonucleotides are produced by T, and pancreatic A ribonuclease digestion of cytoplasmic 5S rRNA from broad beans, dwarf beans, sunflowers, tomatoes and duckweed [26, 27]. It is possible to deduce the probable sequences of the 5S rRNA from these plants as well by alignment of these oligonucleotides with homologous sequences in the rye 5S rRNA (Fig. 3). The sequences are all remarkably alike. There is a slight preponderance of base differences at or near the ends and other

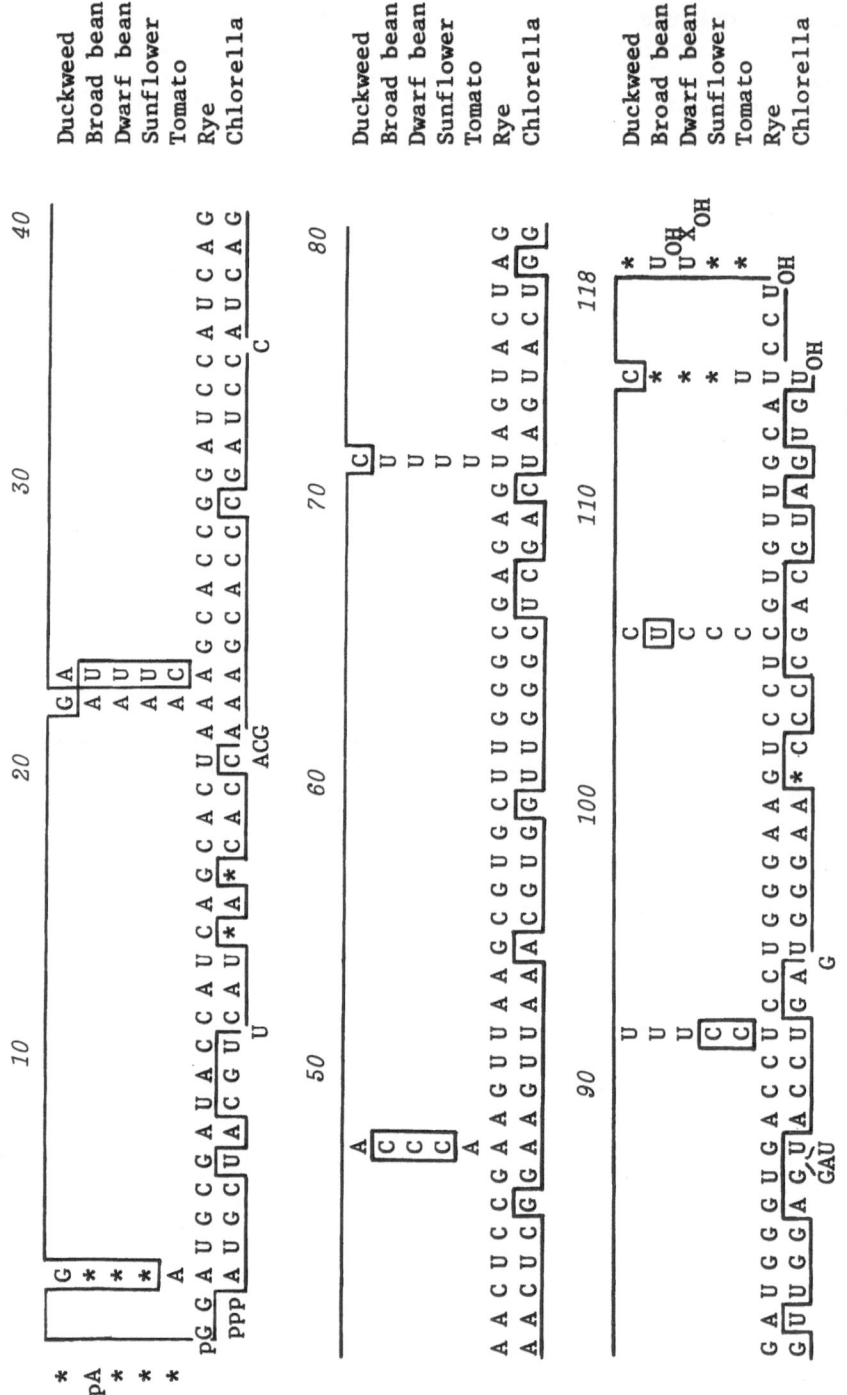

Fig. 3. Comparison of the nucleotide sequence of cytoplasmic 5S rRNA from six flowering plants and from the unicellular green alga *Chlorella pyrenoidosa*.

base differences are dispersed within the molecule. The plants
which were compared are from widely separated taxonomic groups.
Rye and duckweed are monocotyledons and the rest are dicotyledons,
and yet there are only eight base differences between the most
dissimilar (duckweed and broad bean). As the diversification of
the flowering plants probably began over 200 x 10^6 years ago, as
suggested by cytochrome c studies [28], it would seem that in
plants, as in animals, cytoplasmic 5S rRNA is a highly conserved
molecule. Because of this it should also be easy to determine, by
alignment, the sequence of other related plant cytoplasmic 5S rRNAs.
Consequently the study of this molecule may be useful in
establishing the phylogenetic relationship of major plant groups.

The difference in nucleotide sequence between rye 5S rRNA and
that of the green alga *Chlorella pyrenoidosa* [29] is much greater
than that between the flowering plants but there are several long
sequences which are common to all the cytoplasmic 5S rRNAs of
plants studied so far (Fig. 3). Examination of the 5S rRNA of
animals with good fossil records suggests that a nucleotide is
changed in the molecule at intervals of approximately 30 x 10^6
years [14]. If one assumes that the rate of change has been about
the same in plants then the divergence between the ancestors of the
plants and those of the green algae can be calculated to have
occurred about 1400 x 10^6 years ago. The difference between the
rye cytoplasmic 5S rRNA and that of human KB cells is about the
same as that between rye and the green algae and between rye and
the yeast *Torulopsis utilis* [25]. This implies that the divergence
between the major groups of eukaryotes occurred at about the same
time and, also, that diversification of the plants began very early
in their evolution.

In flowering plants the sequence of chloroplast 5S rRNA is
obviously very different from that of cytoplasmic 5S rRNA [27].
This is apparent even from the fingerprints of the oligonucleotides
produced by T, and pancreatic A ribonuclease digestion of duckweed
5S rRNA (Fig. 1 & 2). A special feature of the chloroplast 5S rRNA
is that two oligonucleotides from each end of the molecule have
been found and these are present in equimolar amounts. From this
it appears that there are at least two cistrons coding for its
synthesis. This is of particular interest as hybridization studies
have already indicated that there are two cistrons complementary to
the high-molecular-weight rRNA per chloroplast DNA copy [30]. Like
the cytoplasmic 5S rRNA, the chloroplast molecule contains no
modified bases.

The complete sequence of duckweed chloroplast 5S rRNA is being
determined at present. Preliminary data indicates that certain
sequences of nucleotides in the molecule are very similar to those
in the 5S rRNA of the blue-green alga *Anacystis nidulans* [31]; the

similarity to the cytoplasmic 5S rRNA is much less pronounced. It
appears therefore that the 5S rRNA of chloroplasts and the blue-
green algae diverged in evolution more recently than the 5S rRNA of
the chloroplasts and cytoplasm. This data is consistent with the
endosymbiotic theory for the origin of chloroplasts. However, it
does not show that blue-green algae, as such, were the ancestors of
the chloroplasts of flowering plants - their progenitors may now be
extinct but were almost certainly photosynthetic prokaryotes.

    As would be expected, the dissimilarity between the sequences
of prokaryote and eukaryote 5S rRNA is greater than within the
eukaryotes themselves. Also, all prokaryotes (and the 5S rRNA of
chloroplasts) contain the -pyr-G-A-A-C- sequence which may interact
with the -G-T-Ψ-C-pur- sequence in tRNA [32]. In the cytoplasmic 5S
rRNA of flowering plants the comparable sequence is -pur-G-A-A-C-
and in human KB cells [33] and yeasts [34] it is -pyr-G-A-U-C-.
However, some features of secondary structure may be common to all
5S rRNAs. The nucleotides towards the 5' end are complementary to
those towards the 3' end which suggests that these two segments are
base paired *in situ*. There is probably substantially more base
pairing than this, which may be located in the molecule as suggested
by a recent universal model for its secondary structure [35]. The
cytoplasmic 5S rRNA of flowering plants can be folded as predicted
by this model and our preliminary data suggests that chloroplast
5S rRNA could also be folded in the same way. Furthermore, from an
additional sequencing study of the cytoplasmic 5S rRNA of flowering
plants, it seems that an equivalent segment of this molecule
interacts with ribosomal protein as in *E. coli* Dyer and Zalik,
unpublished results;[36].

    5.8S rRNA. A comparison has been made of the oligonucleotides
produced by T₁ and pancreatic A ribonuclease digestion of the 5.8S
rRNA of several flowering plants [37]. The sequence of this RNA
seems to have been highly conserved during the evolution of
flowering plants (although perhaps not to quite the same extent as
in 5S rRNA). The complete nucleotide sequences of 5.8S rRNA from
Novikoff hepatoma cells and from yeast are now known. There is
about 75% homology between the sequences of the two. In contrast,
only limited homology is evident between the oligonucleotides from
flowering plant 5.8S rRNA and sequences in yeast and hepatoma 5.8S
rRNAs [38]. This finding is somewhat unexpected as there is about
the same degree of similarity between flowering plant and animal
5S rRNA sequences as there is between the sequence of yeast 5S rRNA
and either the flowering plant or animal sequences [25]. The
composition of the flowering plant oligonucleotides demonstrates
that the molecule is very different from 5S rRNA. It seems
probable that the two molecules have not shared a common ancestor
[39]. However, the 5.8S rRNA molecules of all flowering plants
investigated so far have been found to contain a -pyr-G-A-A-C-

sequence and it has been suggested that in cytoplasmic ribosomes of
eukaryotes, the 5.8S rRNA rather than 5S rRNA interacts with tRNA
[34]. One of the most obvious differences between 5.8S rRNA and
the other low-molecular-weight rRNAs is that it contains modified
(probably methylated) nucleotides, as do the high-molecular-weight
rRNAs.

4.5S rRNA. The products of the digestion of 4.5S rRNA from
duckweed with T, and pancreatic A ribonuclease have recently been
investigated [27]. Fingerprints of the oligonucleotides produced
by such digestions (Fig. 1 & 2) show that the molecule is
essentially homogeneous and that it is completely different from
any of the other low-molecular-weight rRNAs studied so far.
Further analysis of the oligonucleotides shows that the 4.5S rRNA
contains no modified nucleotides and it does not have any features
which suggest that it may be like a tRNA or 5S rRNA molecule.

CONCLUDING REMARKS

We have now gained some insight into the structure of the
low-molecular-weight rRNAs in flowering plants and are therefore in
a position to discuss how the RNAs may have evolved. However, we
are still ignorant of their precise role in the synthesis of
proteins. The sequencing studies have indicated which structural
features are common to all 5S rRNAs and may therefore be essential
for the molecule to function. One can only speculate at this stage
as to the nature of 5.8S rRNA and 4.5S rRNA. The 4.5S rRNA could,
for instance, be the chloroplast equivalent of the 5.8S rRNA.
Alternatively, the 4.5S rRNA could be a fragment of the high-
molecular-weight rRNA, but as it is found in extracts made from
plant tissues using methods which rigorously suppress nuclease
activity, fragmentation, if it occurs, must take place *in vivo*. A
further possibility is that the 4.5S rRNA is a piece of transcribed
spacer, perhaps with no particular function in the ribosome.
Whether it has a function or not might be resolved by
reconstituting ribosomes in its presence or absence.

The relationship of the low-molecular-weight rRNAs to each
other and to the high-molecular-weight rRNA will be clarified by
locating the cistrons which code for their synthesis and by
establishing which RNA species share a common precursor molecule.
Some studies of this type have already been reported. For
instance, in maize there is no obvious association between the loci
for cytoplasmic 5S rRNA synthesis and the nucleolus where the
cistrons for the high-molecular-weight rRNAs are located [40].
Furthermore, the results of experiments in which the kinetics of
labelling of cytoplasmic rRNA was studied suggest that the 5.8S
and 25S are part of a precursor with molecular weight of 1.39×10^6

[41]. We may confidently expect that the mapping in the chloroplast genome of cistrons for chloroplast rRNAs will be completed shortly. Eventually, by understanding what controls the transcription of the rRNA and its processing to mature molecules, we may hope to identify factors which regulate plant growth through gene expression.

REFERENCES

1   Blobel, G. and Sabatini, D. (1971) Proc. Natl. Acad. Sci. U.S. 68, 390-394.
2   Dyer, T.A. and Koller, B. (1972) Proc. IInd Int. Congr. Photosynth. Res. pp. 2537-2544. W. Junk N.V., Den Haag.
3   Gaulerzi, C., Janda, H.G., Passow, H. and Stöffler, G. J. Biol. Chem. 11, 3347-3355.
4   Pongs, O., Nierhaus, K.H., Erdmann, V.A. and Wittmann, H.G. (1974) FEBS Lett. Suppl. 40, S28-S37.
5   Dyer, T.A. and Leech, R.M. (1968) Biochem. J. 106, 689-698.
6   Wollgiehn, R., Ruess, M. and Munsche, D. (1966) Flora, Abt. A, 157, 92-108.
7   Ruppel, H.G. (1969) Z. Naturforsch B, 24, 1467-1475
8   Payne, P.I. and Dyer, T.A. (1971) Biochim. Biophys. Acta 228, 167-172.
9   Payne, P.I. and Dyer, T.A. (1971) Biochem. J. 124, 83-89.
10   Lizardi, P.M. and Luck, D.J. (1971) Nat. New Biol. 229, 140-142.
11   Attardi, B. and Attardi, G. (1971) J. Mol. Biol. 55, 231-249.
12   Leaver, C.J. and Harmey, M.A. (1973) Biochem. Soc. Symp. 38, 175-193.
13   Leaver, C.J. and Harmey, M.A. (1976) Biochem. J. 157, 275-277.
14   Hori, H. (1975) J. Mol. Evol. 7, 75-86.
15   Peterman, M.L., Hamilton, M.G., and Pavlovec, A. (1972) Biochem. 11, 2323-2326.
16   Blobel, G. (1971) Proc. Natl. Sci. U.S. 68, 1881-1885.
17   Horne, J.R. and Erdmann, V.A. (1973) Proc. Natl. Acad. Sci. U.S. 70, 2870-2873.
18   Grummt, F., Grummt, I. and Erdmann, V.A. (1974) Eur. J. Biochem. 43, 343-348.
19   Herr, W. and Noller, H.F. (1975) FEBS Lett. 53, 248-252.
20   Azad, A.A. and Lane, B.G. (1975) Can. J. Biochem. 53, 320-327.
21   Payne, P.I. and Dyer, T.A. (1972) Nat. New Biol. 235, 145-147.
22   Bonen, L. and Doolittle, W.F. (1976) Nature 261, 669-673.
23   Brownlee, G.G. (1972) *Determination of Sequences in RNA*, North Holland, Amsterdam.
24   Sogin, S.J., Sogin, M.L. and Woese, C.R. (1972) J. Mol. Evol. 1, 173-184.
25   Payne, P.I. and Dyer, T.A. (1976) Submitted to Eur. J. Biochem.
26   Payne, P.I., Corry, M.J. and Dyer, T.A. (1973) Biochem. J. 135, 845-851.

27 Dyer, T.A. and Bowman, C.M. (1976) In *Genetics and Biogenesis of Chloroplasts and Mitochondria* (Bücher, Th., Neupert, W., Sebald, W. and Werner, S., ed.), North Holland, Amsterdam, in press.

28 Ramshaw, J.A.M., Richardson, D.L., Meatyard, B.T., Brown, R.H., Richardson, M., Thompson, E.W. and Boulter, D. (1972) New Phytol. 71, 773-779.

29 Jordan, B.R., Galling, G. and Jourdan, R. (1974) J. Mol. Biol. 87, 205-225.

30 Thomas, J.R. and Tewari, K.K. (1974) Proc. Natl. Acad. Sci. U.S. 71, 3147-3151.

31 Corry, M.J., Payne, P.I. and Dyer, T.A. (1974) FEBS Lett. 46, 63-66.

32 Richter, D., Erdmann, V.A. and Sprinzl, M. (1973) Nat. New Biol. 246, 132-135.

33 Forget, B.G. and Weissman, S.M. (1967) Science (Wash. D.C.) 158, 1695-1699.

34 Nishikawa, K. and Takemura, S. (1974) FEBS Lett. 40, 106-109.

35 Fox, G.E. and Woese, C.R. (1975) J. Mol. Evol. 6, 61-76.

36 Gray, P.N., Bellemare, G. and Monier, R. (1972) FEBS Lett. 24, 156-160.

37 Woledge, J., Corry, M.J. and Payne, P.I. (1974) Biochim. Biophys. Acta 349, 339-350.

38 Nazar, R.N., Sitz, T.O. and Busch, H. (1975) Biochem. Biophys. Res. Com. 62, 736-743.

39 Cedergren, R.J. and Sankoff, D. (1976) Nature (Lond.) 260, 74-75.

40 Wimber, D.E., Duffey, P.A., Steffensen, D.M. and Prensky, W. (1974) Chromosoma 47, 353-359.

41 Hepburn, A.G. and Ingle, J. (1975) Phytochem. 14, 1157-1160.

THE GENETICS OF THE CHLOROPLAST RIBOSOME IN Chlamydomonas reinhardi

Lawrence Bogorad, Jeffrey N. Davidson and
Maureen R. Hanson
The Biological Laboratories, Harvard University
16 Divinity Avenue, Cambridge, Mass. 02138 USA

INTRODUCTION

Our interest in the general problem of the interaction between
nuclear and organelle genomes in the life and development of eukaryotic cells has led us to study the genetics of proteins of chloroplast ribosomes of the single-celled green alga Chlamydomonas
reinhardi. Specifically, we have dealt with the proteins of the
large, i.e. 52S, subunit of these ribosomes.

The work followed this general rationale and plan: The antibiotic erythromycin blocks protein synthesis in bacteria and
inhibits the growth of Chlamydomonas and we have shown that ^{14}C-
erythromycin binds only to the 52S subunits of chloroplast ribosomes;
not to the small subunit of these ribosomes; not to either of the
subunits of the cytoplasmic ribosomes of this alga (Mets and
Bogorad, 1971). Thus it appeared that mutation of C. reinhardi
to erythromycin resistance might involve genetic alteration of a
protein in the large subunit of the chloroplast ribosome. Cells
were treated with a mutagen and erythromycin-resistant strains
were isolated. After determining that the chloroplast ribosomes
of the resistant mutants had altered erythromycin binding properties
in vitro, the pattern of genetic transmission of the resistance was
examined to determine whether the mutation was in a nuclear or
chloroplast gene. Ribosomal proteins of the 52S subunits of chloroplast ribosomes from wild-type and mutant cells were then compared
to discover the molecular basis of erythromycin resistance (Mets
and Bogorad, 1971, 1972; Davidson et al., 1974; Hanson et al., 1974;
Bogorad, 1975; Davidson, 1976; Hanson, 1976).

In the course of this work, we have characterized the proteins of the large and small subunits of the cytoplasmic and chloroplast ribosomes of <u>Chlamydomonas</u> <u>reinhardi</u> (Hanson <u>et al.</u>, 1974), we have located some genes for ribosomal proteins of the large chloroplast subunit, and have examined a number of aspects of erythromycin inhibition of the growth of <u>Chlamydomonas</u>.

CHLOROPLAST RIBOSOMES OF CHLAMYDOMONAS

Growth of wild-type <u>C</u>. <u>reinhardi</u> in liquid culture is completely inhibited by 10-20 µg/ml erythromycin A. The binding of ^{14}C-erythroymcin A or its derivatives to ribosomes can be estimated by the Millipore filter binding assay of Teraoka (1970). A sample of ribosomes is mixed with the radioactive antibiotic in Mg acetate-KCl-Tris buffer and incubated for one hour (equilibrum is achieved with wild-type ribosomes by 30 minutes). When the suspension is filtered, the ribosomes remain on the cellulose filter; the radioactivity thus retained is a measure of the binding of the antibiotic to the 52S subunits. Each ribosome associates with a single molecule of erythromycin. The erythromycin A binding constant for chloroplast ribosomes of wild-type cells is 3-4 x 10^6 M^{-1} (Hanson, 1976). A binding constant of 8 x 10^{-4} M^{-1} is obtained for the derivative of erythromycin A produced by acid treatment (Mets and Bogorad, 1971). <u>C</u>. <u>reinhardi</u> ribosomes bind erythromycin over a magnesium acetate range of 5 to 30 mM; the optimum is 25 mM. KCl is required for binding erythromycin <u>in vitro</u>; wild-type ribosomes bind the same amount of antibiotic over a KCl concentration range of 25 to 450 mM (Mets, 1973; Hanson, 1976).

For comparison, <u>E</u>. <u>coli</u> MRE 600 ribosomes have a binding constant of 9.2 x 10^6 M^{-1} (Fernandez-Munoz and Vazquez, 1973), while the erythroymcin binding constant of ribosomes from strain Q13 is 6.3 x 10^5 M^{-1} (Mao, 1971). The binding constant for <u>C</u>. <u>reinhardi</u> wild-type ribosomes fits within this range.

A procedure was established for preparing ribosomal subunits by sucrose density gradient centrifugation on a large enough scale for later steps in the work. Ribosomal proteins of each subunit were solubilized with urea and analyzed by two-dimensional gel electrophoresis (Mets and Bogorad, 1972, 1974). The latter procedure involves electrophoresis first in 10 cm long cylindrical, 4% polyacrylamide gels in the presence of urea. The cylindrical gel is then cemented with additional polyacrylamide or agarose on top of a 11.5 x 10 cm 10% polyacrylamide slab. The detergent sodium dodecyl sulfate (SDS) is included in the upper electrode buffer and, when the current is applied, the SDS passes through the first dimension gel, "coating" the ribosomal proteins <u>en route</u>. In this way electrophoretic mobility of pH 5 and the molecular weight (Weber and Osborn, 1969) of each protein in each of the subunits was determined (Hanson <u>et al.</u>, 1974). The results of this analysis are shown in Table 1. Values are calculated from mobilities in SDS gel

Table 1.   Molecular weight estimates for ribosomal proteins of
Chlamydomonas reinhardi

Protein	Large cytoplasmic subunit	Small cytoplasmic subunit	Large chloroplast subunit	Small chloroplast subunit
1	54000	32000	40000	37000
2	53000	32000	34000	29000
3	40000	31000	29500	29000
4	39500	31000	30000	27000
5	31000	32000	27500	27500
6	33000	27500	26000	23500
7	26500	25500	26000	21000
8	26500	21500	26000	21500
9	27000	21500	21500	19500
10	26000	18000	20000	20500
11	27000	18000	22500	18000
12	26000	18000	16500	16000
13	22500	19500	15000	16000
14	23500	17500	18500	18000
15	21500	17500	17500	16500
16	22000	18000	17500	15500
17	23500	18000	15500	16000
18	19000	18500	16000	15500
19	20000	18500	16000	14500
20	19000	16500	17000	15000
21	17500	16000	14500	14000
22	18500	17000	14500	21000
23	20000	14000	15000	
24	17500	15000	14000	
25	17500	15000	13500	
26	19000	12500	14000	
27	19000			
28	19000			
29	16500			
30	17000			
31	16000			
32	16000			
33	17000			
34	18500			
35	15500			
36	13000			
37	15000			
38	14500			
39	16500			
Total	904000	541500	538000	4515000
average	23000	21000	20500	20500

electrophoresis. Each molecular weight is the average, to nearest
500, of determinations from three separate electrophoresis runs.
Molecular weights 29000 and below are approximately ± 1000; deviat-
ion for the proteins above 29000 ranges from ± 2000–4000 (Hanson
et al., 1974).

Our (Hanson et al., 1974) original estimates of the number of
proteins in the ribosomal subunits were: small chloroplast, 22;
small cytoplasmic, 26; large chloroplast, 26; and large cytoplasmic,
39. A photograph of a stained gel slab showing the separation of
the proteins of the 52S ribosomal subunit, and schematic diagram
of the two-dimensional electrophoretogram of these proteins is shown
in Fig. 1. Two pairs of proteins in the small chloroplast and
cytoplasmic subunits and four pairs of proteins in the large subunits had similar electrophoretic mobilities in two dimensions.
These may not be necessarily identical but establish an upper limit
to the number of proteins which can be common to both types of
ribosomal subunits.

As in E. coli the total of the molecular weights of the individual ribosomal proteins does not equal the protein content estimated
for the ribosomes. In E. coli some ribosomal proteins are present
in other than unit amounts. Similar comparisons involving the
total molecular weight of the proteins so far identified in
C. reinhardi ribosomal subunits must await accurate determinations
of the subunit molecular weights as well as molecular ratios of
individual proteins in a subunit.

Isolation methods different from ours may yield higher or lower
numbers of proteins in the subunits (Speiss and Arnold, 1975). For
example, N-H.Chua (personal communication) has observed and we have
confirmed, the presence of an additional high molecular weight
protein in the small chloroplast subunit.

Treatment of 52S chloroplast ribosomal subunits with 6 M urea
using the method of Langer et al., (1975) results in the complete
solubilization of proteins LC19, LC20, LC23, and LC24; these can
be classified as "split" proteins. They are closer to the surface
or more loosely bound than some of the others. LC1, LC12, and
LC13 were partially solubilized. The remaining proteins remained
with the "core" particle (Davidson, 1976). This type of analyses
may reveal the relative locations on the ribosome of classes of
proteins.

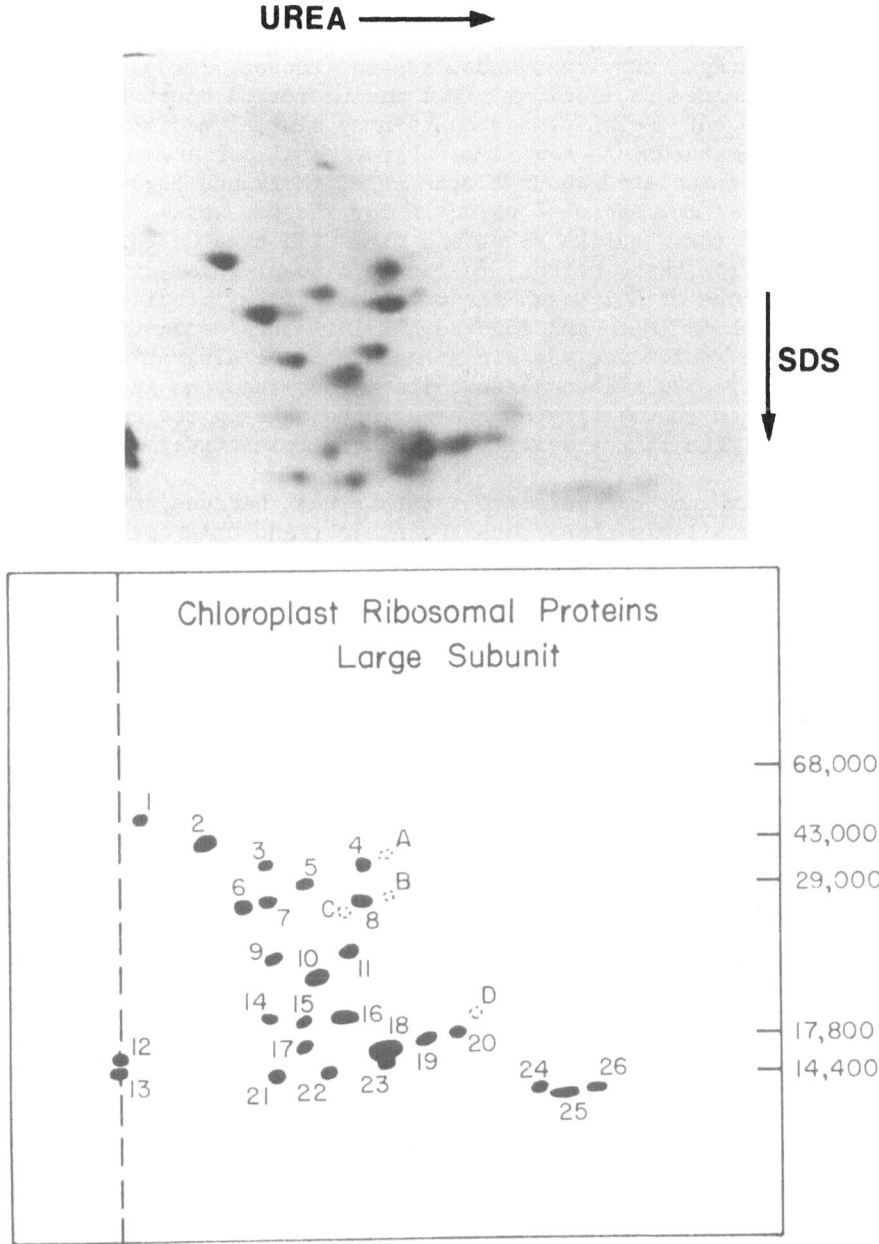

Fig. 1 Upper: Second dimension gel of proteins of the 52S subunit of the C. reinhardi chloroplast ribosome. Lower: Schematic
of two-dimensional electrophoretogram of proteins of the 52S subunit.

Erythromycin Resistant Mutants of C. reinhardi

Wild-type, mating type "+" strain 137c C. reinhardi were treated with the mutagen ethyl methanesulfonate (Loppes, 1968) and after being maintained in liquid minimal medium for 24 hours the cells were spread onto Petri plates containing minimal medium plus 5×10^{-4}M erythromycin (366.5 µg/ml); several erythromycin-resistant mutants were isolated about 1 week later (Mets and Bogorad, 1971). Nine of these mutants were selected for further work. It was established that, unlike ribosomes from wild-type C. reinhardi, the ribosomes from these mutants failed to bind erythromycin in low concentrations of KCl using the cellulose filter binding assay described above (Mets and Bogorad, 1971). This demonstrated that mutation to resistance was a consequence of an alteration in the chloroplast's 52S ribosomal subunits rather than the result of an alteration in the ability of the cells to take up the antibiotic or the acquisition of the ability to render it inactive.

C. reinhardi was selected for this work because genetic information is available about both the nuclear and chloroplast genomes of this organism (Sager, 1972) and transmission patterns of genes in the nuclear and chloroplast genomes are relatively easy to distinguish.

Vegetative cells of C. reinhardi are normally haploid. These cells can be converted into gametes by transfer to a medium lacking nitrogen. The two mating-types are designated "+" and "−". Under appropriate conditions a plus mating-type cell fuses with a minus mating-type cell to produce a zygote. After maturation and meiosis, four or eight haploid zoospores are produced. These progeny can be separated from one another by micro-manipulation and normally each gives rise to a colony of cells. The phenotype of each colony can then be determined.

Mating-type, for example, is transmitted in a Mendelian manner. That is, in a cross half of the offspring will carry the "+" mating-type character and half the "−" mating-type character. Many characters are known to be transmitted in this way and have been mapped to linkage groups in the nuclear genome.

Other genetic characters are transmitted uniparentally. In such cases all of the offspring will carry the character if it is introduced into the cross in the "+" mating-type parent; none will display the trait if it is introduced in the "−" mating-type parent. Genetic determinants carried in the extra-nuclear chloroplast genome are permanently lost if they are introduced in the "−" mating-type parent.

The erythromycin-resistant mutants we obtained were derived from wild-type strain 137c, mating-type "+". Therefore, crossing each of the erythromycin-resistant mutants with "-" mating-type, erythromycin-sensitive (i.e. wild-type) cells would indicate that the gene is nuclear (Mendelian) if half the offspring are resistant. On the other hand, the gene is shown to be in the chloroplast genome if all the offspring carry the resistance marker.

Genetic analyses of this type showed that in eight of the strains the gene for erythromycin-resistance is in the nuclear genome. The ninth strain has a mutation in a chloroplast gene. Thus alterations in either the nuclear or chloroplast genomes were shown to confer erythromycin-resistance on the large subunit of the C. reinhardi chloroplast ribosome; both genomes are involved in chloroplast ribosome function.

As a consequence of crossing the eight Mendelian mutants with one another it was concluded that they fit into two genetically unlinked groups of four each. One group was designated ery-M1a, b, c, d and the other ery-M2a, b, c, d. The uniparentally transmitted character was designated ery-U1a (Mets and Bogorad, 1971).

Ery-U1a. To seek the more immediate consequence of the mutation in the chloroplast genome the proteins from 52S subunits of the erythromycin-resistant mutant ery-U1a were isolated and compared with those from wild-type cells using the two-dimensional gel electrophoretic system. Ribosomal protein LC (large chloroplast) 4 of the ery-U1a mutant strain was found to aggregate more than the wild-type form of LC4. Thus, the mutation to erythromycin resistance assigned to locus ery-U1 in the chloroplast genome may correlate with an alteration in LC4, a 30,000 dalton protein of the 52S subunit (Mets and Bogorad, 1973; Hanson et al., 1974).

Ery-M2. Chloroplast ribosomal proteins from the erythromycin-resistant strain ery-M2d were isolated and examined in one-dimensional β-alanine gels. A protein of the 52S subunit had reduced electrophoretic mobility when compared with wild-type ribosomal proteins (Mets and Bogorad, 1972). This ribosomal protein is small and is often lost during preparation of ribosomal proteins. It has not been identified on the two-dimensional gel system.

Chloroplast ribosomes comprise 35% to 40% of the total ribosomes in wild-type cells but only 13-18% of the total in ery-M2 mutants grown at 25°C. Furthermore, the four erythromycin-resistant mutants in the ery-M2 group are cold-sensitive for growth at 15°C; genetic analyses indicate that the same mutation confers both erythromycin-resistance and cold sensitivity (Hanson, 1976). Finally, the ratio of large to small chloroplast ribosomal subunits in ery-M2 mutants is reduced after a shift of cells from 25°C to 15°C (Hanson, 1976). Thus it seems likely that an

alteration in the protein specified by the locus ery-M2 in these
mutants results in a defect in production or stability of chloroplast ribosomes. This defect could result in a reduced population of chloroplast ribosomes generally and in a specific
reduction in the large chloroplast subunit population at lower
temperatures.

Ery-M1. Detailed genetic analyses of ery-M1 mutants
(Davidson et al, 1974) have revealed that all four of the mutants
map to the same locus. The locus is on linkage Group XI, 12-15
map units to the right of the genetic marker pf2 (paralyzed
flagella-2). If this locus is the structural gene for a chloroplast ribosomal protein, each of the strains in this group should
contain an altered form of the same ribosomal protein. This
proved to be the case.

52S subunits were obtained from each of the four ery-M1
mutants and compared by electrophoresis with proteins from the
same subunit of wild-type cells. All four ery-M1 mutants proved
to have an alteration in the same ribosomal protein, LC6. During
polyacrylamide gel electrophoresis at pH5 in 8 M urea, LC6 of
ery-M1a and -M1c moved behind the wild-type form; LC6 of ery-M1d
moved more rapidly; and LC6 of ery-M1b moved much more slowly than
wild-type LC6. LC6 from ery-M1a is not electrophoretically
distinguishable from LC6 of ery-M1c. Thus there are at least three
altered forms of LC6 among the four mutant strains of the M1 group
(Figure 2). The LC6 of ery-M1a, -M1c and -M1d all have a molecular
weight of 26,000, like the wild-type. However, LC6 of ery-M1b
has a molecular weight of only 18,500, i.e. it is approximately
30% smaller than the normal form of this ribosomal protein. This
work (Davidson et al., 1974) demonstrated that the locus ery-M1 on
nuclear linkage group XI is the site of the structural gene for
chloroplast ribosomal protein LC6. This was the first gene to be
located for any eukaryotic ribosomal protein and for any chloroplast
protein.

The conclusion that the locus ery-M1 is the site of the
structural gene for LC6 has been confirmed in another way. Although
vegetative cells of C. reinhardi are normally haploid, a small
fraction of zygotes fail to undergo meiosis and continue to repro-
duce vegetatively as stable diploids. By the construction of
appropriate strains it is possible to select for such diploids
(Ebersold, 1967; Gillham, 1963). Complementation analyses of
the ery-M1 mutants was carried out to obtain additional information
about the ery-M1 locus. Diploids of the type +/ery-M1 were
selected and the ribosomal proteins of these cells were analyzed
(Hanson, 1976). In each case studied the diploid strain carried
both the wild-type and a mutant from LC6. This is precisely what
would be expected if the locus ery-M1 is the structural gene for

the protein LC6. Although both the wild-type and mutant forms
of the genes are expressed in these diploids, quantitative
analyses of the amounts of the two forms of LC6 in 52S subunits
recovered from the diploids revealed that the ratio of the wild-
type to the altered form was approximately 3:2. One possible
explanation is that both forms of LC6 are produced in equal amounts.
Then the wild-type and mutant forms of the protein compete for a
site on the large chloroplast subunit and the altered forms
have a lower affinity, or form less stable ribosomes. Alternatively,
the difference could result from differential rates of synthesis
and degradation.

Genetic analyses have revealed that mutants ery10, ery11,
and ery14 isolated by Surzycki and Gillham (1971) are allelic
to the ery-M1 mutants. The proteins of the 52S subunits obtained
from these mutants were also analyzed by two-dimensional gel
electrophoresis. LC6 from ery10 and ery11 is indistinguishable
from that of ery-M1b. Thus, in all, seven erythromycin-resistance
mutations which map to the locus ery-M1 have been examined. At
least three alterations in protein LC6 have been found (Davidson
et al., 1974; Davidson, 1976).

Protein LC6 of the 52S subunit of the chloroplast ribosome
of C. reinhardi shares many properties with the 50S ribosomal
protein L4 of E.coli. They have similar molecular weights: LC6,
26,000; L4, 25,800. They also have very similar isoelectric
points. [Data on E.coli are in Garrett and Wittmann, (1973).]
Both proteins are found in the core particles prepared by treatment of ribosomal subunits with 6M urea (Langer et al., 1975).
Finally, alterations in LC6 are found in the erythromycin-
resistant mutants of C. reinhardi and altered forms of L4 are
found in erythromycin-resistant mutants of E.coli (Tanaka
et al., 1968).

Other erythromycin-resistant strains of C. reinhardi.
Genetic analyses were carried out on mutants ery-2y and ery-21,
originally isolated by Surzycki and Gillham (1971). These mutants
are Mendelian and allelic, but are not linked to ery-M1 or
ery-M2 (Davidson, 1976), and these constitute another Mendelian
group, ery-M3.

Er11 is a uniparentally-transmitted factor for erythromycin
resistance and is reported (Gillham, personal communication) to be
allelic with ery-U1a.

Ery-U37 is still another uniparentally-transmitted erythromycin-resistance factor. This is reported to be linked to ery-U1a
but not allelic with it (Gillham, personal communication). Thus,
this appears to be a second uniparental locus for erythromycin-
resistance. Ribosomal proteins of the ery-M3, and ery-U37

strains have not been examined for possible alterations.

Summary. At the present time mutations in three unlinked nuclear genes are known to be able to confer erythromycin-resistance on C. reinhardi. In two cases, ery-M1 and ery-M2, mutation to resistance has been correlated with an alteration in a chloroplast ribosomal protein. Mutations in two uniparental genes, ery-U1a and ery-U37, have been shown to confer erythromycin-resistance on cells of this alga. The mutation to resistance in ery-U1a is correlated with a possible primary change in ribosomal protein LC4.

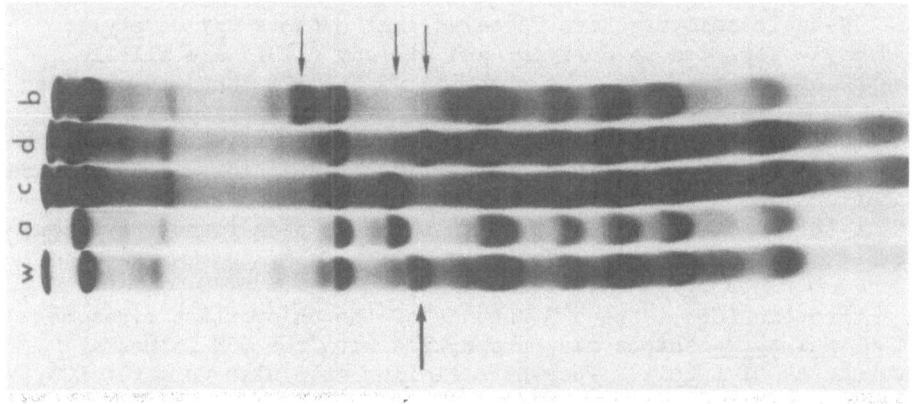

Fig. 2 Portions of polyacrylamide gels showing electrophoretic separation of the 52S ribosomal proteins from wild-type (W), ery-M1a (a), ery-M1c (c), ery-M1d (d), and ery-M1b (b) strains of C. reinhardi. Arrows show positions of wild-type and mutant forms of chloroplast ribosomal protein LC6 (Davidson et al., 1974).

SUPPRESSIVE MUTATIONS

As the result of some of the work described above we had available in the ery-M1 strains mutations which were characterized both with respect to the genetic locus and the alteration of a specific chloroplast ribosomal protein, LC6. Reversion to erythromycin sensitivity could result from additional mutations in the ery-M1 gene or, more probably, mutations in other genes which may affect chloroplast ribosome structure or function. If the revertants should have other altered chloroplast ribosomal proteins, the mapping of genes for components of the 52S subunits could be extended. Consequently ery-M1b cells were mutagenized with ethyl methanesulfonate, plated on minimal medium, and replica plated onto minimal medium containing 350 μg/ml of erythromycin. Colonies which grew on the minimal medium but not on the erythromycin-containing plates were selected and examined for their

erythromycin sensitivity. Three of these erythromycin sensitive
(es) isolates were recovered and analyzed.

Revertants to Erythromycin Sensitivity

Es-M1. C. reinhardi ery-M1b cells are resistant to 500 μg/ml
of erythromycin on agar plates. Cells of the type es-M1ery-M1b
are partially resistant, capable of growing on 100-200 μg/ml erythromycin. Wild-type cells are less tolerant; they cannot grow in a
medium containing 50 μg/ml or more of erythromycin.

Does the mutation es-M1 alter the large subunits of the
chloroplast ribosome? Using the ^{14}C-erythromycin binding assay
the erythromycin binding constant for ribosomes from es-M1eryM1b
was found to be higher than the binding constant for ribosomes
from ery-M1b strains. This shows that the 52S ribosomal subunits
are indeed changed as a result of the mutation. Two-dimensional
polyacrylamide gel electrophoresis of the proteins of 52S ribosomal
subunits from es-M1ery-M1b revealed that the form of LC6 present
is the same as that seen in ery-M1b strains not carrying es-M1.
Using the two-dimensional gel electrophoretic system, so far we
have not detected alterations in any other of the ribosomal
proteins, however, changes could have occurred which are not
detectable in our gel analytical system. Alterations in a protein's
primary sequence can occur which alter its effect on the ribosome
without changing its charge or size significantly.

Detailed genetic analyses have shown the location of the
locus es-M1. It, like ery-M1b, is on nuclear linkage group XI
between pf2 and ery-M1b (Davidson, 1976) as shown:

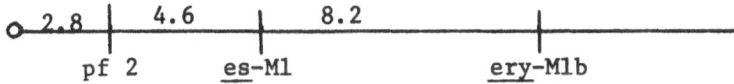

This is the second gene affecting erythromycin binding to chloroplast ribosomes to be mapped to a linkage group in the nuclear
genome.

Es-M2. Like es-M1, es-M2 increases the sensitivity to
erythromycin of cells carrying the ery-M1b or -M1a form of LC6.
There are three bases for concluding that the mutations es-M1 and
es-M2 are different. First, a study of the cross-resistance to a
group of antibiotics has shown that es-M2eryM1b is much more
resistant to oleandomycin and lincomycin, both antibiotic inhibitors
of protein synthesis, than es-M1eryM1b. Second, the erythromycin
A binding constant of ribosomes from es-M2ery-M1b cells is much
lower than that from es-M1ery-M1b cells. Third, es-M1 is not
genetically linked to es-M2.

Es-M3. Strains carrying both es-M3 and ery-M1b are not
affected by the antibiotic carbomycin whereas ery-M1b strains

carrying either es-M2 or es-M1 exhibit only a low level resistance
to this antibiotic. Also, behavior toward lincomycin sets
es-M3 in a separate class from the other two genetic factors for
erythromycin sensitivity. Either es-M2 or es-M3 can partly
suppress erythromycin-resistance of ery-M2d. The masking effects
of es-M1 and es-M3 on ery-M1b are additive. Es-M1ery-M1b can
grow on 200 µg/ml or less of erythromycin; es-M3ery-M1b can grow
on 100 µg/ml; but es-M1es-M3ery-M1b is more sensitive, it can only
grow on erythromycin concentrations up to 50 µg/ml. Es-M3 may be
linked or allelic to es-M2.

Summary. In addition to ery-M1, -M2, and -M3, three other
nuclear genetic factors which alter the erythromycin sensitivity
of Chlamydomonas chloroplast ribosomes have been identified. One
of these, es-M1, has been mapped to nuclear linkage group XI. The
other es mutants are not linked to es-M1. In total, five or six
unlinked loci in the nuclear genome are so far known to affect
the erythromycin sensitivity of the 52S subunit of the chloroplast
ribosome of C. reinhardi.

INTERGENOMIC EFFECTS

The data presented indicate that both chloroplast and nuclear
genes can specify chloroplast ribosomal proteins. Furthermore,
the nuclear genes affecting chloroplast ribosome function are
scattered among the linkage groups of that genome. Research with
es mutants shows that the effect of a nuclear gene which confers
erythromycin-resistance on the 52S subunit can be masked by an
alteration of another nuclear gene, presumably via a change in
another ribosomal protein. These phenotypic reversion mutants
demonstrate that one gene in a genome can affect the expression
of another component of the ribosome specified by another gene
of the same genome. But can a nuclear gene for a chloroplast
ribosomal component, for example, mask, or otherwise alter, the
effect of a chloroplast gene for a different chloroplast ribosomal
protein? This seems likely and a case of this sort would be a
model for some stages in the evolution of organelle ribosomes
(Bogorad, 1975). It could have important implications for understanding barriers to the natural or experimental introduction of
organelles from one individual, one variety or one species into
another.

In fact we have observed some cases where the expression of
a nuclear gene affects the expression of a chloroplast gene. In
the examples that follow, an organism carrying a nuclear gene
mutation plus a chloroplast gene mutation has a phenotype different
from an organism carrying either mutation alone. For example, the
double mutant ery-M1aery-U1a is more sensitive to lincomycin than
strains carrying either gene alone (Davidson, 1976). Another
example is the double mutant ery-M2dery-U1a, which is more

Table 2

RESISTANCE OF SINGLE AND DOUBLE MUTANTS

Strain	Erythro-mycin	Oleando-mycin	Carbo-mycin	Linco-mycin	Cleocin
wild-type	S	S	S	S	S
es-M1ery-M1b	r	r	r	S	S
es-M2ery-M1b	r	R	r	R	S
es-M3ery-M1b	r	r	R̲	r	S
ery-M1b	R	R	R	R	S
ery-M1a	R	R	R	R	S
ery-M1aery-U1a	R	R	–	S	S
ery-U1a	R̲	R	R	r	S
ery-M2dery-U1a	R̲	R	–	R	R
ery-M2d	R	R	R	R̲	r

Growth was scored on agar plate cultures.
Symbols: r, low level resistance
         R, high level resistance
         S, sensitive
         R̲, growth not affected by antibiotic

resistant to cleocin than either ery-M2d or ery-U1a (Hanson, 1976).

Finally, an increase in sensitivity to erythromycin is conferred on an ery-U37 strain by the nuclear suppressor es-M2 (Davidson, 1976). Table 2 summarizes the behavior of revertants and double mutants on a variety of antibiotics.

EVIDENCE FOR OTHER GENETIC LOCI SPECIFYING CHLOROPLAST
RIBOSOMAL PROTEINS

Table 3 lists the mutations discussed above plus others which may affect Chlamydomonas chloroplast ribosomal proteins. Listed are several mutations which affect the sensitivity of chloroplast ribosomal proteins to antibiotics other than erythromycin.

Three mutations have been shown directly or indirectly to affect the functioning of the large subunit. Sr-3 is a nuclear gene mutation which affects the binding by ribosomes of dihydro-streptomycin and results in the loss (at least from the gel) of a ribosomal protein (Brugger and Boschetti, 1975; Speiss and Arnold, 1975). The mutations car and cle are both in the plastid genome and render the ribosomes relatively insensitive in vitro to the antibiotics carbomycin and cleocin respectively (Schlanger and Sager, 1974).

Evidence has been obtained for alterations in proteins of the 52S subunit only in ery-M1, ery-M2, and ery-U1a but perhaps genes

Table 3

MUTATIONS WHICH MAY AFFECT CHLAMYDOMONAS
CHLOROPLAST RIBOSOMAL PROTEINS

	Designation	In vitro test	Protein	Linkage group	
52S	ery-M1	B	LC-6	M-XI	Mets & Bogorad, 1971, 1972; Davidson, Hanson Bogorad, 1974.
	ery-M2	B	A	M$^+$	Mets & Bogorad, 1971, 1972; Hanson, 1976.
	ery-M3			M	Surzycki & Gillham, 1971; Davidson, 1976.
	es-M1	B		M-XI	Davidson, 1976
	es-M2	B		M	Davidson, 1976
	es-M3	B		M	Davidson, 1976
	sr-3(M)	B	M	M	Brugger & Boschetti, 1975; Speiss & Arnold, 1975.
	ery-U1a	B	LC-4	U	Mets & Bogorad, 1971, 1972.
	ery-U37	P		U	Conde et al., 1975
	car (U)	P		U	Schlanger & Sager, 1974
	cle (U)	P		U	Schlanger & Sager, 1974
37S	sm-2 (U)	P	A*	U	Ohta et al., 1975
	sr-35 (U)	B	A*	U	Brugger & Boschetti, 1975.
	spc-u-1-27-3	B	M	U	Boynton et al., 1973.
	nea	P		U	Schlanger & Sager, 1974.
	sd-3-18			U	Gillham et al., 1970
	cr-1,2,3,4			M	Boynton et al., 1972 Harris et al., 1974

* May be the same protein.
A=altered; M= Protein "missing"; B=Binding; P=Protein synthesis.

specifying as many as 10 of the 26 or so proteins of the large subunit of the chloroplast ribosome may be identified through further analyses of the mutants es-M1,-M2 and -M3; sr-3, ery-U37, car and cle. In all, six genes affecting the 52S subunits have been traced to the nuclear genome and four to the chloroplast genome (Table 3).

Up to nine genes are known which may affect functions of the small subunit of the Chlamydomonas chloroplast ribosome. Cells

carrying sm-2, uniparentally transmitted factor for streptomycin
resistance, contain a chloroplast ribosomal protein with altered
mobility in carboxymethyl cellulose chromatography (Ohta et al,
1975). The ribosomes from this mutant are resistant to streptomycin in an in vitro protein synthesis system (Schlanger and
Sager, 1974). Brugger and Boschetti (1975) have studied sr-35, a
uniparentally transmitted factor for streptomycin resistance. The
ribosomes from this mutant do not bind the antibiotic and electro-
phoretograms show an altered protein in the small subunit. (It
is possible that the protein shown to be altered in sm-2 is the
same as that in sr-35.) Boynton et al.(1973) report that chloroplast
ribosomes of the spectinomycin-resistant mutant spc-ul-27-3 do not
bind the antibiotic in vitro and that a protein is absent from
their gels. Schlanger and Sager (1974) have shown that a neamine
resistant mutant contains ribosomes which are resistant to this
antibiotic in an in vitro protein synthesizing system. Less
well characterized is another uniparental mutation, sd-3-18, which
also affects the small subunits of Chlamydomonas chloroplast
ribosomes (Gillham et al., 1970).

In addition to the above, which are all uniparentally transmitted factors, are the cr mutations, cr-1, -2, -3, -4 (Boynton
et al., 1972; Harris et al., 1974) all of which result in slower
growth of the cells and reduced amounts of chloroplast ribosomes.
The first three of this group carry a higher than normal ratio of
large to small subunits of the chloroplast ribosome. This may be
the consequence of the mutation leading to instability of the small
subunits; the converse of what may occur in ery-M2. Thus, four
or five uniparental genes and perhaps some nuclear mutations may
affect the 22 or so proteins of the small chloroplast ribosome
primarily or secondarily.

If only a few of these possible genes are mapped and found to
be structural genes for chloroplast ribosomal proteins, we will be
very far along from the state of affairs in 1971 when the first of
these kinds of identifications began to be made. But it must be
emphasized that, considering the possibility that secondary change
in a polypeptide can affect its behavior on electrophoresis or
ion exchange chromatography or even the properties of a polypeptide
fragment obtained by digestion of the protein, at the present time
the only firm evidence available is for the location of one
structural gene for one chloroplast ribosomal protein, that is
ery-M1.

REGULATORY POSSIBILITIES

One of the central problems in eukaryotic cell biology is the
mechanism of integration of organelles into the life of the cell.
This is probably a two-way road with signals from the nucleus
affecting the organelle and feedback of information from the

organelle to regulate the expression of nuclear genes. What is
the situation with regard to chloroplast ribosomal proteins in
Chlamydomonas?

Probably the best illustration of possibilities of nuclear
control of chloroplast assembly and function through regulation of
plastid ribosome formation are mutants such as ery-M2 strains with
altered levels of chloroplast ribosomes. The ery-M2 mutants
consistently have reduced amounts of chloroplast ribosomes. As
discussed earlier, this may be a result of the nature of the proteins specified by this locus for the 52S ribosome subunit. The
cr mutations described by Boynton et al., (1973) and by Harris
et al., (1974) and ac-20 (Goodenough and Levine, 1970; Boynton
et al., 1972) are other nuclear mutations which result in reduced
levels of chloroplast ribosomes.

It is possible that further study of mutants of these types
may help reveal whether the expression of nuclear genes for chloroplast ribosomal proteins is constitutive or is modulated to regulate the level of organelle ribosomes. We do not know whether the
failure of the nuclear-cytoplasmic system to produce chloroplast
ribosomal proteins results in the arrest of synthesis by the
chloroplast of rRNAs and those ribosomal proteins made from
information in its genes or whether the chloroplast components
are continuously made and degraded in the absence of a complete
set of ribosomal elements. We do not know whether chloroplast
ribosomal protein(s) or some other protein(s) of chloroplast
origin can affect the expression of genes for chloroplast ribosome
components which are present in the nuclear genome. Integrative
problems like these have to be attacked. Knowledge of the
locations of genes for some ribosomal components should promote
interest and the possibility of experimental analysis.

EVOLUTIONARY CONSIDERATIONS

The data which have been considered indicate that genes which
specify chloroplast ribosomal proteins are scattered among a
number of linkage groups in the nuclear genome of C. reinhardi.
Add to this evidence from molecular hybridization experiments
that show genes for C. reinhardi chloroplast rRNA are in chloroplast DNA (Bastia et al., 1971a,b), then there is no doubt that
the genes for components of chloroplast ribosomes are separated
in two genomes.

How did the organelle genome originate? How have genes for
some chloroplast ribosome components come to be in the nuclear
genome and others in the chloroplast genome? Can we understand
why genes for some chloroplast ribosome components are in the
chloroplast genome and others are in the nuclear genome? To begin

to examine these questions we must first consider possible mechanisms for the origin of orangelles in eukaryotic cells and then
the neglected problems of intracellular evolution (Bogorad, 1975).

   Fundamentally two types of proposals have been made. First,
and most widely discussed, is the endosymbiont hypothesis (e.g.
Margulis, 1970). According to this view two or three cells
combined to make one. A nucleated aerobic cell is visualized
as the host for the blue-green algae (or the progenitors of modern
blue-green algae) from which chloroplasts of contemporary eukaryotic cells have developed. Mitochondria, on the other hand, are
visualized as the decendents of aerobic bacteria (or perhaps,
blue-green algae) which first existed as endosymbionts of the
primitive nucleated aerobic cell. The entire apparatus for photosynthesis of a modern green plant cell would thus have originated
with the endosymbiotic blue-green alga; the oxidative repiratory
apparatus of the modern plant or animal cell would have originated
as a functional unit of the invading bacterium.

   The cluster-clone hypothesis provides another possible
description of the origin of eukaryotism (Bogorad, 1975). According to this view a single cell organized roughly like a modern
bacterium or blue-green alga evolved into the modern eukaryotic
cell through the following series of steps: (1) the formation of
clusters of genes (each cluster is a genome);   (2) the development
of membranes surrounding each cluster of genes and a small amount
of cytoplasm to produce two or more gene-containing structures --
nuclei would have evolved from one of these structures, mitochondria
from another, and chloroplasts from the third -- plus the gene-
free space, i.e., the cytoplasm; and (3) the division and faithful
reproduction of each gene cluster-containing unit to give rise to
clones of nuclei, chloroplasts, and mitochondria which, together
with the cytoplasm, have evolved into the multiple types of
contemporary eukaryotic cells. Suggestions by Raff and Mahler (1972),
Uzzell and Spolsky (1974), Meyer (1973) and Nass (1969) can be
taken as step-wise variants on a basic cluster-clone pattern and
have been summarized by Bogorad (1975).

   If eukaryotism originated by endosymbiosis and if ribosomes
originated only once, both the nucleated and anucleate cells which
became partners would have evolved at some earlier point from a
common ancestral type containing a single kind of ribosome. As the
two types of cells evolved, ribosomes with distinctive properties
could have developed. The two or three types of ribosomes which
evolved in separate cell lines would have been brought into the
same cell when endosymbiosis began -- one type of ribosome coming
in with each endosymbiont and one type present already in the
"host". Each participating endosymbiont would have entered the
pre-eukaryote association with a distinctive ribosomal type and the

genes for all of the components for this ribosome. As we have
seen, at least for Chlamydomonas chloroplast ribosomes, the genes
and the gene products are separated in different compartments.
Genes in the nucleus specify some chloroplast ribosomal proteins,
genes in the chloroplast specify others.

Two mechanisms by which genes coding for organellar ribosomal
proteins could have dispersed have been suggested (Bogorad, 1975).
One possibility is the direct transfer of a gene from a genome in
one compartment to a genome in another membrane-limited cell
compartment. Protein and gene substitution is another possibility.
The latter mechanism envisions the functional loss of a gene and/or
its product from it original compartment and the substitution of
the product of a nuclear gene to permit the organelle ribosome to
function.

     This pair of possibilities is easiest to visual in consider-
ing the endosymbiont hypothesis   but such dispersal mechanisms
are also required for forms of the cluster-clone hypothesis
However, if the progenitor of the eukaryotic cell had only one
copy of each gene, separation of genes could have been one of the
earliest steps in the evolution of the modern eukaryotic cell
rather than coming much later.

     Evidence continues to accumulate that multicomponent systems
such as membranes and multimeric enzymes within an organelle
can be the products of genes in two genomes (e.g. summary by
Bogorad, 1975). There are not examples yet of a cytoplasmic
component which is the product of more than one genome but this
possibility cannot be excluded. It may be that intergenomic
cooperation is a principle of only organelle biology or it
may be more general in eukaryotic cells. Intracellular evolution
and intergenomic cooperation are emerging fields of biological
research. To date the most provocative observations have come
from studies of intracellular organelles of eukaryotes. Most of
the questions raised in the second paragraph of this section remain
to be addressed experimentally.

Acknowledgements

     The work described from our laboratory and the preparation
of this manuscript have been supported in part by research grant
GM 20470 of the National Institute of General Medical Sciences
and in part by the Maria Moors Cabot Foundation of Harvard University.

References

1. Bastia, D., Chiang, K.-S. and Swift, H. (1971) Amer. Soc. Cell Biol. Abstracts p. 25.
2. Bastia, D., Chiang, K.-S., Swift, H. and Siersma, P. (1971b) Proc. Nat. Acad Sci. 68: 1157.
3. Bogorad, L. (1975) Science, 188: 891.
4. Boynton, J.E., Gillham, N.W. and Chabot, J.F. (1972) J. Cell Sci. 10: 267.
5. Boynton, J.E., Burton, W.G., Gillham, N.W. and Harris, E.H. (1973) Proc. Nat. Acad. Sci. 70: 3463.
6. Brugger, M. and Boschetti, A. (1975) Eur. J. Biochem. 58: 603.
7. Conde, M.F., Boynton, J.E., Gillham, N.W., Harris, E.H., Tingle, C.L. and Wang, W.L. (1975) Molec. Gen. Genet. 140: 183.
8. Davidson, J.N. (1976) Genes Affecting Erythromycin Resistance in ery-M1 Mutants of Chlamydomonas reinhardi. Ph.D. Thesis. Department of Biology, Harvard University.
9. Davidson, J.N., Hanson, M.R. and Bogorad, L. (1974) Molec. Gen. Genet. 132: 119.
10. Fernandez-Munoz, R. and Vazquez, D. (1973) J. Antibiotics 26: 107.
11. Garrett, R.A. and Wittmann, H.G. (1973) Adv. Prot. Chem. 27: 277.
12. Gillham, N.W., Boynton, J.E. and Burkholder, B. (1970) Proc. Nat. Acad. Sci. 67: 1026.
13. Goodenough, U.W. and Levine, R.P. (1970) J. Cell Biol. 44: 547.
14. Hanson, M.R. (1976) The Genetics and Biochemistry of Chloroplast Ribosome Mutations of Chlamydomonas reinhardi. Ph.D. Thesis. Department of Biology, Harvard University.
15. Hanson, M.R., Davidson, J.N., Mets, L.J. and Bogorad, L. (1974) Molec. Gen. Genet. 132: 105.
16. Harris, E.H., Boynton, J.E. and Gillham, N.W. (1974) J. Cell Biol. 63: 160.
17. Langer, J.A., Acharya, A.S. and Moore, P.B. (1975) Biochim. Biophys. Acta 407: 320.
18. Mao, J. C-H. (1971) In, Drug Action and Drug Resistance in Bacteria, S. Mitsuhashi, ed. pp. 153. U. of Tokyo Press, Japan.
19. Margulis, L. (1970) Origin of Encaryotic Cells. Yale Univ. Press, New Haven.
20. Mets, L.J. (1973) Genetic Analysis of Chloroplast Ribosome Structure in Chlamydomonas reinhardi. Ph.D. Thesis. Department of Biology, Harvard University.
21. Mets, L.J. and Bogorad, L. (1971) Science, 174: 707
22. Mets, L. and Bogorad, L. (1972) Proc. Nat. Acad. Sci. 69: 379.
23. Mets, L. and Bogorad, L. (1974) Anal. Biochem. 57: 200.
24. Meyer, R.R. (1973) J. Theoret. Biol. 38: 647.
25. Nass, S. (1969) Intl. Rev. Cytol. 25: 55.
26. Ohta, N., Sager, R. and Inouye, M. (1975) J. Biol. Chem. 240: 3655.

27. Raff, R.A. and Mahler, H.R. (1972) Science 177: 575.
28. Sager, R. (1972) Cytoplasmic Genes and Organelles, Academic Press, Inc., New York.
29. Schlanger, G. and Sager, R. (1974) Proc. Nat. Acad. Sci. 71: 1715.
30. Speiss, H. and Arnold, C.G. (1975) Arch. Microbiol. 103: 89.
31. Surzycki, S.J. and Gillham, N.W. (1971) Proc. Nat. Acad. Sci. 68: 1301.
32. Teraoka, H. (1970) J. Mol. Biol. 48: 511.
33. Uzzell, T. and Spolsky, C. (1974) Am. Sci. 62: 334.
34. Weber, K. and Osborn, M. (1969) J. Biol. Chem. 244: 4406.

PROTEIN SYNTHESIS IN PLANTS: SPECIFICITY AND ROLE OF THE CYTOPLASMIC AND ORGANELLAR SYSTEMS

CIFERRI, O., TIBONI, O., MUNOZ-CALVO, M.L. and CAMERINO, G., Istituto di Microbiologia e Fisiologia Vegetale, Università di Pavia, Pavia (Italia)

PLANT CELLS CONTAIN THREE PROTEIN-SYNTHESIZING SYSTEMS

Eukaryotic plants contain three different protein-synthesizing systems, localized in the cell cytoplasm, in the mitochondrion, and in the chloroplast (1,2). Indeed it is possible to isolate from plant cells cytoplasmic, mitochondrial and chloroplastic ribosomes that differ in the sedimentation coefficient and in the rRNA composition (3) as well as in the ribosomal proteins (4). In addition different tRNA's and aminoacyl-tRNA synthetases have been reported to be present in the cytoplasm, the chloroplast and the mitochondrion (5) and evidence has been presented suggesting the existence of cytoplasmic-, chloroplast- and mitochondrion-specific translation factors (6). Thus eukaryotic plants, whether unicellular or pluricellular, represent a unique material for many studies such as the regulation and the interrelations of the three protein-synthesizing systems, the comparative structure of the

155

macromolecules that catalyze the same reactions in the three systems, the localization of the genetic information for these macromolecules, etc.

The isolation and characterization of the different macromolecules involved in protein synthesis (RNA polymerase, ribosomes, translation factors, tRNA, aminoacyl-tRNA synthetases etc.) in the three cellular compartments represent, however, a formidable task for the cell biochemist. This task is especially difficult when one tries to isolate the different components involved in mitochondrial and chloroplastic protein synthesis since the two systems are very similar one to the other, as indicated, for instance, by the size of ribosomes, the mechanism for the initiation of the peptide chains, and the sensitivity to ribosome-specific inhibitors (Table I). Nevertheless sufficient experimental data are now available demonstrating that for each of the macromolecules involved in the translation of genetic information there are three distinct types, one for each protein-synthesizing system. The possibility exists that some overlap may occur, e.g. that some tRNA and aminoacyl-tRNA synthetase may be utilized for protein synthesis in the cytoplasm and in one or both the two types of organelles. However it is clear that there exist tRNA's and synthetases specific for each protein -synthesizing system as there are different ribosomes, RNA polymerases, elongation factors, and possibly initiation and termination factors.

From a functional point of view, protein synthesis in mitochondria and chloroplasts appears to be much more similar to the process occurring in prokaryotes than to that taking place in the cytoplasm of eukaryotes. For instance, all tests performed on ribo-

TABLE I

CHARACTERISTICS OF THE PROTEIN-SYNTHESIZING
SYSTEMS PRESENT IN PLANTS

	S value of ribosome, ca.	Initiator tRNA	Inhibition of protein synthesis in vivo and in vitro by	
			CAP	CHI
Prokaryotes	70	F-met-tRNA	+	−
Eukaryotes				
Cytoplasm	80	Met-tRNA	−	+
Mitochondrion	70	F-met-tRNA	+	−
Chloroplast	70	F-met-tRNA	+	−

CAP, chloramphenicol; CHI, cycloheximide.

somes isolated from organelles indicate that many of the reactions for protein synthesis (such as those for peptide chain initiation and elongation) occur in the presence of translation factors isolated from organelles or from prokaryotes but not in the presence of the factors obtained from the cytoplasm of eukaryotes (Table II). Similarly initiation and elongation factors purified from mitochondria and chloroplasts catalyze the same reactions on bacterial ribosomes but not on the ribosomes occurring in the cytoplasm of eukaryotes. In at least one case a crude preparation of translation factors from chloroplasts catalyzes the incorporation in vitro of amino acids by mitochondrial ribosomes (11).

TABLE II

INTERCHANGEABILITY OF RIBOSOMES AND TRANSLATION
FACTORS FROM PROKARYOTES AND EUKARYOTES

Ribosomes from	Translation factors from			
	Prokaryotes	Eukaryotes		
		Cyt.	Mit.	Chloro.
Prokaryotes	+	−	+	+
Eukaryotes				
Cytoplasm	−	+	−	−
Mitochondrion	+	−	+	+
Chloroplast	+	−	(+?)	+

+; active system; −, inactive system; (+?), not tested but presumably active.

Most of the data have been reported in the case of the reactions for peptide chain elongation (7) but a similar specificity has been demonstrated also for some of the reactions involved in peptide chain initiation (8, 9, 10).

The presence of protein-synthesizing systems in mitochondria and chloroplasts raises the interesting question concerning the site of synthesis of the organellar proteins. This is a very cogent question since genetic and biochemical data demonstrate that many of the soluble enzymes present in these organelles

(e.g those of the Calvin and the Krebs cycles) are synthesized
in the cell cytoplasm and then, through a mechanism still unrav-
elled, exported into the organelles. Even some of the proteins
that constitute the organellar protein synthesizing system (e.g.
the ribosomal proteins), appear to coded in the nuclear rather
than in the organellar DNA (Table III) and the genetic information
for many of these proteins is translated in the cell cytoplasm rather than in the organelles.

TABLE III

SITE OF CODIFICATION OF THE COMPONENTS OF MITO-
CHONDRIAL AND CHLOROPLASTIC PROTEIN-SYNTHESIZING
SYSTEMS

Component	Mitochondrial coded in	Chloroplastic
RNA polymerase	nucleus	nucleus (?)
rRNA's	mitochondrion	chloroplast (only?)
ribosomal proteins	nucleus	nucleus (all?)
tRNA's	mitochondrion (many)	chloroplast (some)
aminoacyl-tRNA synthetases	nucleus (some)	nucleus (some)
transformylase	nucleus	nucleus (?)
elongation factors	nucleus	?

For references see (2).

Experiments performed with intact cells or mitochondria isolated from yeast or N. crassa suggest that no soluble protein is
synthesized in substantial amounts in these organelles. Mitochondria synthesize a few protein components of the mitochondrial
membranes, including some peptides of the cytochrome oxidase
complex, of the oligomycin-sensitive ATP-ase, and cytochrome
b (12). Mitochondria from higher plants most probably synthesize
the same proteins (13).

Chloroplasts appear to synthesize a number of membrane proteins, including some components of the thylakoid membrane, of
the cytochromes and of the coupling factor (14,15). However, in
contrast to mitochondria, at least one soluble protein, namely
the large subunit of ribulose-1,5-diphosphate carboxylase (RuDP
carboxylase), is synthesized in the chloroplast itself (16,17).

Assuming that chloroplast DNA has a molecular weight of approximately 1×10^8 daltons (14,18), the genetic information for
the above mentioned proteins as well as for the other macromolecules known to be coded in the chloroplast DNA (chloroplast ribosomal RNA's and tRNA's) does not account for more than 10% of
the organellar genome. This means that approximately 90% of the
chloroplast DNA has not been accounted for. This amount of DNA
could code for approximately 100 proteins with a molecular weight
of 50,000 daltons. We think that the chloroplast-specific translation factors, that is the soluble proteins that catalyze peptide
chain initiation, elongation and termination in the organelle, represent a group of soluble proteins synthesized in the chloroplast.

CHLOROPLAST ELONGATION FACTORS
ARE PRODUCED IN THE CHLOROPLAST

We have previously reported that from the unicellular alga
Chlorella vulgaris it is possible to isolate three elongation factors endowed with translocase activity (6). One is active only on
80S ribosomes and is elongation factor-2 (EF-2), specific for
protein synthesis in the cytoplasm. The other two elongation factors catalyze translocation only on ribosomes of the 70S type.
One of the latter factors may be isolated from chloroplasts and is
present in detectable amounts only in light-grown cells. This has
been considered to be the chloroplast-specific elongation factor
G ($EF-G_{chl}$) while it has been surmised that the other EF-G is
that associated with the mitochondrion ($EF-G_{mit}$). The two EF-G's
may be separated by column chromatography and shown to be different proteins by immunological tests (19) and by their diffe·ent
sensitivity to the inhibition by fusidic acid (6), an inhibitor of protein synthesis that interacts specifically with EF-G or EF-2.

The synthesis of the two EF-G's is differentially affected
when intact cells of the alga are exposed to antibiotics that selectively inhibit protein synthesis. The synthesis of $EF-G_{mit}$ is, if
any, stimulated when a culture of the alga is treated with chloramphenicol or erythromycin (20), antibiotics known to inhibit mitochondrial or chloroplastic protein synthesis but not protein synthesis in the cytoplasm. On the contrary, the synthesis of $EF-G_{chl}$
is inhibited when cells of the alga are exposed to chloramphenicol
but not to cycloheximide (21). The conclusion that $EF-G_{chl}$ is produced in the chloroplast and not in the cytoplasm is confirmed by

demonstrating that labelled $EF-G_{chl}$ may be isolated from cells supplied with radioactive amino acids in the presence of cyclohe- ximide (22). Experiments performed on chloroplasts isolated from spinach confirm that $EF-G_{chl}$ is one of the soluble proteins that becomes labelled when isolated chloroplasts are incubated with ra- dioactive methionine in the presence of cycloheximide (23). Indeed $EF-G_{chl}$ and probably $EF-T_{chl}$ appear to be the most labelled sol- uble proteins after RuDP carboxylase. Thus while mitochondrial elongation factors appear to be a product of cytoplasmic protein synthesis, the chloroplast-specific ones are synthesized in the organelle itself.

Analysis by disc gel electrophoresis of the soluble protein prepared from isolated spinach chloroplasts has demonstrated that $EF-G_{chl}$ and $EF-T_{chl}$ represent a considerable portion of the soluble protein of the organelle (Fig. 1). We have estimated that $EF-G_{chl}$ and $EF-T_{chl}$ account for at least 20% of chloroplast sol- uble protein. This finding is not too surprising if one considers that, in a bacterial cell such as that of E. coli, EF-G represents approximately 2-5% of the soluble protein (24) and that in a bacte- rial cell there are as many molecules of EF-G and $EF-T_s$ as there are ribosomes(25) while $EF-T_u$, the other component of the EF-T complex, is present in even higher proportions, i.e. 8 to 14 molecules per ribosome (26). Chloroplast ribosomes represent approximately 50% of the total cell ribosomes (14) (e.g. it has been reported that there are 300,000 ribosomes in a chloroplast from Chlamydomonas reinhardii (27). If one assumes that the same ratio elongation factors to ribosomes occurs also in chloroplasts, the number of molecules of these factors in the organelle should be very high (i.e. 300,000 molecules of EF-G and $EF-T_s$ and pos-

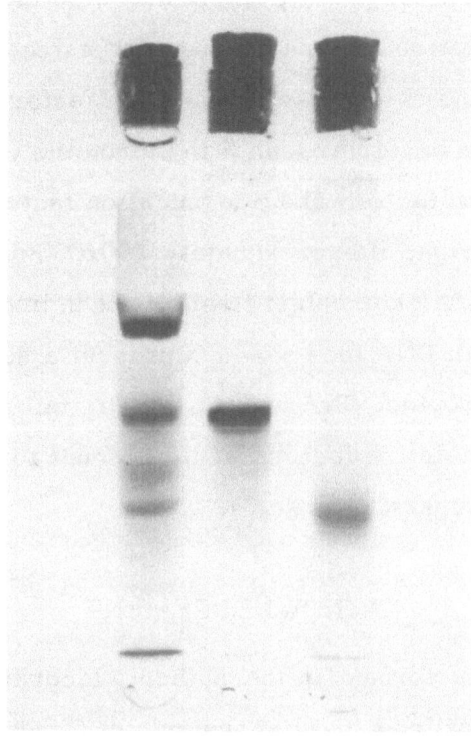

Fig. 1 <u>Separation by disc gel electrophoresis of chloroplast solu-</u>
<u>ble protein from spinach</u>

Left, chloroplast soluble protein (31 μg)

Center, purified EF-T$_{chl}$ (45 μg of protein)

Right, purified EF-G$_{chl}$ (15 μg of protein)

sibly, a few million of EF-T$_u$ molecules per chloroplast). This
would represent a considerable proportion of the soluble proteins
present in the chloroplast and would explain why EF-G$_{chl}$ and
EF-T$_{chl}$ are, after RuDP carboxylase, the most prominent bands
visible when fractionates the soluble protein present in isolated
chloroplasts. If, as it seems, EF-G$_{chl}$ and EF-T$_{chl}$ are produced

in the chloroplast, it is quite likely that also the other organelle-
-specific translation factors may be synthesized in the organelle.
Assuming that the chloroplast translation factors are similar to
those present in bacteria, then 8 to 9 proteins (3 initiation factors,
2 or 3 elongation factors and 3 termination factors) with a combined molecular weight of approximately 400,000 daltons (28) may be
synthesized in the chloroplast itself. If such proteins were coded
in the organellar DNA they would represent a significant proportion of the chloroplast DNA and the genetic information for all the
chloroplast translation factors would account for at least another
10% of the chloroplast genome.

AKNOWLEDGEMENT

The work performed in the authors' laboratory has been supported by a grant from the Laboratorio di Genetica Biochimica ed
Evoluzionistica of the Consiglio Nazionale delle Ricerche. One of
us (M.L.M.C.) is supported by a fellowship of the "Plan de For-
macion de Personal Investigador" of the C.S.I.C., Madrid, Spain.

REFERENCES

1 Boulter, D., Ellis, R.J. and Yarwood, A. (1972) Biol. Rev.47,
   113-175

2 Ciferri, O. (1975) The Chemistry and Biochemistry of Plant
   Proteins (Harborne, J.B. and Van Sumere C.F. eds.), pp. 113-
   -135, Academic Press, New York

3 Avadhani, N.G. and Buetow, D.E. (1972) Biochem. J. 128, 353-
   -365

4 Vasconcelos, A.C.L. and Bogorad, L. (1971) Biochim. Biophys.

Acta 228, 492-502

5 Burkard, G., Guillemaut, P., Steinmetz, A. and Weil, J.H. (1973) Nitrogen Metabolism in Plants (Goodwin, T.W. and Smillie, R.M.S. eds.), Biochemical Society Symposium vol. 38, pp. 43-56

6 Ciferri, O. and Tiboni, O. (1973) Nat. New Biol. 245, 209-211

7 Ciferri, O. and Parisi, B. (1970) Progress in Nucleic Acid Research and Molecular Biology (Davidson, J.N. and Cohn, W.E. eds.), vol. 10 pp. 121-144, Academic Press, New York

8 Sala, F. and Kuntzel, H. (1970) Eur. J. Biochem. 15, 280-286

9 Sala, F., Sensi, S. and Parisi, B. (1970) FEBS Lett. 10, 89-91

10 Tiboni, O., Di Pasquale, G. and Ciferri, O. (1976) Plant Sci. Lett. 6, 419-429

11 Avadhani, N.G. and Buetow, D.E. (1974) Biochem. J. 140, 73-78

12 Schatz, G. and Mason, T.L. (1974) Ann. Rev. Biochem. 43, 51-87

13 Leaver, C.J. personal communication

14 Ellis, R.J. and Hartley, M.R. (1974) MTP International Review of Sciences (Burton, K. ed.), Biochemistry Series, vol. 6, Medical Technical Publishing Co., Lancaster

15 Mendiola-Morgenthaler, L.R., Morgenthaler, J.J. and Price, C.A. (1976) FEBS Lett. 62, 96-100

16 Blair, G.E. and Ellis, R.J. (1973) Biochim. Biophys. Acta 319, 223-234

17 Bottomley, W., Spencer, D. and Whitfeld, R.P. (1974) Arch. Biochem. Biophys. 164, 106-117

18 Kirk, J.T.O. (1972) Subcell-Biochem. 1, 333-361

19 Munoz-Calvo, M.L., Tiboni, O. and Ciferri, O. manuscript in preparation

20 Ciferri, O., Tiboni, O., Lazar, G. and Van Etten, J. (1974) The Biogenesis of Mitochondria (Kroon, A.M. and Saccone, C. eds.), pp. 107-115 Academic Press, New York

21 Ciferri, O., Tiboni, O. and Amileni, A.R. (1975) Molecular Biology of Nucleocytoplasmic Relationships (Puiseux-Dao, S., ed.), pp. 47-51, Elsevier Publishing Company, Amsterdam

22 Ciferri, O. and Tiboni, O. Plant Sci. Lett. in press

23 Tiboni, O. and Ciferri, O. unpublished results

24 Leder, P., Skogerson, L.E. and Nau, M.M. (1969) Proc. Natl. Acad. Sci. U.S., 62, 454-460

25 Gordon, J. (1970) Biochem. 9, 912-917

26 Furano, A.V. (1975) Proc. Natl. Acad. Sci. U.S. 72, 4780--4784

27 Boynton, J.E., Gillham, N.W. and Chabot, J.F. (1972) J. Cell. Sci. 10, 267-305

28 Altman, L. and Dittmer Katz, D. (eds.), (1976) Biological Handbooks I. Cell Biology, pp. 302-303, FASEB Bethesda, Md.

FUNCTIONAL CHARACTERIZATION OF THE INITIATION FACTORS OF WHEAT GERM

Samarendra N. Seal, Marcella Giesen, Ruth Roman and
Abraham Marcus
The Institute for Cancer Research, The Fox Chase Cancer
Center, Philadelphia, Pennsylvania 19111

A. Introduction

    Protein biosynthesis is one of the first biological processes
"activated" by exposure of wheat embryos to water (1). A major
facet of the "activation" process is the attachment of ribosomes to
preformed mRNA (2), and the reaction can, in part, be reproduced
in vitro (3). In further studies it was found that the ribosome
attachment reaction could be carried out with a number of eucaryotic
mRNAs, in particular with plant viral RNAs (4, 5, 6). Utilizing
particularly TMV-RNA, we have studied aspects of the initiation
reaction in vitro and the subsequent sections summarize the current
status of these studies.

B. Characteristics of TMV-RNA Catalyzed Amino Acid Incorporation

    With isolated ribosomal subunits (7) and an initiation inhibitor,
aurintricarboxylic acid (8), it can be shown that the initiation
reaction requires ATP, GTP and mRNA as well as at least two initiation factors (9). The process occurs on the 40S ribosomal subunit
and results in the formation of a 50S complex containing an initiator
species of methionyl-tRNA. This complex subsequently combines with
a 60S ribosomal subunit forming an 80S ribosome-Met-tRNA initiation
complex. Amino acid incorporation into protein requires, in addition,
two elongation factors.

C. Resolution of the Wheat Protein Chain Initiation System and
    Analysis of the Functions of the Resolved Factors

    One of the more controversial questions with regard to the
initiation process is the ordering of the sequence of attachment of

167

the mRNA and the initiator tRNA to the 40S ribosomal subunit. In
procaryotic systems, the binding of fMet-tRNA to ribosomes has an
almost absolute requirement either for the trinucleotide AUG or for
an mRNA containing the AUG codon (10). Furthermore, binding of
radioactive mRNA to 30S ribosomal subunits can be carried out in
the absence of fMet-tRNA. Thus the procaryotic system seems to
function by initially binding the mRNA which then provides an
aligning site for the binding of the initiator tRNA[1]. In early
studies of Met-tRNA binding to ribosomes with the wheat system, a
strong requirement for mRNA was noted (Table 1) suggesting either
the simultaneous binding of the mRNA and the initiator tRNA or a
sequential process with the mRNA attaching first. Similar results
were obtained by Anderson and coworkers (13, 14) with a fractionated
system obtained from the ribosomal wash of rabbit reticulocytes.
Studies from a number of other laboratories (15-21), however,
described systems from rabbit reticulocytes and mouse fibroblasts
in which a 40S-Met-tRNA complex could be formed in the absence of
an mRNA. Furthermore, a factor functioning in all of these systems

Table 1. Requirements for ribosomal binding of methionyl-tRNA

| Substrate | Conditions | Met-tRNA bound (pmol) | | |
		+ TMV-RNA	− TMV-RNA	Δ
Met-tRNA	No factors	0.20		
	C + D	1.44	0.26	1.18
	C alone	0.17		
	D alone	0.20	0.18	
	C + D (ATP omitted)	0.35	0.11	0.24
	C + D (GTP omitted)	0.70	0.11	0.59
Met-tRNA$_i$	C + D	1.45	0.18	1.27
Met-tRNA$_m$	C + D	0.42	0.22	0.20

The reaction mixture contained in a volume of 0.34 ml: 1.1 mM
Met, 30 mM Tris acetate, pH 8, 1.1 mM ATP, 60 µM GTP, 10 µg of TMV
RNA, 2.6 mM dithiothreitol, 1.3 mM MgAc$_2$, 51 mM KCl, ribosomes (220
µg of RNA), 38 pmol of unresolved [14C]Met-tRNA or 34 pmol of [14C]-
Met-tRNA$_i$ or [14C]Met-tRNA$_m$ (1 pmol = 330 c.p.m.) and initiation
factors: C, 170 µg and/or D-final, 125 µg (ref. 12). After incubating 10 min at 20°C the [14C]Met-tRNA retained by nitrocellulose
filters (i.e. [14C]Met-tRNA bound to ribosomes) was determined.

could bind Met-tRNA in a form adsorbable to nitrocellulose filters.
In the partially fractionated wheat system (see Table 1) no such
reactions were observed. Subsequent experiments, however, involving the further resolution of the wheat factors, have resulted in
very different findings. As seen in Table 2, a factor (C3β) can
be isolated that does indeed bind Met-tRNA in a reaction independent

Table 2. <u>Aminoacyl-tRNA binding to factor C3β</u>

| | | Aminoacyl-tRNA bound | |
	Conditions	Met-tRNA$_1^{Met}$ pmol	Phe-tRNA pmol
1	complete system	2.24	
	−C3β	0.08	
	−GTP	0.61	
	−MgAc$_2$	2.02	
	−C3β + EF$_1$	0.04	
2	complete system	2.75	2.14
	+ D	1.34	0.18

The complete system contained in a volume of 0.17 ml: 1 mM ^{12}C amino acid (corresponding to the radioactive aminoacyl-tRNA used), 30 mM tris acetate pH 8.0, 2.6 mM dithiothreitol, 60 μM GTP, 70 mM KAc, 5 mM KCl, 1.3 mM MgAc$_2$, 7 mM phosphoenolpyruvate, 5.6 μg pyruvate kinase, 20 μg C3β and 19.6 μg tRNA containing either 3.3 pmol ^{3}H Met-tRNA$_1^{Met}$ (2300 cpm/pmol) or 14.3 pmol ^{14}C Phe-tRNA (666 cpm/pmol). Where indicated, 90 μg D-final were added or C3β was replaced by 45 μg EF1 (C3γ). Data similar to that presented for EF1 (exp. 1, line 5) were obtained when C3β was replaced either by C3α, A, D-final, or A2, all in amounts as used in Table 3, exp. 1. After 10 min incubation at 20°C, aminoacyl-tRNA binding was determined by the nitrocellulose filter assay. The preparation of the various factors is described in ref. 22.

of mRNA. The reaction requires GTP and is unaffected by variation of the Mg^{++} concentration from 0.2 to 5 mM. The resolution leading to the preparation of factor C3β is shown in fig. 1 and was developed on the basis of a requirement for TMV-RNA catalyzed transfer of amino acids into protein from radioactive aminoacyl tRNA (see Table 3). A clue to the inability to detect the ribosome and mRNA-independent Met-tRNA binding reaction in the cruder fractions is provided by experiment 2 of Table 2. The presence of factor D, a component absolutely required for the overall initiation sequence (see figs. 2 and 3) results in a considerable decrease in the binding of Met-tRNA. A substantially greater effect of this type in crude fraction C would reduce the Met-tRNA binding to an essentially insignificant level.

A further divergence with the more fractionated preparations was obtained in studies of the formation of a 40S-Met-tRNA complex. As seen in fig. 2, the reaction is independent of mRNA, clearly in contrast to the earlier results. The explanation here, however, is clear. The 40S-Met-tRNA complex is fixed with glutaraldehyde which

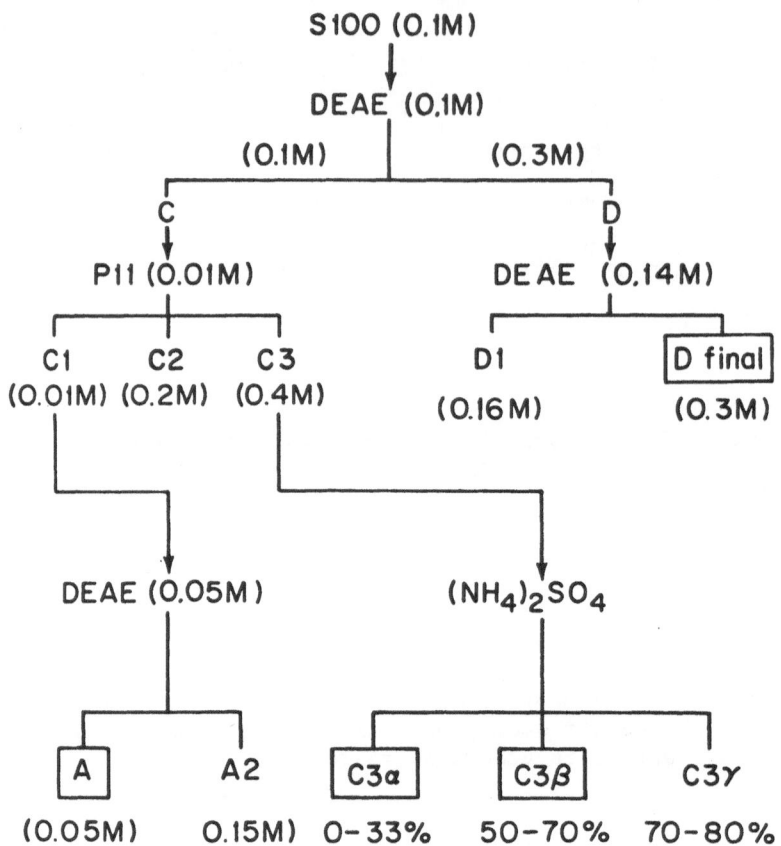

FIG. 1. Schematic representation of the fractionation of wheat germ supernatant. A detailed description is presented in ref. 22. The components surrounded by a box are those required for amino acid polymerization (see Table 2).

stabilizes an otherwise essentially nondetectable complex. Formation of this complex requires GTP, C3β (the Met–tRNA binding factor), and factor D. The function of factor D may be related to its ability to lessen the affinity of factor C3β for the Met–tRNA, thereby perhaps allowing the Met–tRNA to bind to an appropriate ribosomal site. Another possible function of factor D might be to insure the specificity of the C3β reaction. As noted in Table 2, experiment 2, the presence of factor D, while only halving the binding of Met–tRNA, almost completely eliminates the binding of Phe–tRNA.

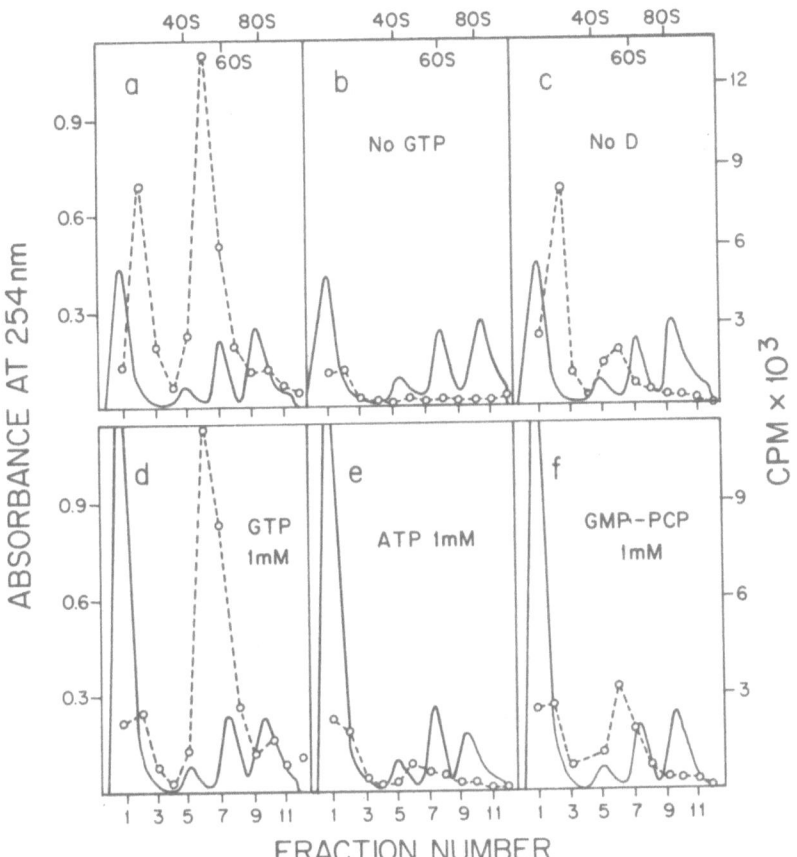

FIG. 2. Requirements for the formation of a 40S·Met-tRNA$_f^{Met}$ complex. a) The complete reaction mixture (0.34 ml) contained 1 mM unlabeled methionine, 30 mM tris acetate pH 8.0, 2.6 mM dithiothreitol, 60 μM GTP, 70 mM KAc, 5 mM KCl, 1.3 mM MgAc$_2$, 7 mM phosphoenolpyruvate, 11.2 μg pyruvate kinase, 39 μg tRNA containing 6.6 pmol [CH$_3$-^{3}H]Met-tRNA$_f^{Met}$ (2300 cpm/pmol), 25 μg C3β, 157 μg D final and 115 μg K$_{100}$M$_2$ ribosomes. In b) GTP was omitted. In c) fraction D was replaced by 155 μg bovine serum albumin. In d,e,f), phosphoenolpyruvate, pyruvate kinase and the low 60 μM GTP were omitted, and 1 mM GTP, ATP, or GMP-P(CH$_2$)P (guanylyl-5'-methylene diphosphonate) were added as indicated. After incubation for 10 min at 20°, 38 μl of 10% glutaraldehyde in 40 mM triethanolamine were added and the mixtures were kept on ice for 5 min. The samples were then centrifuged through a 5 ml 13-21% sucrose gradient in an SW50.1 Spinco rotor for 70-80 min at 230,000 x g. 0.45 ml fractions were collected and analyzed for bound aminoacyl tRNA by nitrocellulose filtration. The solid line traces absorbance while radioactivity is denoted by the open circles.

Table 3. <u>Factor requirements for TMV-RNA catalyzed amino acid polymerization</u>

	Conditions	Amino acid incorporation pmol
1	complete system	8.6
	$-C_3\beta$	0.4
	$-C_3\alpha$	1.9
	$-A$	2.7
	$-D$	0.4
2	complete system	8.6
	$-A$	0.9

The complete system contained in a volume of 0.2 ml: 2.2 mM dithiothreitol, 50 µM GTP, 1 mM ATP, 25 mM Tris-acetate pH 8.0, 3.6 mM Mg-acetate, 70 mM K-acetate, 5 mM KCl, 8 mM creatine phosphate, 8 µg creatine phosphokinase, 158 µg $K_{100}M_2$ ribosomes, 5 µg TMV-RNA, 7.2 µg tRNA containing 29 pmol 8-^{14}C-aminoacyl-tRNAs (450 cpm/pmol), and 8 ^{12}C-amino-acids (corresponding to the ^{14}C-amino-acids; 1 mM each). In experiment 1 the following factors were added in the complete system: 20 µg C3β, 20 µG C3α, 27 µg A, 94 µg D-final, 4.5 µg EF1, and 42 µg A2 as a source of EF2. In experiment 2, 600 µg D were added in place of D-final (this level of D contains a saturating level of EF2) and A2 was omitted. Under these conditions, a clearer requirement for fraction A can be demonstrated. After 15 min at 30°C, the radioactive material insoluble in hot trichloroacetic acid was determined.

Fig. 3a describes the further function of the resolved fractions in forming an 80S-Met-tRNA$_i^{Met}$ complex. The reaction is carried out in two steps. In the first step, the 40S-Met-tRNA$_i^{Met}$ complex is made at 1.3 mM Mg^{++} with factors C3β and D. In the second step the Mg^{++} concentration is adjusted to 3.6 mM and mRNA, factor C3α, and ATP are added. In contrast to the 40S-Met-tRNA$_i^{Met}$ complex, the 80S-Met-tRNA$_i^{Met}$ complex is stable without glutaraldehyde so that this reagent is not added. Consequently, the 40S-Met-tRNA$_i^{Met}$ complexes that are present in the various control incubations (fig. 3 b-f) are not seen. Of particular interest are the requirements for mRNA (fig. 3c) and ATP (fig. 3d). The latter requirement is consistent with the observations made previously with the unfractionated wheat system, and has also been reported by Schreier and Staehlin, for the reticulocyte system (15), and Kramer <u>et al</u>. for a mixed <u>Artemia</u> reticulocyte system (23). With

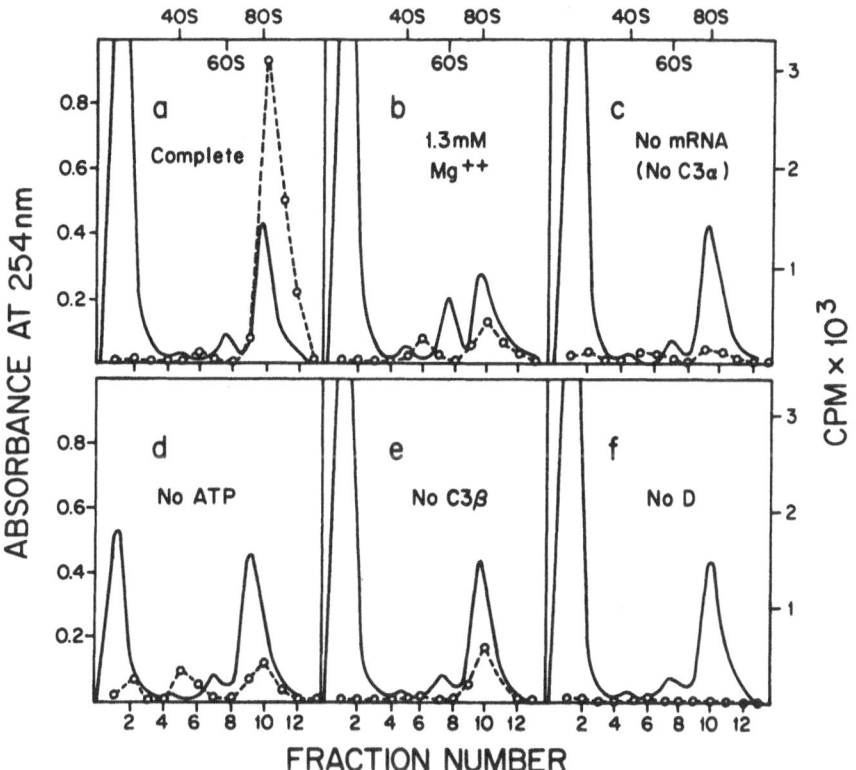

FIG. 3. Requirements for the formation of an 80S·Met-tRNA$_1^{Met}$ complex. The reaction was carried out in two steps. The first incubation contained in 0.34 ml, the complete system as in Fig. 2 except that 52 μg tRNA containing 28.1 pmol [CH$_3$-^{3}H]Met-tRNA (2300 cpm/pmol), and 189 μg fraction D-final were used. After 2 min at 20°C, 35 μg C3α, 6 μg alfalfa mosaic virus RNA-top a (AMV-RNA), ATP (1 mM), MgAc$_2$ (3.6 mM), and KAc (70 mM) were added to a volume of 0.40 ml, and the incubation was continued for 10 min at 20°C. The samples were centrifuged without glutaraldehyde fixation through sucrose gradients and analysed as in Fig. 2. In (b) Mg^{2+} was maintained at 1.3 mM, and in (c)-(f) either AMV-RNA, ATP, fraction C3β, or fraction D-final were omitted. In experiments where C3α was omitted a profile similar to that of Fig. 3c was obtained.

regard to the function of ATP, it should be noted that a GTP-generating system is added in the initial step of the incubation so that the role of ATP does not appear to be related to the regeneration of GTP. Such a role for ATP, utilizing the nucleoside diphosphate kinase reaction, has recently been suggested (24).

The requirement for mRNA for the formation of the 80S-Met-tRNA$_i^{Met}$ complex suggests that mRNA participates in this reaction. Fig. 4 shows a sucrose gradient analysis of an 80S-reaction with ^{14}C-labeled TMV-RNA. Although there is considerable retention of radioactivity throughout the 40S region, a distinct peak of radioactivity can be seen in the position of the 80S-Met-tRNA$_i^{Met}$ complex. The specificity of the reaction is indicated by the absence of the peak in an incubation in which factor C3α is omitted (fig. 4b). Two other points relevant to this experiment suggest that the mRNA attachment reaction may be more complex. 1) The amount of ^{14}C-radioactivity transferred to the 80S-complex (based on a molecular weight of 2 x 10^6 for TMV-RNA) is only 1/5 that of the Met-tRNA$_i^{Met}$ transferred. 2) In preliminary experiments, it was observed that a shift of ^{14}C-TMV-RNA radioactivity similar to that of fig. 4a could be obtained in the absence of both factor C3β and Met-tRNA$_i^{Met}$. Only ATP and factors C3α and D were required.

In explanation of the stoichiometric balance two points are relevant. It has been reported that TMV-RNA has 2 initiation sites (25). Furthermore, the mRNA component present finally in the 80S region may only be a fragment of the mRNA that was initially attached. The lack of requirement for C3β and Met-tRNA$_i^{Met}$ for mRNA attachment suggests the possibility that mRNA attachment may be a distinct reaction, independent of Met-tRNA$_i^{Met}$ binding, perhaps even preceding it. Such a scheme would still allow that the Met-tRNA$_i^{Met}$ anticodon would align the mRNA codon, but the initial attachment of the mRNA would be relegated to a site on the mRNA other than the initiating triplet. Two possibilities for such binding sites are polypurine-rich stretches preceding the AUG codon (26) and the m^{7}G5'ppp 5'-termini which have been shown to facilitate the translation of many eucaryotic mRNAs (27).

D. Analysis of the Sequence of Reactions Leading to the Formation of the 80S Complex

The studies described above suggest the following sequence of reactions:

(1) 40S + mRNA → 40S-(mRNA)

(2) C3β + Met-tRNA$_i^{Met}$ → C3β-Met-tRNA$_i^{Met}$

(3) 40S-(mRNA) + C3β-Met-tRNA$_i^{Met}$ → 40S-(mRNA)-C3β-Met-tRNA$_i^{Met}$

(4) 40S-(mRNA)-C3β-Met-tRNA$_i^{Met}$ → 40S-mRNA-C3β-Met-tRNA$_i^{Met}$

(5) 40S-mRNA-C3β-Met-tRNA$_i^{Met}$ + 60S → 80S-mRNA-Met-tRNA$_i^{Met}$

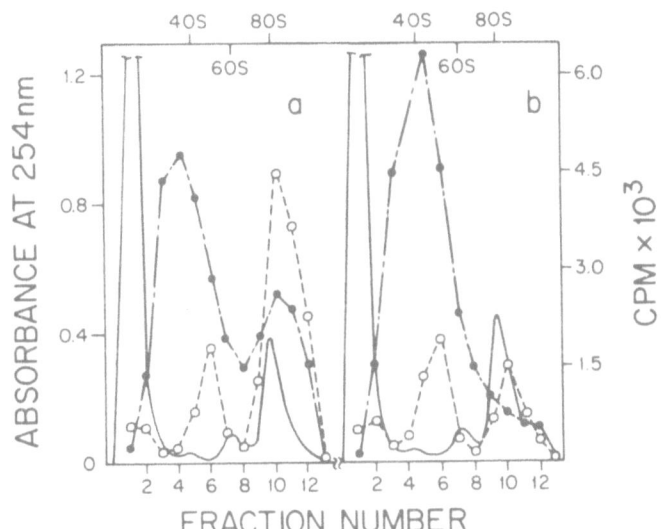

FIG. 4. Binding of ^{14}C-TMV-RNA to an 80S-initiation complex. The incubation conditions were those of Fig. 3 except that the second incubation contained 2 µg ^{14}C-TMV-RNA (4550 cpm/µg) in place of AMV-RNA. This level of viral RNA is below saturation as assayed by Met-tRNA$_i^{Met}$ transfer to the 80S-complex. Higher levels of ^{14}C-TMV-RNA, however, obscured the peak in the region of the 80S complex. In (b) fraction C3α was omitted. The solid line represents absorbance at 254 mµ. The closed circles (●——●) show ^{14}C-radioactivity from TMV-RNA and the open circles (o---o) denote ^{3}H-radioactivity from Met-tRNA$_i^{Met}$.

It appeared possible to test the role of the C3β-Met-tRNA$_i^{Met}$ complex and the 40S-Met-tRNA$_i^{Met}$ complex as intermediates in reactions 3 and 5 respectively by preparing each of these complexes with radioactive Met-tRNA$_i^{Met}$ and then carrying out their respective reactions in the presence of excess unlabeled Met-tRNA$_i^{Met}$. The identical experiments with either no unlabeled Met-tRNA$_i^{Met}$ added or with the unlabeled diluent added prior to the reaction in which the

presumed intermediates are made, would serve as controls. When
such an experiment was performed with C3β in reaction 3 there was
no greater transfer of radioactivity from the preformed C3β-Met-
$tRNA_i^{Met}$ than when the radioactive complex was not preformed. This
result suggests either that the C3β-Met-$tRNA_i^{Met}$ complex is not an
intermediate in the forming of the 40S-complex or that the rate of
exchange of Met-$tRNA_i^{Met}$ out of the C3β complex is considerably
greater than its rate of transfer to the 40S subunit. When the
same type of experiment was done to test the role of the 40S-Met-
$tRNA_i^{Met}$ complex in reaction 5, a positive result was obtained. As
seen in Table 4, experiment 1, when the 80S reaction is terminated
by glutaraldehyde fixation after 3 min, the added unlabeled
Met-$tRNA_i^{Met}$ has little effect and the radioactive Met-$tRNA_i^{Met}$ is
transferred essentially to the same extent as in the control lacking the diluent Met-$tRNA_i^{Met}$. These data are consistent with the
40S-Met-$tRNA_i^{Met}$ complex as an intermediate in the formation of an
80S complex. If the 80S reaction is run for 10 min (experiment 2)

Table 4. Transfer of Met-$tRNA_i^{Met}$ from the 40S complex to an 80S
complex

	Components Added		^{3}H-Met-tRNA Bound (pmol)	
	Incubation 1	Incubation 2	40S complex	80S complex
1	2 min 20°	3 min 20°		
	^{3}H-Met-tRNA		1.4	0.66
	^{3}H-Met-tRNA + Met-tRNA		0.21	0.13
	^{3}H-Met-tRNA	Met-tRNA	0.99	0.57
2	2 min 20°	10 min 20°		
	^{3}H-Met-tRNA		0.49	1.4
	^{3}H-Met-tRNA + Met-tRNA		0.17	0.26
	^{3}H-Met-tRNA	Met-tRNA	0.19	0.37
	^{3}H-Met-tRNA[a]	–	0.56	

[a]Reaction terminated at end of first incubation.

The reaction conditions were those of Fig. 3 except that in
experiment 1 the second incubation was terminated after 3 min. 28
pmol of [CH_3-^{3}H]Met-$tRNA_i^{Met}$ (52 μg of tRNA), and 140 pmol of unlabeled Met-$tRNA_i^{Met}$ (260 μg of tRNA) were added to the particular
incubations as indicated, and the reactions were terminated by the
addition of glutaraldehyde as in Fig. 2. The data presented are a
calculated sum of the nitrocellulose bound radioactivity in the
single fractions corresponding to the specified regions of the
gradient.

Table 5. ^{14}C-Leucine incorporation catalyzed by STNV-RNA in the presence and absence of S-Adenosylhomocysteine (SAH)

Additions	^{14}C-Leucine Incorporated cpm
-	1,200
STNV-RNA (2.5 µg)	24,331
" + SAM (4 µM)	26,752
" + SAH (320 µM)	25,250
reovirus RNA (2.5 µg)	8,873
" + SAM (4 µM)	13,944
" + SAH (320 µM)	2,480

A 0.1 ml incubation containing 20 mM HEPES-KOH pH 7.6, 1 mM ATP, 20 µM GTP, 8 mM creatine phosphate, 40 µg/ml creatine phosphokinase, 2.5 mM dithiothreitol, 2 mM MgAc$_2$, 80 µM spermine, 20 mM KCl, 120 mM KAc, 30 µM 19 amino acids-leucine, 0.625 µCi/ml ^{14}C-L-leucine, 15 µg/ml wheat tRNA, 25 µl wheat germ S23, and viral RNA as indicated, was incubated for 60 min at 25° and the hot TCA insoluble radioactivity was determined. Wheat germ S23 was prepared by grinding 5 g wheat germ (General Mills Inc.) first with 5 g of sand and then with 25 ml 1 mM MgAc$_2$, 2 mM CaCl$_2$, 90 mM KCl using a chilled mortar and pestle. After centrifuging at 23,000 x g for 10 min (pellet discarded), adjusting to 20 mM Tris acetate pH 7.6, 2 mM MgAc$_2$ and recentrifuging, the supernatant was passed through a column of Sephadex G-25 and eluted with 1 mM Tris acetate pH 7.6, 50 mM KCl, 3 mM MgAc$_2$, 4 mM β-mercaptoethanol. The turbid fraction was collected, centrifuged at 23,000 x g for 10 min (pellet discarded), and stored in small aliquots at -70°C. Reovirus RNA having a 5'-terminus of 75% ppGp and 25% GpppG was a gift of M. Morgan and A. Shatkin of the Roche Institute.

there is considerably greater dilution by the added Met-tRNA$_f^{Met}$. This is due in part to the fact that in the 10 min incubation about 3 x as much Met-tRNA$_f^{Met}$ appears in the 80S complex as was present initially in the 40S-complex. Thus at least 2/3 of the 80S-Met-tRNA$_f^{Met}$ complex derives from 40S-Met-tRNA$_f^{Met}$ complexes made with fully diluted Met-tRNA$_f^{Met}$. The longer incubation also allows equilibration of the initially formed 80S-Met-tRNA$_f^{Met}$ with the 40S-Met-tRNA$_f^{Met}$ pool, again resulting in lowered radioactivity in the 80S-complex. The short-term incubation (experiment 1) indicates clearly, however, that the Met-tRNA$_f^{Met}$ present in the 40S complex can be transferred to the 80S complex with little dilution.

E. <u>The 5'-7-Methylguanosine "cap" and mRNA Translation</u>

An important aspect of the initiation process, as noted in section C, is the attachment of mRNA to the 40S ribosomal subunit. Both, Shatkin, and coworkers (28) have shown that the attachment of reovirus mRNA to ribosomes and the subsequent translation of this mRNA requires the presence of an m^7G^5 ppp "cap" on the mRNA. As a test of the universality of this conclusion, we studied the translation of satellite tobacco necrosis virus RNA (STNV-RNA). This viral RNA can be translated authentically in the wheat system (5, 29) although its 5' terminus has been reported to be either ppApGpUp$\cdots$ (30) or pppApGpUp$\cdots$ (31). If one could ascertain that the translational system was not adding the 5'-"cap", this would indicate that STNV-RNA did not require the "cap" for translation. Table 5 demonstrates exactly this point showing that the addition of S-adenosyl homocysteine (SAH), a methyl transfer inhibitor known to prevent the "capping" reaction (32), has no effect on the translation of STNV-RNA. Further evidence supporting the idea that STNV-RNA functions without a 5'-"cap" was obtained in experiments using the inhibitor, 7-methyl-guanosine-5' phosphate, $m^7G^5{}'p$ (33). As shown in Table 6, this inhibitor has no effect on the formation of the 40S-Met-tRNA$_1^{Met}$ complex, but is strongly inhibitory to the

Table 6. <u>Effect of $m^7G^5{}'p$ on the formation of 40S-Met-tRNA$_1^{Met}$ and 80S-Met-tRNA$_1^{Met}$ complexes</u>

	Addition (.25 mM)	Met-tRNA$_1^{Met}$ bound cpm	Inhibition %
1) 40S reaction	–	1701	
	$m^7G^5{}'p$	1640	3.6
2) 80S reaction (TMV-RNA)	–	1706	
	$m^7G^5{}'p$	342	80.0
	$m^7G2(3')p$	1468	14.0
	$G^5{}'p$	1397	18.1
3) 80S reaction (STNV-RNA)	–	1953	
	$m^7G^5{}'p$	2147	
4) 80S reaction (AMV-RNA)	–	5297	
	$m^7G^5{}'p$	1363	74.3

The reaction systems were those of Figs. 2 and 3, with the viral RNAs added at 6 μg per 0.34 ml total volume. The data are a calculated sum of the nitrocellulose bound radioactivity; in exp. 1 for the region of the 40S complex and exp. 2-4 for the region of the 80S complex. The data in exp. 2-4 are corrected for a control in which mRNA was omitted.

mRNA-dependent reaction in which an 80S-Met-tRNA$_i^{Met}$ complex is formed. The effect, however, is strikingly dependent on the specific mRNA added with the reaction with TMV-RNA and AMV-RNA being strongly inhibited and that with STNV-RNA unaffected. Clearly, the most compatible explanation of these data is that STNV-RNA translation is independent of a 5'-"cap".

F. Summary

Protein chain initiation in extracts of wheat germ proceeds first by the mRNA-independent formation of an unstable 40S ribosome·Met-tRNA$_i^{Met}$ complex that is detected in sucrose gradients only after fixation with glutaraldehyde. Two soluble factors (C3β and D) and GTP are required for this reaction. One of the factors (C3β) binds Met-tRNA$_i^{Met}$ suggesting that a C3β·Met-tRNA$_i^{Met}$ complex may be an intermediate in the formation of the 40S complex. When ATP, mRNA, magnesium acetate (final concentration 3.6 mM), and a third factor (C3α) are added, a stable 80S·Met-tRNA$_i^{Met}$ complex is formed. Radioactive TMV-RNA binds to this complex suggesting that the mRNA is a component. Evidence that the 5'-cap of the mRNA is a participant in the 80S-forming reaction is obtained in experiments with m^{7}G$^{5'}$p. The reaction catalyzed by either TMV-RNA or AMV-RNA is inhibited by the "cap" analogue, whereas when STNV-RNA (an RNA lacking the 5' cap) is used, the analog is without effect. Preformed 40S·Met-tRNA$_i^{Met}$ transfers Met-tRNA$_i^{Met}$ to the 80S complex in preference to free Met-tRNA suggesting that the 40S complex is an intermediate in the 80S reaction.

Analyses of the factor requirements for TMV-RNA catalyzed amino acid polymerization show that in addition to the three factors already described and to two elongation factors there is a requirement for a fourth factor suggesting that the initiation process may be more complex than thus far indicated.

REFERENCES

1. MARCUS, A., AND FEELEY, J. (1964) Proc. Natl. Acad. Sci. U.S. 51, 1075-1079.
2. MARCUS, A. (1969) Symp. Soc. Dev. Biol. 23, 143-160.
3. WEEKS, D. P., AND MARCUS, A. (1970) Biochim. Biophys. Acta 232, 671-684.
4. MARCUS, A., LUGINBILL, B., AND FEELEY, J. (1968) Proc. Natl. Acad. Sci. U.S. 72, 1189-1193.
5. KLEIN, W. H., NOLAN, C., LAZAR, J. M., AND CLARK, J. M. JR. (1972) Biochem. 11, 2009-2014.
6. DAVIES, J. W., AND KAESBERG, P. (1974) J. Gen. Virol. 25, 11-20.

7. WEEKS, D. P., VERMA, D. P. S., SEAL, S. N., AND MARCUS, A.
    (1972) Nature 236, 167-168.
8. MARCUS, A., BEWLEY, J. D., AND WEEKS, D. P. (1970) Science
    167, 1735-1736.
9. MARCUS, A., WEEKS, D. P., AND SEAL, S. N. (1973) in
    Nitrogen Metabolism in Plants (Smellie, R. M. S. and Goodwin,
    T. W., eds.), Biochem. Soc. Symp. 38, 97-109.
10. HASELKORN, R., AND ROTHMAN-DENES, L. B. (1973) Ann. Rev. of
    Biochem. 42, 397-438.
11. JAY, G., AND KAEMPFER, R. (1975) J. Biol. Chem. 250,
    5742-5748.
12. SEAL, S. N., BEWLEY, J. D., AND MARCUS, A. (1972) J. Biol.
    Chem. 247, 2592-2597.
13. SHAFRITZ, D. A., AND ANDERSON, W. F. (1970) Nature 227,
    918-920.
14. SHAFRITZ, D. A., PRICHARD, P. M., GILBERT, J. M., MERRICK,
    W. C., AND ANDERSON, W. F. (1972) Proc. Natl. Acad. Sci. 69,
    983-989.
15. SCHREIER, M. H., AND STAEHLIN, T. (1973) Proc. 24th Mosbacher
    Colloq. Springer Verlag., 335-349.
16. LEVIN, D. H., KYNER, D., AND ACS, G. (1973) Proc. Natl.
    Acad. Sci. 70, 41-45.
17. LEVIN, D. H., KYNER, D., AND ACS, G. (1973) J. Biol. Chem.
    248, 6416-6425.
18. DETTMAN, G. L., AND STANLEY, W. M. JR. (1972) Biochim.
    Biophys. Acta 287, 124-133.
19. CASHION, L. M., AND STANLEY, W. M. JR. (1974) Proc. Natl.
    Acad. Sci. 71, 436-440.
20. GUPTA, N. K., WOODLEY, C. L., CHEN, Y. C., AND BOSE, K. K.
    (1973) J. Biol. Chem. 248, 4500-4511.
21. GUPTA, N. K., CHATTERJEE, B., CHEN, Y. C., AND MAJUMDER, A.
    (1975) J. Biol. Chem. 250, 853-862.
22. GIESEN, M., ROMAN, R., SEAL, S. N., AND MARCUS, A., J. Biol.
    Chem., in press.
23. KRAMER, G., KONECKI, D., AMADEVILLA, J. M., AND HARDESTY, B.
    (1976) Arch. Biochem. 174, 355-358.
24. WALTON, G. N., AND GILL, G. N. (1976) Biochim. Biophys.
    Acta 418, 195-203.
25. KNOWLAND, J., HUNTER, T., HUNT, T., AND ZIMMERN, D. (1975)
    in In Vitro Transcription and Translation of Viral Genomes
    (Haenni, A. L. and Beandt, G., eds.), Inserm, Paris. p. 211-
    216.
26. SHINE, J., AND DALGARNO, L. (1974) Proc. Natl. Acad. Sci.
    71, 1342-1346.
27. FURUICHI, Y., MORGAN, M., SHATKIN, A. J., JELINEK, W.,
    DALDITT-GEORGIEV, M., AND DARNELL, J. E. (1975) Proc. Natl.
    Acad. Sci. U.S. 72, 1904-1908.
28. BOTH, G. W., FURUICHI, Y., MUTHUKRISHNAN, S., AND SHATKIN,
    A. J. (1975) Cell 6, 185-195.

29. ROMAN, R., BROOKER, J. D., SEAL, S. N., AND MARCUS, A. (1976)
    Nature 260, 359–360.
30. LESNAW, J., AND REICHMANN, M. (1970) Proc. Natl. Acad. Sci.
    U.S. 66, 140–145.
31. HORST, J., FRANKEL-CONRAT, H., AND MANDELES, S. (1971)
    Biochem. 10, 4748–4752.
32. BOTH, G. W., BANERJEE, A. K., AND SHATKIN, A. J. (1975) Proc.
    Natl. Acad. Sci. U.S. 72, 1189–1193.
33. HICKEY, E. D., WEBER, L. A., AND BAGLIONI, C. (1976) Proc.
    Natl. Acad. Sci. 73, 19–23.

Footnote

[1]Evidence for the alternative sequence in which fMet-tRNA
first binds to the 30S subunit and then directs the attachment of
the mRNA is reported by Jay and Kaempfer (11).

Acknowledgement

This work was supported by U.S.P.H.S. grants GM15122, CA-06927
and RR-05539 from the National Institutes of Health, and by an
appropriation from the Commonwealth of Pennsylvania. Marcella
Giesen was a recipient of postdoctoral fellowships from NATO and
the Deutsche Forschungsgemeinschaft, and Ruth Roman was supported
by a postdoctoral fellowship from the Universidad Nacional Autonama
de Mexico.

PHENOTYPIC MARKERS FOR CHLOROPLAST DNA GENES IN HIGHER PLANTS

AND THEIR USE IN BIOCHEMICAL GENETICS

Kevin Chen, Sarjit Johal, and S. G. Wildman
Department of Biology
Molecular Biology Institute
University of California
Los Angeles, California 90024

Mode of Inheritance of Chloroplast and Nuclear DNA Phenotypic Markers in Reciprocal, Interspecific Hybrids

Plants belonging to the genera Nicotiana, Avena, and Triticum appear to transmit chloroplast DNA exclusively by the maternal parent. In Nicotiana this mode of inheritance has been demonstrated for a variegation caused by the presence of normal and defective chloroplasts contained within the same cells of N. tabacum leaves (1). The defective chloroplasts have a DNA which is slightly different in physical structure from the DNA of the normal chloroplasts (2). Maternal inheritance also governs transmission of coding information for the large subunit of Fraction I protein (3) as well as the site of its ribulose diphosphate carboxylase catalytic activity (4). In Avena, Steer (5) has found that the electrophoretic mobility of Fraction I protein is controlled by coding information transmitted via the maternal line. In Triticum, genetic information controlling the isoelectric points of the three polypeptides comprising the large subunit of Fraction I protein is also inherited exclusively from the maternal parent (6). In all three genera, the demonstration of chloroplast DNA phenotypic markers for genes controlling the primary structure of proteins has arisen because F_1 plants can be created by interspecific hybridization, often in reciprocal fashion. The mode of inheritance of the coding information for the change in primary structure determines the location of the genetic information. If maternal, the information is most probably contained in chloroplast DNA. If maternal and paternal, the information is either also, or exclusively, contained in the DNA of the nucleus (7).

The fact that different species of plants can be hybridized to form an F_1 generation of new plants provides an additional approach

to the biochemical genetics of chloroplast DNA beyond that already
available with the unicellular organisms, Euglena and Chlamydomonas.
Euglena has no known mode of sexual reproduction whereas hybridizable
"species" of Chlamydomonas are limited to + and - strains of the same
species. The ability to hybridize species maximizes the opportunity
to find differences in the primary structure of proteins and to trace
out the source of genetic coding information for the difference.

Stability of Chloroplast DNA

Chloroplast DNA in higher plants has evolved at an exceedingly
slow rate as shown by two recent investigations. In the case of the
large subunit of Fraction I protein, there have been only four differences so far encountered in the isoelectric points of the three
polypeptides among 63 out of the 65 species of plants composing the
genus Nicotiana (8). (Plants of two species were not available for
analysis). It is suspected that the cluster of three polypeptides
is a chloroplast DNA genetic unit and therefore, $2^x=4$ differences
translates into the probability that only 2 mutations in the chloroplast DNA coding for the large subunit of Fraction I protein have
survived during the entire life span of the genus Nicotiana.
Atchison, Whitfeld and Bottomley (9) have investigated the specific
fragmentation pattern of Nicotiana chloroplast DNAs produced by EcoRI
endonuclease. They have found only 5 differences which is a rather
remarkable correlation with the 4 differences in chloroplast DNA
coding for Fraction I protein large subunits. The 20 Australian
species of Nicotiana have been separated from the 43 Western Hemisphere species for about 100 million years (7). There are no differences among the Australian species in either chloroplast DNA fragmentation patterns or Fraction I protein large subunit polypeptide
patterns. Consequently, there have been no discernable mutations
in chloroplast DNA in 100 million years by the two tests applied.

Lack of Effects of Plant Breeding on Chloroplast DNA Genes
in Cultivars of Nicotiana

Man has been experimenting with one species of Nicotiana, viz.,
N. tabacum, the source of smoking tobacco, for many hundreds of years
As a plant species, N. tabacum evolved as a consequence of interspecific hybridization. In the process, genetic information for one
small subunit polypeptide of Fraction I protein was provided by N.
tomentosiformis as the male parent whereas N. sylvestris contributed
the chloroplast DNA genes for the large subunit as well as a nuclear
gene for the second small subunit polypeptide in N. tabacum (12).
The time of domestication of N. tabacum by the aborigines of the
Western Hemisphere is unknown but the plant's presence was noted as
early as 1492 by an entry in the diary of Columbus. In the nearly
500 years since, the N. tabacum genome has come under constant

attention and manipulation by plant breeders with the intent of
changing this or that character as it might effect disease resistance, smoking quality of the leaf, climatic adaption, etc. The
consequence has been that about 1000 distinct cultivars of this
species have been created.

     About 30 of the N. tabacum cultivars have been analyzed for
their Fraction I protein composition. In no instance has a change
been observed in the isoelectric points of the two polypeptides constituting the small subunit coded by nuclear genes. Included were
cultivars with haploid, diploid, triploid and tetraploid chromosome
complements. Thus, genetic manipulations which have completely
changed the morphological characteristics of the tobacco plant have
had no recognizable, gratuitous effect on the primary structure of
either polypeptide of the small subunit of Fraction I protein as
tested by isoelectric points (10), or by Kawashima et al. (11) using
chymotryptic peptide fingerprints. Of interest is the absence of
either dominance or recessiveness in regard to expression of the
nuclear genes coding for the small subunit. In all 30 cultivars,
the ratio of the two polypeptides appears to be 1:1 as judged by
intensity of staining of the two protein bands. However, Kung and
Marsho (13) have found that the specific enzymatic activity of
Fraction I protein in one cultivar was reduced by one-half in the
case of a double recessive mutant although there was no detectable
change in the ratio, tryptic peptide composition, or isoelectric
points of the two polypeptides of the small subunit or the three
polypeptides of the large subunit. It is clear that a precise knowledge of the sequences of amino acids in the two small subunit polypeptides of appropriate N. tabacum cultivars could provide important
new insight into the molecular biological meaning of the terms dominance and recessiveness as applied to Mendelian inheritance. Some
progress has already been made in sequencing the small subunit polypeptides by laboratories in Denmark and England (14).

             Phenotypic Markers for Multiple Chloroplast DNA Genes
                       Among Cultivars of N. tabacum

     Cultivars of N. tabacum display a considerable array of dif-
ferent characteristics which are inherited only via the maternal
line. The characters are listed in Table 1 where it can be seen
that some cultivars contain as many as three phenotypic markers to
distinguish their extranuclear DNA genes from those in another
cultivar.

     As for the chloroplast DNA genes coding for the large subunit
of Fraction I protein all of the 4 differences in polypeptide isoelectric points encountered among the 63 species of the genus
Nicotiana are also present among the cultivars of N. tabacum as
shown in Table 1 but with a surprising correlation. The cultivars

Table 1

Phenotypic markers for chloroplast DNA genes present in different cultivars of _Nicotiana tabacum_

Cultivar	Type of Fraction I protein Large Subunit				Type of Variegation		Type of Male Sterility	
	Australian Glauca	Glauca	Tabacum	Glutinosa	White	Green	No Anthers	Defective Anthers
Turkish Samsun	−	−	+	−	−	−	−	−
Var. T. Sam (W)	−	−	+	−	+	−	−	−
Var. T. Sam (G)	−	−	+	−	−	+	−	−
Burley 21 MS	+	−	−	−	−	−	+	−
Var. Burley 21 MS	+	−	−	−	+	−	+	−
Nadwislanski MS	−	+	−	−	−	−	+	−
Kupchunos MS	+	−	−	−	−	−	−	+
Burley 21 MS (Plumbaginifolia)	−	−	−	+	−	−	−	+

are either male fertile (MF), that is the pollen fertilized the
ovules within the same flower to produce seeds, or male sterile (MS).
The ovules in MS flowers require pollen from the flowers of another
MF N. tabacum cultivar to achieve fertilization. Cultivars which
are MF contain the same three large subunit polypeptides with respect
to isoelectric points. The surprising result is that cultivars which
are MS also contain the usual three large subunit polypeptides, but
one or more differ in isoelectric points from the three in the MF
cultivars. The MS cultivars analyzed are indicated by name in Table
2 together with the type of large subunit polypeptides they contain.
The presence of Australian type large subunits in the Fraction I
in the MS Burley 21 cultivar was previously shown to have resulted
from an Australian species of Nicotiana having served as the female
component during hybridization with N. tabacum as the male component
(10). The MS Burley 21 cultivar was then produced by backcrossing
the F$_1$ hybrid with MF Burley 21 pollen and repeating the backcrosses
until all of the nuclear genes which regulated the phenotypic expression of the Australian species' characters were eliminated. The
backcrossing also removed the genetic information for the small subunit polypeptides present in the original female parent. However,
the chloroplast DNA genes for the large subunit derived by maternal
inheritance in the first interspecific hybrid remained unchanged
throughout the many breeding steps which finally transformed the
hybrid plant back into an authentic cultivar of N. tabacum. In 6
out of 7 of the MS cultivars listed in Table 2, the difference in
large subunit polypeptide composition from that of MF N. tabacum
can be traced to an original interspecific F$_1$ hybrid where the female
species contained chloroplast DNA genes coding for Fraction I protein
which were different from those in MF N. tabacum. The seventh MS
cultivar was discovered by Berbec (15) as a spontaneous MS mutant

Table 2
Correlation between male sterility in cultivars of N. tabacum
and a change in the isoelectric points of the three large subunit polypeptides of Fraction I protein coded by chloroplast DNA

Male Sterile Cultivar	Origin	Kind of Large Subunit
Burley 21	Megalosiphon ♀	Australian
BP 210 Glauca MS	Glauca ♀	Glauca
BP 210 Exigua MS	Exigua ♀	Australian
BP 210 Debneyi MS	Debneyi ♀	Australian
Nadwislanski Maly	Spontaneous	Glauca
BP 210 MS	N. M. ♀	Glauca
Burley 21	Plumbaginifolia ♀	Glutinosa

of the otherwise MF Nadwislanski Maly commercial cultivar of N. tabacum grown in Eastern Europe. The discovery was made because the mutation resulted in flowers which contained normal female structures but the total absence of male structures. That is, a normal ovary, pistil and stigma were present in each flower but stamens were completely missing. Otherwise, there was no other discernable difference in the appearance of the MS and MF plants. However, the mutation which produced MS was accompanied by a mutation in the chloroplast DNA genes coding for Fraction I protein. The large subunit polypeptides were transformed to the type found in N. glauca. The correlation between these two phenomena therefore makes more persuasive the probability that the genetic information for MS is contained in chloroplast DNA. In summary, we now have numerous phenotypic markers to investigate the behavior of chloroplast DNA genes as shown in Table 1. The question is whether anything useful can be done with them that might constitute the beginning of the science of chloroplast DNA genetics of higher plants. The analysis of Fraction I proteins contained in MS N. tabacum cultivars has shown how inaccessible chloroplast DNA genes are to alteration by conventional techniques used by plant breeders. Therefore, the announcement by Carlson, Smith and Dearing (16) that parasexual hybrids could be created by fusion of protoplasts between two different species of plants was of great interest. The opportunity existed to see whether chloroplast DNA genes could be transferred from plant to plant by this novel technique of fusion of protoplasts derived from single leaf cells.

Expression of Chloroplast DNA Genes Coding for Fraction I Protein in Parasexual Hybrids

The first parasexual hybrid was produced by bringing N. langsdorffii and N. glauca protoplasts into physical contact. Fraction I protein was extracted from the leaves of this first self-fertile new plant species created in a test tube. Electrofocusing showed the protein to contain three small subunit polypeptides, one derived from the genes in the nucleus of the N. glauca protoplast and the other two from the nuclear DNA of N. langsdorffii. However, the three large subunit polypeptides were only of the N. glauca type (17). Had the N. langsdorffii polypeptides also been present, a cluster of four polypeptides should have been present. The same communication reported the analysis of the Fraction I protein composition contained in another remarkable plant that Dr. P. S. Carlson created by bringing N. tabacum protoplasts into physical contact with N. suaveolens isolated chloroplasts. The leaves of this unique plant contained Fraction I protein with five large subunit polypeptides. Since the large subunits of the two proteins in question have only one of their three polypeptides in common in regard to isoelectric points, the presence of five polypeptides was a clear indication the chloroplast DNA genes derived from the N. suaveolens chloroplasts were being translated together with the chloroplast DNA genes in the N. tabacum protoplasts. There was a remarkable difference in the phenotypic appearance between the plants created by the two different schemes of genetic engineering. The plant created by fusion of protoplasts, but which contained only one species of chloroplast DNA, was robust, displayed hybrid vigor, was self-fertile and therefore fully capable of self-perpetuation as a new species. The plant created by uptake of foreign chloroplasts into N. tabacum protoplasts and contained a mixture of chloroplast DNA's was a monster. It survived only briefly under intensive care conditions because it failed to grow appreciably in length and did not form sexual organs. The small number of leaves which did develop were deformed and variegated. The contrast between the two kinds of test-tube created plants was so extreme as to evoke the question as to whether the deformed plant could have been the consequence of two, dissimilar chloroplast DNAs having undergone recombination. Perhaps reshuffling genes coding for RuDPCase activity had produced deleterious effects in the structure of the crucial enzyme which makes photoautotrophic life possible in the first place.

Capitalizing on Kao, Constabel, Michayluk and Gamborgs' discovery of polyethylene glycol as an agent which induces fusion of leaf protoplasts with a high degree of frequency (18), Smith, Kao and Combatti (19) have created an additional 15 parasexual plants by fusion of N. langsdorffii and N. glauca protoplasts some of whose characters are listed in Table 3. As determined by chromosome number of each parasexual hybrid, some plants arose by fusion of single protoplasts of one species with multiple protoplasts of the other species. In some instances, (e.g., 8-3 #1 and #2) the callus formed after fusion of protoplasts was further divided so that more than one parasexual hybrid plant could be generated from the same fusion product. Crystalline Fraction I protein was obtained from all plants shown in Table 3 and electrofocused. In each plant, the Fraction I protein small subunit consisted of three polypeptides, one derived from N. glauca and the other two derived from N. langsdorffii. Therefore, there was no doubt that the nuclear genes coding for this portion of Fraction I protein from both species of plants were being expressed in the protein extracted from the parasexual hybrid plants. However, in 15 out of 16 parasexual hybrids, only one kind of chloroplast DNA coding for the large subunit of Fraction I protein was expressed. As shown in Table 3, it was apparently a random circumstance as to whether the N. glauca or the N. langsdorffii chloroplast DNA genes were expressed. The plant which arose from fusion 30-8 contains two complements of N. glauca chromosomes to one of N. langsdorffii but the chloroplast DNA expressed is the N. glauca type. Conversely, the plant from fusion 31-6 contains two N. langsdorffii to one N. glauca nuclear genomes and yet still also expresses only N. glauca type chloroplast DNA. These 14 parasexual hybrid plants,

Table 3
Type of Fraction I Protein Large Subunit Polypeptides
Found in Parasexual Hybrid Plants Produced by Fusion of
N. glauca (2N=24) and N. langsdorffii (2N=18) protoplasts

Fusion Number	Chromosome Number	Probable Genome Composition	Type of Large Subunit
7-2	62		L
8-3 #1	60	G+2L	L
8-3 #2	–	–	L
23-2	60	G+2L	L
30-6 #2	56	?	L+G
30-8	64	2G+L	G
31-6	60	G+2L	G
32-1	60	G+2L	G
35-1	58	?	L
35-6 #1	63	?	L
35-6 #2	60	G+2L	L
35-6 #3	60	G+2L	L
35-6 #4	–	–	L
35-6 #6	–	–	L
41-6	58	?	L
44-3	–	–	L
46-4 #2	–	–	G
46-6	58	–	G
47-1	60	G+2L	G
51-6	63	2G+L	G

as in the case of the predecessor formed by physical fusion, are
vigorous in appearance and a number have produced F_2 progeny whose
Fraction I proteins were found to be identical to the parasexual
parents in regard to both large and small subunit polypeptide composition.

One fusion, 30-6 #2 (#1 did not survive), produced a plant whose
Fraction I protein consists of four large subunit polypeptides,
some being coded by N. glauca and the others by N. langsdorffii
chloroplast DNA. This plant is a freak which has remained dwarfed,
distorted, and unable to produce flowers. So, in the only instances
where two dissimilar species of chloroplast DNAs have been present
in the same plant, the result seems to have been a catastrophe as
far as survival of the plant as a new species is concerned. Could
it be that nature evolved the system of exclusive maternal inheritance of chloroplast DNA in higher plants for the expressed purpose
of preventing recombination of dissimilar DNA genes? Has the barrier
to paternal transmission of chloroplast DNA evolved so as to prevent

the kind of catastrophes now observed in the two instances where chloroplast DNA mixtures have been made?

In the quest to improve plants by exploiting these new techniques of genetic engineering, the large subunit of Fraction I protein should prove to be an uncommonly useful genetical marker for determining the survival and functioning of a foreign chloroplast DNA after it has been introduced into a new environment of living protoplasm. Fraction I protein has so far been found in every organism which contains chlorophyll a. A survey of Fraction I proteins throughout the Plant Kingdom has shown them all to contain three large subunit polypeptides (20).

Phenotypic Markers for Chloroplast DNA Genes Coding for
Fraction I Proteins in the Plant Kingdom

Fig. 1 is a diagram showing the polypeptide compositions of Fraction I proteins obtained from plants representing all but one of the Phyla of the Plant Kingdom. The diagram also includes protein from plants belonging to the major divisions of the Angiosperms. It also includes Fraction I protein from plants which have evolved an additional system for fixing CO_2 via the Hatch-Slack, C_4 pathway of carbon assimilation. In each instance, the protein dissociates into large and small subunits in the presence of 8M urea and resolves into individual polypeptides upon isoelectric focusing. Thus every Fraction I protein so far encountered appears to be built on the same principle of large and small subunits in a 3:1 proportion in regard to mass. They are alike also in regard to a fixed number of large subunit polypeptides and a variable number of small subunit polypeptides.

Each Fraction I protein resolves into a cluster of three large subunit polypeptides separated by about 0.1 pH unit from each other. A similar cluster of three polypeptides was found for Fraction I protein isolated from a prokaryotic, blue-green algae. It was not included in the diagram because of uncertainty in regard to the number of polypeptides constituting the small subunit. As for proteins obtained from different genera of plants belonging to the same family, the subunit polypeptides may be identical or different in isoelectric points. For example, the protein from tomato (Lycopersicum esculentum) has both large and small subunit polypeptides with isoelectric points the same as those of Nicotiana petunioides. Sequence analysis could establish whether the notion is tenable that tomato and Nicotiana Fraction I proteins are very closely related in their evolution. In contrast, none of the isoelectric points of the large subunit polypeptides of protein from eggplant (Solanum melongena) belonging to another genus of the Solanacea correspond in charge with any other representatives of the same family.

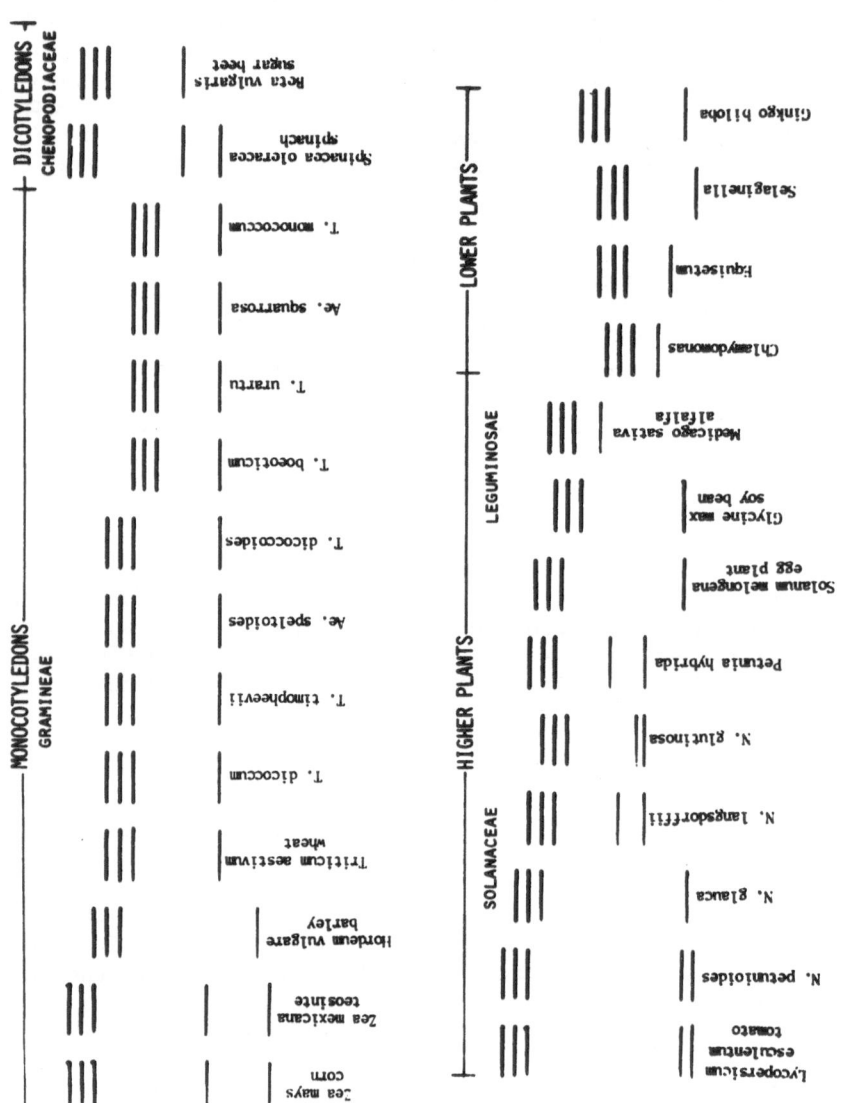

Fig. 1. Number of large and small subunit polypeptides and relative position of isoelectric points of Fraction I proteins isolated from representatives of the Plant Kingdom.

In the Chenopodiaceae, proteins from spinach and <u>Beta vulgaris</u> have non-identical large subunit polypeptides but share one small subunit polypeptide in common. The presence of two small subunit polypeptides in spinach would argue for this plant being a product of interspecific hybridization. Perhaps, <u>Beta vulgaris</u> was the male parent of spinach.

The chloroplast DNA coded component of Fraction I proteins from the Monocotyledons is of interest because the large subunit has served as a marker for determining the maternal ancestors of modern hexaploid wheats (6). The seven species of Triticum and two species of Aegilops are distinguished by two different clusters of large subunit polypeptides. In contrast to the <u>Nicotiana</u> large subunit clusters, the two wheat clusters share only one polypeptide of the same isoelectric point in common. The question remains therefore as to whether broadening the analysis to include still more species might uncover a species with a large subunit cluster of three polypeptides, two of which would have isoelectric points in common with the two groups shown in Fig. 1. In the case of the <u>Nicotianas</u>, it appears unlikely that the chloroplast DNA genes coding for the large subunit polypeptides have undergone single mutational events sufficient to change the isoelectric points of two out of three of the polypeptides at once.

Both the three large and two small subunit polypeptides of corn (<u>Zea mays</u>) and Teosinte (<u>Zea mexicana</u>) have identical isoelectric points. Thus, Teosinte has the correct chloroplast DNA genes to have been a maternal ancestor of modern day corn. The presence of two small subunit polypeptides suggests that Teosinte may have been the product of interspecific hybridization. The presence of two small subunit polypeptides in corn protein identical in isoelectric points to the two in Teosinte is consistent with the argument advanced by Beadle (21) that corn arose by a very small number of mutations of Teosinte.

ACKNOWLEDGEMENT

Research supported by Contract E (04-3) 34, Project 8, U.S. Energy Research and Development Administration, U.S. National Science Foundation grant 75-07368, and a grant from Monsanto Commercial Products Corporation.

REFERENCES

1. Wildman, S.G., Liao, C.L. and Wong-Staal, F. (1973) Planta
    113: 293-312.

2. Wong-Staal, F. and Wildman, S.G. (1973) Planta, 113: 313-26.
3. Chan, P.H. and Wildman, S.G. (1972) Biochim. Biophys. Acta, 277: 677-80.
4. Singh, S. and Wildman, S.G. (1973) Molec. gen. Genet., 124: 187-90.
5. Steer, M.W. (1975) Can. J. Genet. Cytol., 17: 337-344.
6. Chen, K., Gray, J.C. and Wildman, S.G. (1975) Science, 190: 1304-6
7. Wildman, S.G., Chen, K., Gray, J.C., Kung, S.D., Kwanyuen, P. and Sakano, K. (1975) in Genetics and Biogenesis of Mitochondria and Chloroplasts, C.W. Birky, P.S. Perlman, T.J. Byers, Eds., Ohio State University Press, Chap. 9.
8. Chen, K., Johal, S. and Wildman, S.G. (1976) in Genetics and Biogenesis of Chloroplasts and Mitochondria, Th. Bucher, W. Neupert, W. Sebald, S. Werner, Eds., North Holland Publishing, In press.
9. Atchison, B.A., Whitfeld, P.R. and Bottomley, W. (1976) Submitted to Molec. gen. Genet.
10. Chen, K., Kung, S.D., Gray, J.C. and Wildman, S.G. (1975) Biochem. Genet., 13: 771-778.
11. Kawashima, N., Tanabe, Y. and Iwai, S. (1974) Biochim. Biophys. Acta., 371: 417-31.
12. Gray, J.C., Kung, S.D., Wildman, S.G. and Sheen, S.J. (1974) Nature, 252: 226-7
13. Kung, S.D. and Marsho, T.V. (1976) Nature, 259: 325-326.
14. Gibbons, G.C. et al. (1975) Experientia, 31: 1040-41.
    Haslett, B. et al. (1976) Biochim. Biophys. Acta., 420: 122-132
15. Berbec, J. (1974) Z. Pflanzenzucht, 73: 204-216.
16. Carlson, P.S., Smith, H.H. and Dearing, R.D. (1972) Proc. Natl. Acad. Sci. U.S.A., 69: 2292.
17. Kung, S.D., Gray, J.C., Wildman, S.G. and Carlson, P.S. (1975) Science, 187: 353-354.
18. Kao, K.N., Constabel, F., Michayluk, M.R. and Gamborg, O.L. (1974) Planta, 120: 215-227.
19. Smith, H.H., Kao, K.N. and Combatti, N.C. (1976) J. Hered., In press.
20. Chen, K., Gray, J.C., Kung, S.D. and Wildman, S.G. (1976) Plant Science Letters, In press.
21. Beadle, G.W. (1939) J. Hered., 30: 245-47.

THE SYNTHESIS OF CHLOROPLAST PROTEINS

R. JOHN ELLIS

DEPARTMENT OF BIOLOGICAL SCIENCES
UNIVERSITY OF WARWICK
COVENTRY, UNITED KINGDOM

INTRODUCTION

It is a fundamental feature of the organisation of eukaryotic cells that they contain organelles possessing genetic systems additional to the one located in the nucleus. In the case of chloroplasts, the fact that these organelles contain both DNA and ribosomes was demonstrated first in 1962. It soon became apparent that both these components are present in significant quantities. Thus the chloroplast genome has the potential capacity for encoding about 125 proteins, each of molecular weight 50 000, whilst chloroplast ribosomes can represent up to 50% of the total ribosomal complement of leaves [1,2]. The existence of such quantities of chloroplast DNA and ribosomes prompts the question as to their roles in the formation of chloroplasts. Which genes are encoded in chloroplast DNA? Which proteins are synthesised by chloroplast ribosomes? I believe it is necessary to answer such simple direct questions before it is possible to tackle meaningfully the more interesting but far more complex question as to the molecular basis of chloroplast development.

THE USE OF INTACT ISOLATED CHLOROPLASTS

When I entered this field in 1970, it was clear that little progress had been made in identifying either the genes located in chloroplast DNA, or the proteins which are synthesised by chloroplast ribosomes. The chloroplast genome occurs in multiple copies in each chloroplast. This makes conventional genetic analysis difficult, because a mutation in one copy will be swamped for a long time by all the other wild-type copies. The other question,

195

as to the function of chloroplast ribosomes, had been tackled by
two methods. The first method was to determine the effect of
treating cells making chloroplasts with antibiotic inhibitors
believed to be specific for chloroplast ribosomes. The results of
such experiments carried out in different laboratories were often
conflicting; I believe this is partly due to the lack of specificity
of action of antibiotics in some tissues [2]. The second method
was to identify the products of protein synthesis by isolated
chloroplasts. Preparations of isolated chloroplasts were known to
incorporate labelled amino acids into proteins, but no convincing
identification of any protein had been reported; most papers showed
that the products tended to be polydisperse rather than discrete in
nature. Nevertheless, it seemed to me in 1970, that the identification of protein and RNA molecules made by isolated chloroplasts
was, in principle, the most direct and unambiguous method for
determining the function of the chloroplast genetic system. If a
method could be found for preparing isolated chloroplasts in which
translation is coupled to transcription, and in which both processes
occur with fidelity, it should be possible to determine both the
structural genes present in chloroplast DNA and the function of
chloroplast ribosomes. Since 1970, we have developed in my
laboratory a system in which isolated chloroplasts from pea, spinach,
barley and maize leaves use light energy to make discrete RNA and
protein molecules, some of which we have identified. The coupling of
transcription to translation however, still eludes us, so we cannot
conclude that because a protein is made inside the chloroplast it
is therefore encoded in chloroplast DNA.

        The rationale of our approach is that in order to produce
discrete, identifiable protein and RNA molecules, conditions must
be used in which correct elongation, termination, and release of
chains occurs. It seemed to me that such conditions were more
likely to be met in intact chloroplasts, rather than in the lysed
systems that were commonly used, and we therefore adopted methods
that were currently in vogue for preparing intact chloroplasts
capable of high rates of photosynthetic carbon dioxide fixation.
It is characteristic of such intact chloroplasts that they will use
light as a source of energy in the absence of added cofactors. By
using light as the source of energy for protein and RNA synthesis,
it is thus possible to ensure that the incorporation of labelled
precursors is taking place solely in intact chloroplasts. It is
therefore not necessary to use purified preparations of chloroplasts
for these studies, because other organelles, as well as broken
chloroplasts, cannot use light to synthesise ATP. Once the pattern
of protein synthesis by intact chloroplasts had been established,
it provided a base-line which enabled us both to fractionate the
system, and to assess the results of in vivo inhibitor experiments.
This approach has been very successful in establishing the basic
patterns of protein synthesis in chloroplasts isolated from several

TABLE 1

IN VITRO CHLOROPLAST PROTEIN SYNTHETIC SYSTEMS

System	Energy source	Species	Reference
1. Intact chloroplasts	Light or ATP	Pea, spinach, barley, maize	[5,6,7]
2. Lysed chloroplasts	ATP	Pea	[8]
3. Free ribosomes	ATP	Pea	[8]
4. Bound ribosomes	ATP	Pea	[8]
5. Chloroplast mRNA + E. coli ribosomes	ATP	Pea, spinach	[9,10]
6. Etioplasts	ATP	Pea	[11]

higher plants, and it has led to the detection of one of the first specific messengers for a plant enzyme. The general applicability of this approach is testified by the subsequent confirmation of our results in two other laboratories [3,4]. Table 1 lists the in vitro chloroplast protein synthetic systems now available. Highlights of studies with these systems will now be summarised; the references in Table 1 should be consulted for experimental details. These studies represent the combined work of Eric Blair, Allan Eaglesham, Martin Hartley, Peter Highfield, Ken Joy, Stuart Siddell, Annabel Wheeler and the author.

1. Intact Chloroplasts

Suspensions of chloroplasts, prepared as described in Fig. 1, contain 40-50% intact chloroplasts, and incorporate labelled leucine, phenylalanine and methionine into protein when illuminated (Fig. 1). The rate of incorporation falls rapidly to zero after about 20 min; this falling rate is not accompanied by lysis of the chloroplasts. A vital component of the incubation medium is the presence of K^+ ions. When chloroplasts are incubated in media containing sucrose or sorbitol as the osmoticum, protein synthesis is greatly reduced, while replacement of KCl by NaCl prevents all light-dependent incorporation. If chloroplasts are first lysed by resuspension in

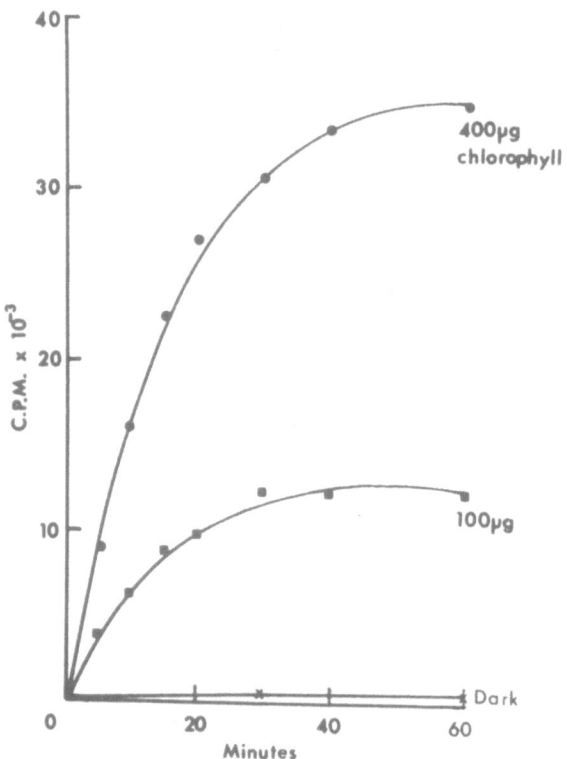

Fig. 1. Time course of light-driven chloroplast protein synthesis.
Pea seeds (Pisum sativum var. Feltham First) were grown in compost
for 7-10 days under a 12 h photoperiod of 2,000 lux white light.
The apical leaves (15 g) were homogenised for 4 sec in a Polytron
homogeniser in 100 ml of a sterile ice slurry containing 0.35 M
sucrose, 25 mM HEPES-NaOH, 2 mM EDTA and 2 mM sodium isoascorbate
(pH 7.6). The homogenate was immediately strained through 8 layers
of muslin and centrifuged at 2,500 x g for 1 min at 0°C. The pellet
was resuspended by cotton-wool in 1-5 ml of sterile 0.2 M KCl, 66 mM
tricine-KOH, 6.6 mM $MgCl_2$ (pH 8.3). Chloroplasts (300 μl) were
incubated in a final volume of 500 μl with 0.5 μCi of either [^{14}C]
leucine (3 μM) or [^{35}S] methionine (36 nM). Tubes were illuminated
at 20°C with filtered red light at 4,000 lux. Protein was extracted
and counted by liquid scintillation spectrometry in toluene - 0.5%
PPO scintillant at 70% counting efficiency.

25 mM tricine-KOH, 10 mM $MgSO_4$ (pH 8.0), addition of KCl to 0.2 M
does not result in any light-dependent incorporation of amino acids
into protein, but incorporation can be driven by added ATP. These
results suggest that KCl is acting both as an osmoticum to preserve
the intactness of the chloroplasts, and as a specific cofactor for
protein synthesis. In the case of pea and spinach, similar rates
of incorporation are achieved when 0.33 M sorbitol is used as the
osmoticum, but only if K^+ ions are present at an optimum concentration of about 30 mM (Table 2). The specific requirement of
bacterial ribosomes for K^+ or NH_4^+ ions is well known, but the
importance of K^+ ions does not seem to have been appreciated by
previous workers on chloroplast protein synthesis. In the case of
barley and maize, light-driven incorporation occurs only when KCl
is the osmoticum, and not when sorbitol is the osmoticum (Table 2).
The presence of Mg^{2+} ions also has different effects on chloroplasts from different species. I recommend for general use the
medium containing 0.2 M KCl and 6.6 mM $MgCl_2$, although it should be
emphasised that in such a medium the chloroplasts do not carry out
carbon dioxide fixation and thus may be in an abnormal state.

   Some characteristics of protein synthesis by pea chloroplasts
have been studied to establish the nature of the process [5].
Light can be partially replaced as an energy source by added ATP
and an ATP-generating system, while addition of ATP as well as
light gives only a slight stimulation. Inhibitors of photophosphorylation, such as CCCP and DCMU, inhibit protein synthesis, as do
antibiotics specific for 70S ribosomes, such as lincomycin and D-
threo chloramphenicol. Added ribonuclease does not inhibit protein
synthesis significantly, and I regard this as additional evidence
that incorporation is proceeding in intact chloroplasts only. We
have shown that addition of ribonuclease causes RNA to be hydrolysed
to a percentage equal to the percentage of broken chloroplasts [12].
Actinomycin D at 10 μg/ml does not inhibit protein synthesis; we
have found that this concentration inhibits light-driven incorporation of [H-3] uridine into RNA by the same chloroplast preparation
by 85%. Incorporation of leucine into protein is not stimulated by
the plant hormones indole-3-acetic acid or gibberellic acid, or by
inorganic phosphate, cyclic AMP, $NADP^+$ or phenazine methosulphate.
Addition of poly (U), which stimulates phenylalanine incorporation
by intact mitochondria from <u>Xenopus</u>, does not have this effect in
this chloroplast system.

   Incorporation of labelled amino acids by other components of
the incubation medium can be excluded. Bacterial contamination of
the chloroplast preparation is minimised by using sterile glassware
and media. If the preparations are solubilised in 2% Triton X-100
at the end of the incubation, less than 0.1% of the radioactivity
incorporated into protein is present in the pellet spun down at
10 000 x g for 10 min. This indicates that incorporation is not
due to bacteria, nuclei or whole leaf cells [13]. The large

TABLE 2

RESPONSE OF ISOLATED CHLOROPLASTS TO CATIONS

| Incubation Media (mM) | | | | Rate of Light-Driven Protein Synthesis | | |
KCl	NaCl	MgCl$_2$	Sorbitol	Pisum Spinacia	Zea	Hordeum
200	0	6.6	0	100	100	100
30	0	0	330	100	0	0
0	200	6.6	0	0	0	0
0	30	0	330	0	-	-
200	0	0	0	100	50	45
30	0	6.6	330	10	40	0

Chloroplasts were prepared from four species, resuspended in incubation media of varying composition, and incubated with light and [S-35] methionine. The amount of label incorporated into protein in 40 min for each species incubated in 200 mM KCl and 6.6 mM MgCl$_2$ has been called 100, and the other values expressed as a percentage. All media contained in addition 50 mM tricine, buffered to pH 8.3 with either KOH or NaOH.

stimulation of incorporation by light, and the sensitivity to inhibitors of photophosphorylation argues for chloroplast, rather than mitochondrial, protein synthesis. The activity of the crude chloroplast suspensions decreases by about 50% if stored for 1 h at 0°C, so we have not attempted to purify the chloroplasts from other cellular components present in the suspension. Other workers have shown recently that spinach chloroplasts purified on silica sol gradients show light-driven protein synthesis with similar characteristics [4].

We conclude from all these characteristics that protein synthesis is proceeding in intact chloroplasts only, is being driven by photophosphorylation, and is probably using messenger RNA synthesised before the chloroplasts are isolated. The products of protein synthesis have been analysed by electrophoresis on sodium

dodecylsulphate (SDS) polyacrylamide gels. After staining to reveal
the protein bands, the gels are cut into 1 mm slices, and each slice
counted for radioactivity. This method gives a quantitative estimate of the amount of label in each product. Other workers have
presented their results as scans of optical density from autoradiographs [3], but such a procedure can give a misleading impression
of the relative amount of label in each product unless it is care-
fully calibrated.

Isolated intact chloroplasts from pea synthesise at least six
major discrete proteins as revealed by SDS gel electrophoresis
(Fig. 2). Similar results have been obtained with chloroplasts from
barley, spinach, and maize, except that in the latter case, peak B
is missing. When the chloroplasts are lysed and the thylakoids are
spun down, it is seen that only one of these proteins (peak B in
Fig. 2) is soluble. This protein has been identified by tryptic
peptide analysis as the large subunit of Fraction I protein [5].
I regard this as the first definitive identification of a protein
that is synthesised by chloroplast ribosomes. The high proportion
of chloroplast ribosomes found in leaves may thus be required, not
because they make a wide range of different proteins, but because
Fraction I protein is one of the most abundant proteins [14].
Traces of other labelled soluble products can be detected in the
gels beside the large subunit of Fraction I protein, but these are
labelled so poorly as to defy analysis. These other soluble products
have been detected by autoradiography [3]. The small subunit of
Fraction I protein is not synthesised by isolated chloroplasts, but
it has been detected as the product of protein synthesis by isolated
cytoplasmic ribosomes from bean leaves [15].

The failure of chloroplasts isolated from maize leaves to
synthesise the large subunit of Fraction I protein is consistent
with the observation that most of these chloroplasts are derived
from mesophyll cells. Maize plants possess the C4 pathway of
photosynthesis, a feature of which is the absence of ribulose
bisphosphate carboxylase (or Fraction I protein) from the mesophyll
cells. Instead this protein occurs in the bundle-sheath cells, and
can be readily labelled in vivo. So far we have not succeeded in
demonstrating the synthesis of the large subunit in chloroplasts
isolated from bundle-sheath cells. These results raise the interesting question as to the nature of the mechanism which ensures
that the large subunit gene is not expressed in mesophyll chloroplasts.

Reports from both laboratories which have repeated our work on
isolated chloroplasts have suggested that the large subunit labelled
in vitro is integrated into the Fraction I molecule as judged by
analysis on non-denaturing polyacrylamide gels [3,16]. The essential
observation is that the large subunit labelled in vitro migrates
with Fraction I protein when run on gels containing 5% acrylamide.

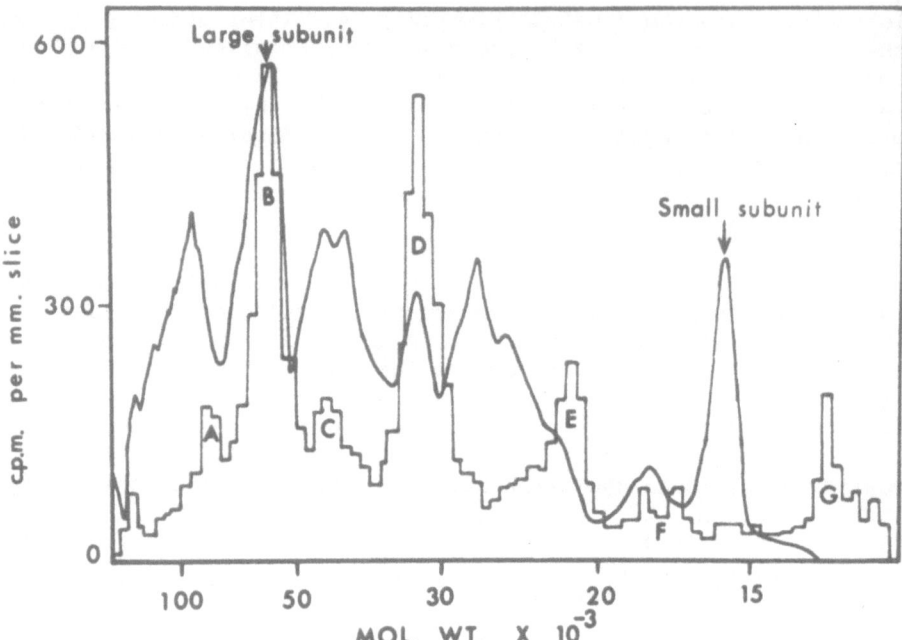

Fig. 2. SDS gel electrophoresis of labelled whole chloroplasts.
Isolated pea chloroplasts were incubated with [^{35}S] methionine
as described in Fig. 1. dissolved in SDS, and run on 15% gels
(20 μg chlorophyll per gel). The smooth line represents the
absorbance at 620 nm after staining with Amido Black, and the
histogram the radioactivity in each 1 mm gel slice. The letters
A - G mark the discrete labelled peaks. Large subunit and small
subunit refer to the subunits of fraction I protein.

However this coincidence is purely fortuituous, as can be shown by
using gels of higher or lower percentage acrylamide [5].
In fact the large subunit synthesised in vitro migrates on nondenaturing gels as an aggregate with an apparent molecular weight
of over 700 000 [17]. The failure of the large subunit to enter
Fraction I protein may reflect either a lack of small subunits in
isolated chloroplasts or the decay of the assembly process.

Recent studies from Wildman's laboratory have shown that the
large subunit of Fraction I protein from many species shows three
bands when analysed on electrofocussing gels [18]. The nature of
the differences between these bands is not clear, but the band
pattern is useful as a genetic marker for taxonomic purposes. We
have shown that the large subunit made by isolated pea chloroplasts

also separates as three bands on electrofocussing gels, so whatever
process produces this pattern is still operative in vitro.

The other major proteins made by isolated pea chloroplasts
are all membrane-bound, and have so far resisted identification.
They do not include cytochrome f, the coupling factor ATPase, RNA
polymerase, or photosystem I and II proteins. It is clear that
they are minor components of the thylakoids because fractionation
of the mebranes with digitonin shows that none of these labelled
peaks coincides with the staining bands [6]. The identification
of these major membrane-bound products of chloroplast protein
synthesis, which contain about 75% of the total protein label, is
the most pressing problem in this field.

Recent work has shown that there are several minor membrane-
bound products of chloroplast protein synthesis. A procedure for
isolating the chloroplast envelope was first devised by Mackender
and Leech [19]. By using this procedure we have shown that chloroplast envelopes isolated from pea leaves have a protein composition
distinct from that of the thylakoids, and that at least two of these
proteins contain a small amount of label when prepared from chloroplasts labelled in vitro [7]. Another minor product of chloroplast
protein synthesis is the coupling factor CF_1 or ATPase. This ATPase
is removed from thylakoids by washing with EDTA. Since our earlier
work showed that treatment with EDTA did not alter the pattern of
thylakoids labelled in vitro, we concluded that none of the labelled
peaks A - G could be a component of the ATPase [6]. We were therefore most interested in a recent report that three of the five
subunits of the ATPase are labelled in isolated spinach chloroplasts
[16]. We have confirmed that isolated intact pea chloroplasts also
label these three subunits, and have shown that our earlier failure
to detect this stems from the fact that these labelled subunits
contain only about 1% of the total label in membrane protein. It
is tempting to speculate, by analogy with Fraction I protein, that
the other two subunits are made by cytoplasmic ribosomes. This may
be incorrect, because the failure to label these two subunits in
isolated chloroplasts may merely reflect that these chloroplasts
already possess their full complement of the ATPase protein, and
are merely turning over the other three subunits. More detailed
inhibitor experiments with intact cells should resolve this point.
In any event, it is clear that chloroplasts are different from
mitochondria with regard to the site of synthesis of at least some
of their coupling factor subunits. I venture to predict that those
coupling factor subunits which are synthesised inside the chloroplasts will be shown to be encoded in the chloroplast genome.

Our experience with isolated intact pea chloroplasts confirms
the prediction that elongation, termination and in the case of the
large subunit of Fraction I protein, release of polypeptide chains
occurs within them. We have recently asked the question as to

TABLE 3

INITIATION OF PROTEIN SYNTHESIS IN ISOLATED CHLOROPLASTS

Energy source	Inhibitor (100μg/ml)	CPM x 10^{-6} S-35 in:-		
		f-Met puromycin		Protein
None	None	0.08		0.53
Light	None	3.1		20.05
Light	D-threo chloramphenicol	0.4		0.57

Isolated pea chloroplasts were incubated with 1 mM puromycin and [S-35] methionine in the presence and absence of light and chloramphenicol. f-Met puromycin was extracted and counted. The cpm in protein was measured in parallel incubations when puromycin was omitted.

whether initiation is also proceeding at a significant rate in these chloroplasts. Peter Highfield has shown that these chloroplasts incorporate [S-35] methionine into N-formylmethionylpuromycin; this incorporation is dependent upon light and is inhibited by chloramphenicol (Table 3). If several assumptions are made, it is possible to calculate that each messenger RNA that is being translated when the chloroplasts are isolated undergoes initiation at least twice during the incubation. It must be emphasised that the assumptions are the most unfavourable that are still feasible [20]. These results suggest that initiation is proceeding at a significant rate in isolated chloroplasts, and we must therefore withdraw an earlier suggestion of mine that protein synthesis in isolated chloroplasts represents only run-off of preformed polysomes due to a lack of initiation factors [21].

2. Lysed Chloroplasts

About one-third of the ribosomes in isolated pea chloroplasts are tightly bound to the membranes and cannot be removed by washing in hypotonic buffer. The remainder are readily released on osmotic lysis. It seemed possible to us that these two classes of ribosome could have distinct roles; for example the free ribosomes might synthesise the large subunit of Fraction I protein, while the bound ribosomes might synthesise the membrane-bound proteins. Using the knowledge obtained with intact chloroplasts, we have devised a method of persuading lysed and fractionated chloroplasts to synthesise discrete proteins.

The method consists of isolating pea chloroplasts as described in Fig. 1. but then resuspending them in a hypotonic buffer containing 25 mM tricine-KOH, 10 mM $MgSO_4$, and 5 mM 2-mercaptoethanol (pH 8.0). This causes immediate lysis as judged by phase-contrast microscopy. The lysed preparation will incorporate [S-35] methionine into protein if given ATP and GTP, but in contrast to intact chloroplasts, it will not use light energy [8]. The incorporation is sensitive to ribonuclease and D-threo chloramphenicol and requires 100 mM KCl for optimal rates. The rate of incorporation by lysed chloroplasts is similar to that by intact chloroplasts. Analysis on SDS polyacrylamide gels of the products of such lysed systems shows that the pattern is very similar to that given by intact chloroplasts using light energy. The most highly labelled peak runs coincident with the large subunit of Fraction I protein. A lower-molecular weight peak, running near the position of peak D (Fig. 2) is also seen, but is never as clearly discrete as peak D. These results suggest that lysed chloroplasts make the same products as intact chloroplasts, but that incomplete termination of polypeptide chains occurs more frequently in the lysed situation. The high activity of lysed chloroplasts provides a sensitive assay for testing the effect of compounds on chloroplast protein synthesis in the absence of the chloroplast envelope, which may otherwise act as a permeability barrier.

3. Free Ribosomes

When lysed chloroplasts are fractionated by centrifuging at 38 000 x g for 5 min, both the colourless supernatant fraction and the green membrane fraction incorporate [S-35] methionine into protein when supplied with ATP and GTP (Table 4). The sum of the incorporations by the supernatant and membrane fractions is only 50% of that given by the unfractionated lysed chloroplasts. The products of protein synthesis by free chloroplast ribosomes have been analysed on SDS polyacrylamide gels [8]. The chief product is a protein which runs coincident with the large subunit of Fraction I protein; its identity has been confirmed by tryptic peptide analysis. The free ribosomes also make some lower molecular weight products which run in the same positions as the membrane-bound products made by intact chloroplasts. In this case, these products are largely soluble and do not centrifuge down at 200 000 x g for 3 h. The identities of these products are unknown. They are not breakdown products of the large subunit, or the products of contaminating cytoplasmic ribosomes. One possibility is that they are soluble forms of membrane proteins which fail to attach to membranes in lysed chloroplasts.

TABLE 4

PROTEIN SYNTHESIS BY FRACTIONS PREPARED

FROM LYSED PEA CHLOROPLASTS

Fraction	Incorporation
Total lysate	100
38 000g Supernatant	31
38 000g Pellet	18
Washed 38 000g Pellet	2.6
200 000g Supernatant	0
Washed pellet + 200 000g Supernatant	10

Chloroplasts were lysed and fractionated as described in the text, and incubated with [S-35] methionine, 2 mM ATP, 0.2 mM GTP and 100 mM KCl for 40 min at 25°C. Results are expressed relative to the incorporation given by lysed, unfractionated chloroplasts.

4. Bound Ribosomes

If the membrane fraction is washed by resuspension in lysis medium, and recentrifuged, its activity is much reduced. This is expected because washing will remove soluble factors such as tRNAs and amino acid-activating enzymes. However, the activity of the washed membrane fraction can be partially restored by supplementing with a supernatant prepared by centrifuging lysed chloroplasts for 3 h at 200 000 g (Table 4). This high-speed supernatant has no protein synthetic activity of its own because it lacks ribosomes. Analysis of the products of supplemented bound ribosomes reveals a rather heterogenous pattern with at least four major peaks. All these products are firmly bound to the chloroplast membranes.

The results of these experiments with free and bound ribosomes

suggest that there is a division of labour in the chloroplast.
The large subunit of Fraction I protein is made by the free ribosomes, but not by the bound ribosomes. Both types of ribosome may
be implicated in the synthesis of the membrane-bound proteins.

5. Bacterial Ribosomes Programmed
with Chloroplast Messenger RNA

It is my view that one of the most important problems for
biochemists to study today is the molecular basis of development.
Most developmental processes are likely to involve messenger RNA,
and thus we must have the technology to measure specific messengers
[22]. Our discovery that the large subunit of Fraction I protein
is a major product of synthesis by chloroplast ribosomes prompted
us to see whether we could detect the messenger for this subunit
by adding chloroplast RNA to a heterologous protein-synthesising
system. Martin Hartley and Annabel Wheeler has shown that spinach
chloroplast RNA added to a cell-free ribosomal system from
Escherichia coli causes the production of a very similar set of
proteins to that labelled in isolated chloroplasts [9]. The
similarity between the products of isolated chloroplasts and of
the chloroplast RNA-directed heterologous system suggests that the
chloroplast messengers are being correctly translated. Current
work is aimed at purifying the messenger RNA for the large subunit
of Fraction I protein, but is hampered by the lack of poly (A) in
this messenger. In fact, all chloroplast messengers that are
translated by bacterial ribosomes appear to lack poly (A) as judged
by their failure to bind to poly (U) or to oligo d(T) columns [10].

6. Etioplasts

The question can be asked whether the products of light-driven
protein synthesis reflect the total spectrum of proteins that is
made by chloroplast ribosomes. We have designed our experiments so
that only chloroplasts with functioning photosystems can incorporate
amino acids into protein. It is therefore possible that chloroplasts
at an earlier stage of development - before they can utilise light
energy - may make a different range of proteins. We have examined
this possibility by characterising the products of ATP-driven protein
synthesis by etioplasts isolated from pea plants grown in the dark
[11]. The principal product is the large subunit of Fraction I
protein. This result is consistent with the well-known observation
that the synthesis of Fraction I protein does not require light in
many plants. The other products of etioplast protein synthesis are
membrane-bound, and present the same pattern on SDS polyacrylamide
gels as the products from chloroplasts. There is thus no evidence
that etioplasts synthesise a different range of proteins to chloroplasts. The same conclusion is reached from studies of the

products of protein synthesis by plastids isolated at intervals
from greening etiolated pea plants [11].

THE USE OF INHIBITORS OF PROTEIN
SYNTHESIS ON INTACT CELLS

It is important to try to confirm conclusions reached from
in vitro chloroplast experiments by means of inhibitor studies on
intact cells. This is because studies on isolated chloroplasts are
open to the objection that controls by nuclear and cytoplasmic
factors may not be operative. The validity of results from experiments with any inhibitor depends absolutely on the specificity of
its action in intact cells, and there is evidence that both chloramphenicol and cycloheximide have effects on systems other than
protein synthesis in some higher plants [13]. Systems such as ion
uptake, oxidative phosphorylation and photophosphorylation are
inhibited by all four stereoisomers of chloramphenicol, whereas the
inhibition of protein synthesis by isolated chloroplast ribosomes
is specific for the D-threo isomer [24]. This stereospecificity
provides a means of establishing for any particular tissue whether
chloramphenicol is inhibiting protein synthesis directly at the
ribosomal level, or in addition, is affecting some other process
such as the energy supply. It is strongly recommended that only if
an inhibition is produced specifically by the D-threo isomer should
an interpretation directly involving protein synthesis be invoked.
Another problem is that in most of the inhibitor experiments on
chloroplast ribosomal function, increases in specific proteins
have been measured as enzymic activities rather than as amounts of
protein. Failure to observe an effect by a particular inhibitor
might therefore mean that the increase in enzymic activity is due
to activation of a precursor protein, rather than to de novo synthesis by either chloroplast or cytoplasmic ribosomes. Bearing these
difficulties in mind, I interpret the bulk of the published inhibitor experiments as suggesting that most of the chloroplast proteins
are synthesised on cytoplasmic ribosomes [12,21]. In all the studies
reported on several algae and higher plants, the synthesis of
Fraction I protein was found to be inhibited by 70S ribosomal inhibitors. In most cases, the synthesis of the other soluble enzymes
of the photosynthetic carbon dioxide reduction cycle appears to
occur on cytoplasmic ribosomes; the same is true for ferredoxin and
the chloroplast RNA polymerase. Besides Fraction I protein, the
only other proteins that appear to be synthesised by chloroplast
ribosomes are some of the chloroplast ribosomal and lamellar proteins, including the photosynthetic cytochromes. In my laboratory
we have carried out inhibitor experiments on pea shoots with regard
to both the chloroplast membrane proteins and the chloroplast
envelope proteins. The results of these experiments confirm completely the results of the experiments with isolated chloroplasts
[7,21]. It must be emphasised, however, that these inhibitor

experiments are never more than suggestive. Strictly interpreted,
they never say more than that the activity of a particular group
of ribosomes is required for a particular protein to accumulate in
the chloroplast. It is not the same as saying that these ribosomes
actually synthesise that protein, because it is possible that the
apoenzyme is synthesised by one class of ribosomes but requires
for its appearance in the chloroplast in an active state, additional protein(s) which are synthesised by another class of ribosomes.

CO-OPERATION BETWEEN CHLOROPLAST AND NUCLEAR GENOMES

It is clear from the available evidence that the chloroplast
genome requires for its expression the co-operation of the nuclear
genome. The best information about the details of this co-operation
concerns the synthesis of Fraction I protein. By combining the
results of Wildman's genetic studies with those of the in vitro
studies of protein synthesis, a model for the synthesis of Fraction
I protein can be constructed (Fig. 3). In this model, the large
subunit is both encoded and synthesised within the chloroplast,
while the small subunit is both encoded and synthesised outside the
chloroplast. This model is tidy, and requires protein, but not
nucleic acid, to cross the chloroplast envelope. This transport of
protein must be on a large scale because it involves not only the
small subunit of Fraction I protein but all the other proteins
which, from the inhibitor evidence, are made on cytoplasmic ribosomes.
The mechanism of this transport is unknown, but it must be able to
distinguish between different proteins. Our suggestion is that a
protein exists in the outer envelope of the chloroplast which recognises a site common to all those proteins that are made on cytoplasmic ribosomes, but which are destined to function in the
chloroplast. Another possibility is that cytoplasmic ribosomes
synthesising chloroplast proteins are attached to the outside of
the chloroplast envelope, and feed the growing polypeptide chain into
the chloroplast through a hole created by a special 'signal' sequence
at the N-terminus of the polypeptide [25].

Besides structural and enzymic proteins, it is likely that
regulatory proteins also cross the chloroplast envelope. How light
regulates chloroplast development is not known, but it would be
reasonable to speculate in terms of proteins entering the organelle
to trigger nucleic acid and protein synthesis. I regard this
question of protein transport across the chloroplast envelope as a
crucial one to study if we are to understand how the chloroplast and
nuclear genomes interact; almost no research has been carried out in
this area. The movement of nucleic acids across the envelope
remains a possibility, but no compelling evidence to suppose it
occurs has been published.

Another problem that needs to be studied is the nature of the

mechanism which integrates the synthesis of the two subunits of
Fraction I protein in different cellular compartments. The dashed
lines in Fig. 3. illustrate one possibility; the small subunit is
postulated to be required as a positive factor for the initiation
of either transcription or translation of the large subunit messenger [21]. Inhibitor experiments with pea shoots suggest that a
product of cytoplasmic ribosomes is required for the continual
synthesis of the large subunit [21]. An obvious candidate for this
protein is the small subunit. This idea should be testable once
the messenger RNA for the large subunit has been purified.

PROSPECTS

     The following areas of research on the synthesis of chloroplast
proteins seem ripe for exploitation:
1.   The analysis of the products of protein synthesis by isolated
chloroplasts in a high resolution two-dimensional system. The

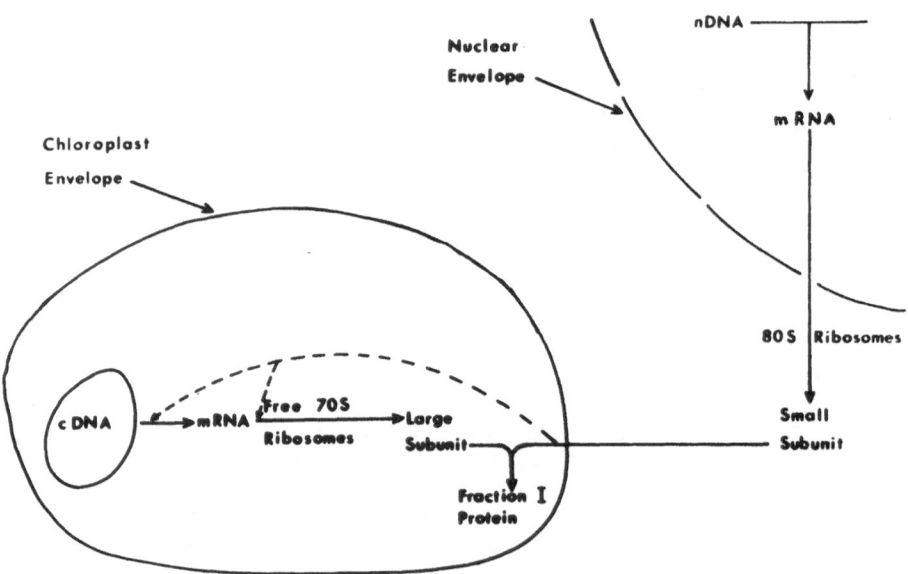

Fig. 3. Model for the co-operation of nuclear and chloroplast
genomes in the synthesis of Fraction I protein. cDNA and nDNA
stand for chloroplast and nuclear DNA respectively.

discovery and confirmation that isolated chloroplasts incorporate
a small amount of labelled amino acid into three subunits of the
coupling factor raises the possibilitv that other soluble and
membrane-bound proteins may be similarly labelled. Such products
would not be detected by the one-dimensional electrophoretic system
of analysis used so far, because they would be masked by the much
more heavily labelled major products. Analysis on a two-dimensional system may reveal that there are many more components being
labelled to a small degree in isolated chloroplasts. If this turns
out to be the case, and the assumption is made that these proteins
are also encoded inside the chloroplast, the potential coding
capacity of the order of 100 genes in chloroplast DNA may be accounted for.
2.   More determined efforts to isolate chloroplasts in which
translation is coupled to transcription. The key to success here
may be to supply cytoplasmic factors to isolated chloroplasts. If
it is correct that proteins enter the chloroplast in the manner
envisaged by the signal hypothesis [25], it may even be worth
trying the addition of leaf cytoplasmic messengers plus a wheat
germ protein-synthesising system to isolated chloroplasts. In this
way, the macromolecular synthetic activity of isolated chloroplasts
may be prolonged to a point where translation depends on newly-
synthesised RNA.
3.   Studies of the patterns of protein synthesis by isolated proplastids. The conversion of the proplastid to the chloroplast is
the most commonly encountered developmental sequence of the plastid
in vivo, and until the range of proteins synthesised on the ribosomes undergoing this differentiation is known, our knowledge of the
function of plastid ribosomes cannot be regarded as complete. It
is possible that the unidentified genes in chloroplast DNA encode
proteins that are synthesised during the development of chloroplasts
from proplastids. If this is the case, studies of isolated developing proplastids may reveal that they synthesise a different
spectrum of proteins from either chloroplasts or etioplasts. The
problem here is to find a tissue from which it is possible to
isolate proplastids readily.
4.   Establishment of a quantitative assay for the messenger RNA
of the large subunit of Fraction I protein, so that variations in
its amount in different developmental situations can be measured.
The light-stimulated synthesis of Fraction I protein in etiolated
plants will be of especial interest in this regard.

REFERENCES

1. Ellis, R.J. and Hartley, M.R. (1974) Nucleic Acids (K. Burton, ed.), MTP International Review of Science Series in Biochemistry, vol. 6, 323 - Medical and Technical Publishing Co., Lancaster, and Butterworth, London.
2. Boulter, D., Ellis, R.J. and Yarwood, A. (1972) Biol. Rev. 47, 113-175.
3. Bottomley, W., Spencer, D. and Whitfeld P.R. (1974) Arch. Biochem. Biophys. 164, 106-117.
4. Morgenthaler, J.J. and Mendiola-Morgenthaler, L. (1976) Arch. Biochem. Biophys. 172, 51-58.
5. Blair, G.E. and Ellis, R.J. (1973) Biochim. Biophys. Acta 319, 223-234.
6. Eaglesham, A.R.J. and Ellis, R.J. (1974) Biochim. Biophys. Acta 335, 396-407.
7. Joy, K.W. and Ellis, R.J. (1975) Biochim. Biophys. Acta 378, 143-151.
8. Ellis, R.J. (1975) Membrane Biogenesis 8, 247-278 (Tzagoloff, A., ed). Plenum Publishing Co., New York.
9. Hartley, M.R., Wheeler, A. and Ellis, R.J. (1975) J. Mol. Biol. 91, 67-77.
10. Wheeler, A.M. and Hartley, M.R. (1975) Nature 257, 66-67.
11. Siddell, S.G. and Ellis, R.J. (1975) Biochem. J. 146, 675-685.
12. Ellis, A.J., Blair, G.E. and Hartley, M.R. (1973) Biochem. Soc. Symp. 38, 137-162.
13. Parenti, F. and Margulies, M.M. (1967) Plant physiol. 42, 1179-1168.
14. Ellis, R.J. (1973) Comment. Plant Sci. 4, 29-38.
15. Gray, J.C. and Kekwick, R.G.D. (1974) Eur. J. Biochem. 44, 491-500.
16. Mendiola-Morgenthaler, L.R., Morgenthaler, J.J. and Price, C.A. (1976) FEBS Letts. 62, 96-100.
17. Ellis, R.J. (1973) Trans. Biochem. Soc. 1, 13-16.
18. Kung, S. (1976) Science 191, 429-434.
19. Mackender, R.O. and Leech, R.M. (1970) Nature 228, 1347-1348.
20. Highfield, P.E. and Ellis, R.J. (1976) Biochim. Biophys. Acta. (in press).
21. Ellis, R.J. (1975) Phytochem. 14, 89-93.
22. Ellis, R.J. (1976) Perspectives in Experimental Biology Vol. 2, 283-298 (N. Sunderland, ed.) Pergamon Press, Oxford and New York.
23. Ellis, R.J. and MacDonald, I.R. (1970) Plant Physiol. 46, 227-232.
24. Ellis, R.J. (1969) Science 163, 477-478.
25. Blobel, G. and Dobberstein, B. (1975) J. Cell Biol. 67, 835-851.

BIOSYNTHESIS OF PLANT MITOCHONDRIAL PROTEINS

C. J. LEAVER and P. K. POPE

DEPARTMENT OF BOTANY, UNIVERSITY OF EDINBURGH

EDINBURGH, EH9 3JH, SCOTLAND

I   INTRODUCTION

The number of recent reviews on the area of mito-chondrial biogenesis is evidence of the impressive amount of research devoted to this subject. The predominant organisim in the field continues to be yeast (Saccharo-myces cerevisiae) although there have been important contributions from those working with Neurospora and animal cells. The cogent review by Schatz and Mason[1] gives a critical evaluation of the experimental appro-aches to the subject, while articles on the biogenesis of mitochondria in HeLa (Attardi[2] et al) and fungi (Mahler[3] et al and Tzagoloff[4] et al) provide a more detailed discussion of results.

In the early 1960's it was shown that both mito-chondria and chloroplasts contained DNA. Subsequent studies were directed towards characterisation of the means of replication and expression of the organellar genome independently of the rest of the cell. It was then realised that these organelles, while containing the macromolecular components for autonomy, namely a specific DNA, DNA polymerase, DNA-dependent RNA poly-merase, and an indepednent protein synthesising system, were not autonomous in any meaningful sense.

It emerged that the biogenesis of a functional mitochondrion or chloroplast required the co-operative expression of both organellar and nuclear genetic in-formation. In the case of the mitochondrion consider-

able evidence suggests that about 90% of mitochondrial
proteins, specifically those of the outer membrane, the
soluble matrix and a large proportion of the proteins
of the inner membrane, are encoded in nuclear DNA, synthesised on cytoplasmic ribosomes and subsequently transferred into the mitochondrial structure by an, as yet,
unknown mechanism.

Following the detailed characterisation of the
genetic, transcriptional and translational systems of
the mitochondrion (see Borst[5]for review), attention has
now turned towards three main areas of investigation:
(1) the construction of a complete physical and genetic
map of mitochondrial (mt)DNA, (2) the identification of
mitochondrial genes and their translation products, and
(3) finally, and perhaps of most interest, is the study
of mitochondrial growth and differentation, and the
mode of integration of mitochondrial genome expression
with the overall process of cell division, growth and
differentiation.

In this review we shall briefly describe the properties of mtDNAs and RNAs, then turn to their participation in mitochondrial protein synthesis and devote
most of our attention to the characterisation of the
products of protein synthesis by higher plant mitochondria. Finally we shall look to the future and those
unique aspects of plant development where a study of
the co-ordination of the synthesis and assembly of
certain mitochondrial proteins, should advance our know-
lege of the physiological and other factors controlling
their biogenesis and turnover. In turn, this may help
our understanding of the way in which their main metabolic function of respiration and the generation of
energy, is geared to the development of the plant cell.

II. THE MITOCHONDRIAL GENOME

Since the first convincing demonstration of the
presence of DNA in mitochondria by Nass[6], a considerable body of information concerning the characterisation of a range of mtDNAs has become available. The
mtDNA of all organisms so far studied, with the exception of Tetrahymena, exists in the form of closed circles, which vary in contour length and kinetic complexity from 5 μm in animals[5] to 20-26 μm in ascomycete
fungi[7] and at least 30 μm in higher plants[8] (cf. 40 μm
circles of chloroplast DNA). The relatively rapid re-

naturation kinetics of denatured mtDNA suggests that the
complexity is low and that the 2-10 circles of mtDNA
found in one mitochondrion are similar if not identical
to the mtDNA in the remaining mitochondria of the organism. The base composition and therefore buoyant density
of mtDNAs differs from that of the homologous nuclear
DNAs in the organisms so far studied. In a range of
higher plants the density of mtDNA is remarkably constant around 1.706-1.707 g/cm3[9,10,11] in contrast to
nuclear DNAs which vary in density between 1.691 and
1.702 g/cm3.

The total information content of the small animal
mtDNA (mol. wt. 10[7]) corresponds to about 15,000 base
pairs. This is quite sufficient to code for the structural mitochondrial ribosomal (rRNA) and transfer (tRNA)
RNAs which have been shown by RNA-DNA hybridisation
studies, to be encoded in mtDNA, in addition to accom-
modating the cistrons for all the mitochondrial translation products[1]. Recently, Borst suggested that 5 μm
is sufficient to code for the structural genes required
for all mitochondria, and that the increased size of
mtDNA in ascomycetes and higher plants is likely to be
due to regulatory sequences rather than protein coding
sequences[12]. As to why some organisms need more control sequences than animals, it has been suggested by
Slonimski that the extra sequences are found in facul-
tative aerobes such as yeast, which might need more
control sequences than obligate aerobes such as animals,
or in green plant cells where the mitochondrion interacts metabolically with the chloroplast.

Some confirmation for the suggestion that the 5 μm
circle of mtDNA may contain enough structural information for the whole range of mitochondria found in nature,
has been gained from the recent extensive use of bacterial
nucleases and endonucleases, coupled with DNA-RNA hybridisation, in the analysis of mtDNA. It has been possible
to map the position of the rRNA cistrons on the yeast
and animal mt genome[2] and in addition allow the formula-
tion of a model for the yeast genome which suggests
that it consists of interspersions of A+T-rich and G+C-
rich sequences. These sequences are thought to correspond to spacers and genes[13] leading to the interesting conclusion that perhaps a maximum of 50% of the
genes has an informational content.

Mitochondrial Genome Replication and Transcription

Relatively little is known about mitochondrial
replication, although several groups have purified and
partially characterised a mtDNA polymerase which is apparently coded for by the nuclear genome[5]. We know a little
more of the details of mitochondrial transcription, and
a mtDNA-dependent RNA polymerase has been isolated from
yeast[14,15] and Neurospora[16]. Perhaps the most unexpected
observation, is that of Attardi[2] who has shown in HeLa
cell mitochondria, that both light (L) and heavy (H)
strands of mtDNA are transcribed into functional tRNAs.
If symmetrical transcription is a general phenomenon, as
seems to be the case, this will necessitate a reappraisal
upwards of our estimates of the coding capacity of the
mitochondrial genome.

In recent years several discrete RNA components coded
for by mtDNA have been identified. Kuriyama and Luck[17]
demonstrated the synthesis of a high molecular weight rRNA
which is subsequently processed into two stable mtrRNAs,
in mitochondria from Neurospora crassa. While Attardi[2]
et al, using low doses of actinomycin-D, to selectively
suppress the synthesis of cytoplasmic rRNA in HeLa cells,
have shown the synthesis of specific mtrRNAs and tRNAs.

To date, there is no direct proof that mtDNA is
transcribed into mt-messenger RNA (mRNA). There are
conflicting results which suggest that the mitochondria
of certain animal cells[2,18] and yeast[19] contain distinct
Poly (A)-containing RNAs which have properties consistent
with those of mRNA. The length of the Poly (A) segment
is in the range of 50-80 nucleotides, compared to values
of 100-200 in the nucleo-cytoplasmic system. In contrast
to these reports Groot[20] et al and Eggitt and Scragg[21]
failed to detect any significant quantity of Poly (A)
material in yeast mtRNA. The later authors did however
demonstrate that total yeast mtRNA will direct the synthesis of a limited range of discrete polypeptides when
injected into Xenopus laevis oocytes or added to an E.
coli cell-free protein synthesising system. Scragg[22]
also attempted to identify mitochondrial gene products
of yeast synthesised from mtDNA in a coupled transcrip-
tion-translation system of E. coli, and found that some
of the labelled polypeptides could be precipitated with
antisera against an insoluble subfraction derived from
yeast mitochondria.

III. THE MITOCHONDRIAL TRANSLATION SYSTEM

The progressive decrease in mt genome size from higher plants to animals appears to be associated with or involve the ribosomal cistrons, which have been reduced in animal mitochondria to an unprecedented small size[5]. The reported sedimentation coefficients of animal mt-ribosomes range from 55-60S, while 72-78S mt-ribosomes have been isolated from several ascomycete fungi[5]. We have shown that a range of higher plant mitochondria contain ribosomes which sediment at 77-78S[11] and this has been independently confirmed in Maize[23]. Higher plant mt-ribosomes can be dissociated into large and small subunits which contain discrete rRNA species sedimenting at 24S and 18.5S. These have estimated molecular weights of $1.12-1.26 \times 10^6$ and $0.69-0.78 \times 10^6$, depending on the plant species (Fig. 1B)[11,24]. Estimates of the size of animal and fungal mt-rRNAs have been complicated by the low G+C content (from as low as 20% in yeast to as high as 48% in some animals) which leads to anomalous behaviour on gel electrophoresis. Bearing this in mind, animal mt-rRNAs sediment at 12-13 S and 16-17S and have calculated molecular weights of $0.3-0.36 \times 10^6$, and $0.35-0.56 \times 10^6$. Fungal mt-rRNAs are larger and sediment at 14-18S and 21-25S corresponding to molecular weights of approximately $0.03-0.72 \times 10^6$ and $1.23-1.30 \times 10^6$(5). Although mt-ribosomes differ in size and physical properties from the cytoplasmic 80S and bacterial and chloroplast 70S ribosomes, they do have several functional similarities with the procaryotic (70S) ribosome; e.g. they require a formylated methionine for initiation of protein synthesis, there is an interchangeability of protein factors required for protein synthesis[25] and a selective sensitivity to certain antibiotics[1,3]. Several studies have shown that animal and ascomycete mitochondria contain at least 19 transfer RNAs (so called 4S RNA with a molecular weight of ca 25,000) which are distinct from their counterparts in the cytoplasm and which hybridise to both the H and L strands of homologous mtDNA[5,2]. In addition, mitochondria also contain aminoacyl-tRNA synthetases which show some specificity with respect to the mt-tRNA which they will acylate[26]. In higher plants we have shown the presence of mitochondria-specific 4S RNA and in addition a 5S rRNA (38-40,000 mol. wt.) component (Fig. 1E and F)[27]. This 5S rRNA is a constituent of the large ribosomal subunit and an equivalent mole-

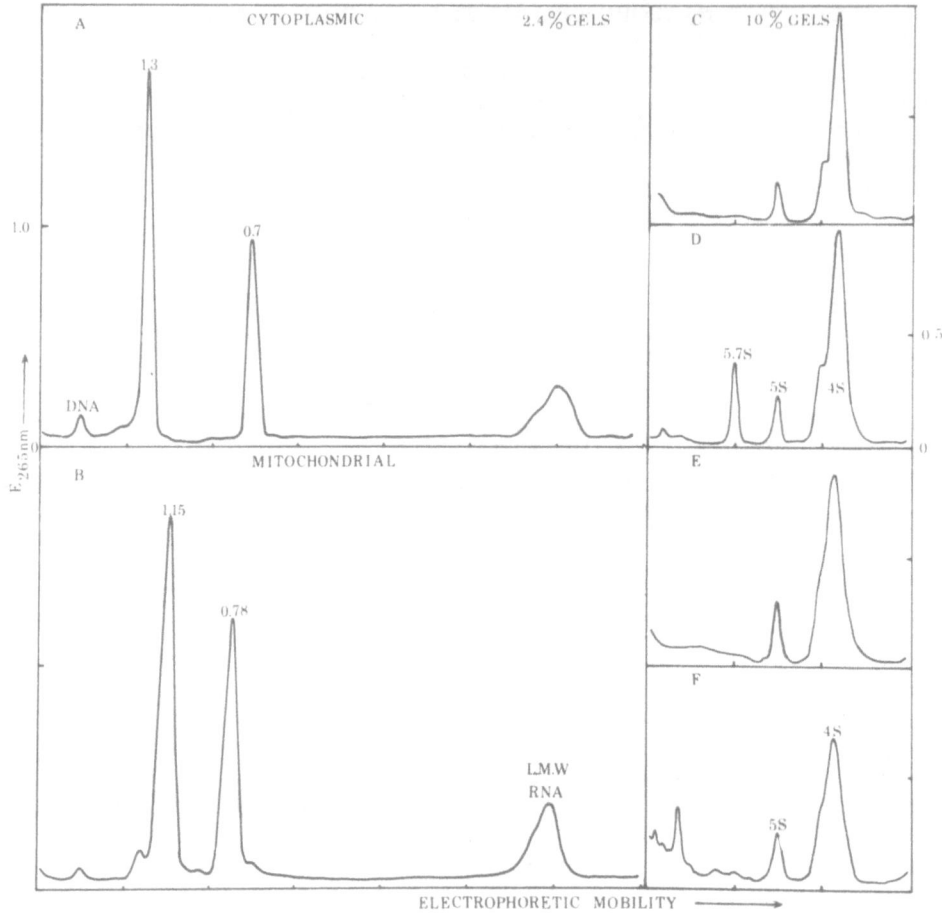

Fig. 1   Polyacrylamide-Gel Electrophoresis of Nucleic
              Acids from Mung Bean Hypocotyls

(A) Total cytoplasmic nucleic acids and (B) total mitochondrial nucleic acids were fractionated on 2.4% (W/V)
polyacrylamide-gels for 3.5 h at 50 V (3mA/9 cm gel) and
at 5°C.

Aliquots of the same samples were fractionated on 10%
(W/V) polyacrylamide-gels for 5 h at 50 V to further
resolve the low-molecular-weight nucleic acid components. (C) total cytoplasmic nucleic acid (D) as (C)
but heated to 65°C for 10 min. and cooled by rapid
immersion in liquid N_2 prior to fractionation, (E) total
mitochondrial nucleic acid, (F) as (E) but heated to
65°C before fractionation.

cule has been detected in the ribosomes of all procaryotic and eucaryotic ribosomes studied[28] except those from animal and fungal mitochondria[5]. Higher plant mitochondrial ribosomes, in common with the ribosomes of prokaryotes, chloroplasts and other mitochondria, do not contain an RNA equivalent to the 5.8S rRNA (50,000 mol. wt.). In eucaryotes this molecule is bonded to the larger of the cytoplasmic rRNAs (cf Fig. 1D and F).

IV. MITOCHONDRIAL PROTEIN SYNTHESIS

In terms of total cellular protein synthesis the mitochondrial contribution is very low, and in higher plants is almost certainly less than 1%. This is in contrast to the situation in chloroplasts where organelle ribosomes may represent as much as 50% of the total cellular ribosome complement and are responsible for the synthesis of a large proportion of the protein of the leaf[29].

There are several means by which the products of mitochondrial translation can be studied and these have been critically reviewed recently[1]. The most direct method is the _in vitro_ approach, in which isolated mitochondria are allowed to incorporate radioactive amino acids into protein, in a process dependent upon oxidative phosphorylation or an added ATP-generating system. There are however, several drawbacks to this potentially useful approach. The first is the problem of contaminating bacteria and cytoplasmic ribosomes, necessitating the use of sterile or nearly sterile mitochondrial preparation. Second, the rates of incorporation are slow, of relatively short duration and may be markedly affected by seemingly minor variations in the incubation medium. Third, many of the presumed controls imposed by cytoplasmic "partner proteins" may be lost, since isolated mitochondria probably contain only limited pools of these 'partner' proteins necessary for the assembly of stable mitochondrial complexes.

The most widely used approach to the study of mitochondrial protein synthesis is to analyse proteins labelled _in vivo_ in the presence of cycloheximide[30,1]. This method depends on the demonstration that protein synthesis on cytoplasmic ribosomes is specifically inhibited by cycloheximide, whereas protein synthesis on mitochondrial ribosomes is specifically inhibited by chloram-

phenicol[31]. This system has the advantage that the
newly synthesised polypeptide products are labelled to
a high specific activity and are able to combine with
cytoplasmically-synthesised 'partner' proteins which
apparently are present in excess in the cycloheximide
treated cells. In addition certain aspects of the
physiological control exerted by the cytoplasm seem to
be operative even under conditions when the two translation systems are 'uncoupled' from each other. The
main caution which must be applied in these experiments
is that the inhibitor must specifically inhibit cytoplasmic protein synthesis and not cause any serious
side effects.

Another means of analysis of the products of mitochondrial protein synthesis is a genetic approach, which
unfortunately is not obviously applicable to plant mitochondria[32]. This approach can be best illustrated by
work based on the observation that under normal conditions of growth, a small percentage of yeast cells are
produced which have respiratory deficient mitochondria[33].
These so called 'petite' mutants have aberrant mtDNA although they still contain a 'mitochondria like' organelle[34]. The 'petite mutant' is incapable of mitochondrial protein synthesis[30] thus making it possible to
analyse for missing polypeptide products of the mitochondrial system and conversely confirming the cytoplasmic origins of other mitochondrial proteins which
are still synthesised.

The systems described above have all proved useful
in studying mitochondrial protein synthesis, the major
problem until the late 1960's was the availability of
good methods for separating and characterising the relatively insoluble protein products. This situation
improved with the introduction and application of two
powerful techniques: the first was the use of SDS-
polyacrylamide gel electrophoresis, which solubilised
the polypeptides of the mitochondrial membrane and
fractionated them on the basis of size[35,36]; the second was the purification of certain important protein
complexes from the inner mitochondrial membrane, their
subsequent fractionation into discrete polypeptide subunits and the preparation of monospecific antibodies to
these subunits[37]. These antibodies were then used to
specifically precipitate radioactive polypeptides from
mitochondria labelled under the conditions already described.

A combination of the experimental approaches outlined above has shown that the only products of mitochondrial protein synthesis are 8-10 relatively insolu-
uble polypeptides which form part of the inner mitochondrial membrane. In yeast and Neurospora these polypeptides vary in molecular weight from 7,8000-45,000
and have been identified as subunits of three enzyme
complexes - namely cytochrome oxidase[1,38], cytochrome
b[39], and the oligomycin-sensitive F₁-ATPase[4]. These
complexes are all involved in electron transport and
oxidative phosphorylation. The system basically consists of a redox chain of electron-transferring proteins and carriers concerned with the transport of
reducing equivalents from the TCA cycle intermediates
for the reduction of molecular oxygen. Intimately
associated with the membrane-bound electon transport
chain is the F_1-ATPase complex, which can be seen under
the electron microscope as stalk-like structures lining
the cristae and protruding into the soluble matrix of
the mitochondrion. This enzyme complex is responsible,
in an as yet unknown manner, for the coupling of electron
transport with the phosphorylation of ADP to produce ATP.
The electron transport chain consists of four functional
complexes; complex IV, contains cytochrome oxidase
which consists of seven polypeptide subunits of which
the three largest have been shown to be synthesised on
mitochondrial ribosomes[40,38,4]. The four small polypeptides are made on cytoplasmic ribosomes. A component of cytochrome b in complex III is also synthesised
in the mitochondrion[39]. The F_1-ATPase enzyme complex
consists of ten polypeptides, four of the most hydro-
phobic of which are embedded in the mitochondrial inner
membrane and are synthesised by mitochondrial ribosomes[4].
These subunits subtend a single polypeptide, the OSCP
(oligomycin sensitivity-conferring protein) stalk protein, which in turn supports the five polypeptides which
constitute the F_1-ATPase. These six subunits are all
synthesised on cytoplasmic ribosomes and are nuclear
coded.

The products of mitochondrial protein synthesis by
animal mitochondria are less well characterised, although
Constantino and Attardi[41] using HeLa cells, have described ten polypeptides which fulfil the criteria of products of mitochondrial translation. They go on to note
that the gross correspondence in the number and size
range between these discrete mitochondrially synthesised
proteins and the number and size of Poly (A)-containing
RNA components transcribed from mtDNA, suggests that

these proteins may be specified by mitochondrial genes[2].

V. MITOCHONDRIAL PROTEIN SYNTHESIS IN HIGHER PLANTS

Studies on protein synthesis by higher plant mitochondria are limited to one by Moore[42] et al who characterised a mitochondrial amino acid incorporating system from etiolated soyabean hypocotyls, a short report
by Goswami[43] et al using Vigna sinensis mitochondria,
and our own work with a range of higher plant mitochondria[44,45].

A. In Vitro Protein Synthesis by Plant Mitochondria

In view of the encouraging results obtained with
in vitro chloroplast protein synthesis[29] and despite
the limitations already discussed, we decided to establish optimal conditions for amino acid incorporation
by isolated plant mitochondria. Mitochondria were prepared from 5-day-old etiolated Mung bean (Phaseolus aur-
eus) hypocotyls grown under sterile conditions or ster-
ilised artichoke (Helianthus tuberosus) tuber tissue.
The isolation of mitochondria was carried out under sterile conditions by a modification of the method of Douce[46]
et al.

An important prerequisite to the study of protein
synthesis by isolated organelles, is that they should
be intact, free from bacterial and cytoplasmic contamination, with their physiological integrity preserved.
Therefore all preparations were rountinely checked for
purity and respiratory competence by a combination of
electron microscopy, oxygen electrode polorography,
RNA and DNA analysis and bacterial counts.

The time course of incorporation of labelled amino
acid into hot-trichloroacetic acid-insoluble material by
isolated mung bean mitochondria incubated in a simple
buffered medium is shown in Fig. 2. After an initial
slight lag the rate of incorporation is essentially linear for the first 30 to 40 min., followed by a gradual
decrease to 60 min. The incorporation is dependent
upon the supply of an oxidisable substrate, adenine
nucleotides, phosphate and coupled mitochondria (Fig.
2 and Table I). Incorporation is reduced to near background levels when malate is replaced by a non-oxidisable
substrate, acetate. When some of the commoner TCA cycle

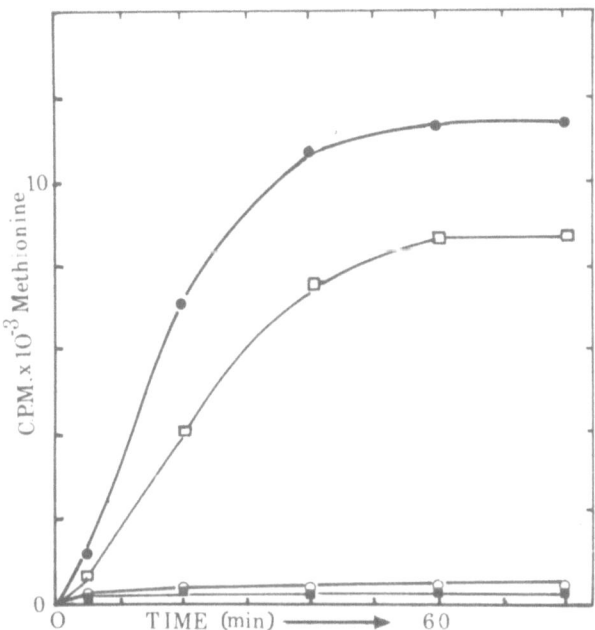

Fig. 2. Time Course of [35S]-methionine Incorporation into the Protein of Isolated Mung Bean Mitochondria.

Mitochondria were isolated from 5-day old etiolated mung bean hypocotyls as previously described[11]. Mitochondria (1.5 mg protein) were incubated at 20°C in a final volume of 500 μl containing: 200 mM-mannitol, 12.5 mM-KCl, 15 mM-MgCl2, 5 mM-KH2PO4, 12.5 mM-Tris-HCl pH 7.2, 1 mM-ADP, 0.025 mM-amino acid mix minus methionine, 0.5 μCi [35S]-methionine (280 Ci/mmole) and (—●—) + 20 mM-malate; (—□—) + 20 mM-succinate; (—○—) + 20 mM-malate–ADP; (—■—) − 20 mM-malate.

Table I. Characteristics of ^{14}C-amino Acid
Incorporation by Isolated Mung Bean Mitochondria

System	Incorporation as % of Complete System
Complete	100
– malate	4
– ADP	8
– phosphate	11
– malate + succinate (10 mM)	71
– malate + acetate (20 mM)	5
– ADP + ATP (1 mM)	84
– ADP + ATP (1 mM) + PEP + Pyruvate kinase	28
+ Dinitrophenol (0.1 mM)	6
+ KCN (1.0 mM)	2
+ Antimycin (0.4 µg)	19
+ Triton-X100 (0.05%)	0

Mitochondria were incubated as described in Fig. 2
except that the [35S]-methionine and amino acid mix
were replaced with 2.5 µCi of [14C]-amino acid mix
(54 mCi/m atom carbon). Incorporation by the complete
system was 8202 cpm/mg mt protein/60 min.

intermediates were used as substrates, their efficiency
in promoting amino acid incorporation reflected the measured P/O values and suggested that the incorporation
rates were due to the energy yield of these substrates
(Table I). This emphasises the close coupling of the
amino acid incorporation to the endogenous phosphory-
lating systems. The dependence upon oxidative phosphorylation can be replaced by the substitution of ADP
by an energy generating system in the reaction mixture
(Table I), although the levels of incorporation were
considerably lowered. This result is consistent with
the observation that the addition of a phosphoenol pyruvate – pyruvate kinase energy generating system inhibited
in vitro amino acid incorporation by rat liver mitochondria[47]. This presumably being due to the stimulation of
the ATP-translocase system thus reducing the ATP levels
available for incorporation.

Inhibitors and uncouplers of respiration inhibit
amino acid incorporation by isolated mitochondria (Table
I), their effect apparently being directly related to
their effect on respiration.

The effect of inhibitors of nucleic acid and protein synthesis on amino acid incorporation in this system is shown in Table II.

The addition of ribonuclease had no effect suggesting that incorporation is only proceeding in intact mitochondria. Actinomycin-D at concentrations shown to inhibit RNA synthesis in the same mitochondrial preparations by over 90%, has no effect on protein synthesis. This suggests that in vitro protein synthesis is not dependent on continued RNA synthesis, and probably uses mRNA synthesised before the mitochondria were isolated. In common with other mitochondrial protein synthesising systems, the active D-threo isomer of chloramphenicol markedly inhibits amino acid incorporation, while having little effect on respiration. The L-threo isomer was a much less effective inhibitor of incorporation and had a greater effect on respiration. Lincomycin, in contrast to reports from other systems[29], had a marginal effect on incorporation by plant mitochondria. This may be a measure of the insensitivity of the mitochondrial protein synthetic system, or merely impermeability of the plant mitochondrial membrane to this inhibitor. Cycloheximide, a potent inhibitor of cytoplasmic protein synthesis, had no effect on this system.

Table II. The Effect of Inhibitors of RNA and Protein Synthesis on ^{14}C-amino Acid Incorporation by Isolated Mung Bean Mitochondria

Additions	Incorporation as % of Control
Control	100
Ribonuclease (30 μg/ml)	98
Actinomycin D (10 μg/ml)	96
L-threo-Chloramphenicol (160 μg/ml)	73
D-threo-Chloramphenicol (160 μg/ml)	7
Lincomycin (100 μg/ml)	81
Cycloheximide (20 μg/ml)	98

Incubation conditions were as described in Table I except that 5 μCi of [^{14}C]-amino acid mix were added. The control incorporated 18,386 cpm/mg mt protein/ 60 min.

The rate of incorporation of radioactive leucine by isolated mung bean mitochondria was calculated to be 150–180 pmoles/mg protein/hr and compares favourably with the rates reported by other authors[48,49].

An important consideration which has been overlooked in some cases is that amino acid incorporation into mito-chondrial protein involves not only the formation of the peptide bond, but also the transport of amino acid across the mitochondrial membrane and other energy-dependent processes. Under normal conditions these processes are closely coupled with ATP synthesis by oxidative phosphor-ylation[50] and each step is potentially rate limiting. It should also be borne in mind that any limitations imposed by experimental conditions, such as ionic or osmotic composition of the medium, on any of these associated processes may drastically affect protein synthesis itself

The results discussed above demonstrate that isolate(plant mitochondria are capable of incorporating amino acids into protein and that the process is dependent upon coupled oxidative phosphorylation and displays character-istic sensitivity towards inhibitors of respiration and protein synthesis.

B. Products of In Vitro Protein Synthesis

To examine the products of in vitro protein syn-thesis, isolated mung bean or artichoke tuber mitochon-dria were incubated with [^{35}S]-methionine for one hour. The mitochondria were then solubilised in SDS and frac-tionated on 15% SDS-polyacrylamide gels. At least eight distinct radioactive peaks, labelled A to I (Fig. 3), were resolved by tube gel electrophoresis followed by estimation of the radioactivity in the sliced gel. Using the improved resultion obtained by slab gel electrophor-esis followed by autoradiography of the dried gel, it was possible to detect at least nineteen labelled poly-peptides (Fig. 4). The same labelled peaks were ob-served if an ATP-generating system was substituted for malate as a source of energy. The labelled peaks are absent if the mitochondria are labelled in the pres-ence of D-threo-chloramphenicol, and are digested by a prior incubation in protease.

It would be premature to conclude that all the labelled polypeptides detected represent unique gene products; polypeptide aggregation, specific and/or

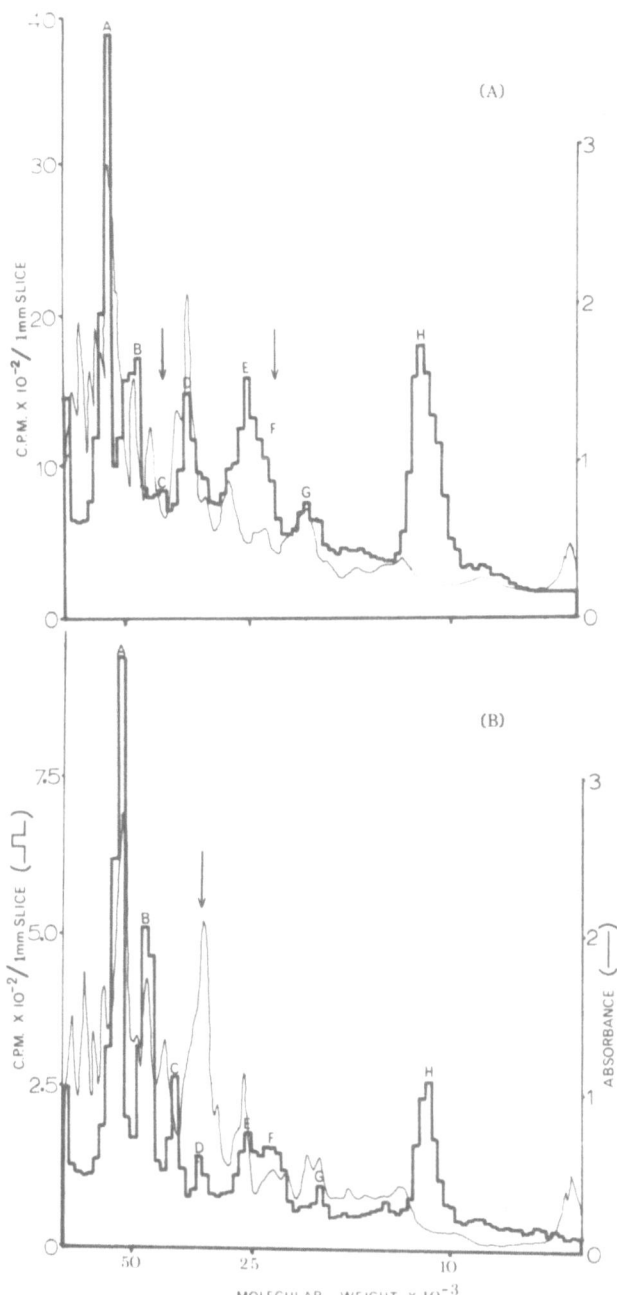

Fig. 3 <u>Sodium Dodecylsulphate (SDS) Polyacrylamide</u>
<u>Gel Electrophoresis of the Polypeptides Synthesised</u>
<u>(A) In Vitro and (B) In Vivo by Artichoke</u>
<u>Tuber Mitochondria</u>

[Fig. 3. (A) isolated mitochondria were incubated with
50 µC [^{35}S]-methione as described in Fig. 2. (B) 100g
artichoke tuber tissue was incubated in 20 µg/ml cycloheximide for 45 min., followed by the addition of 250
µCi [^{35}S]-methionine for 6 h. Mitochondria solubilised
with SDS were fractionated by SDS-PAG-electrophoresis
on 15% gels for 6 h at 3 mA/gel.]

non-specific cleavage products cannot be ruled out, particularly for the minor products.

    The apparent molecular weights of the major label-
led polypeptides were estimated from gels calibrated
with standards of known molecular weight. Table III
compares both the number and molecular weights of the
polypeptides synthesised by artichoke mitochondria
with those polypeptides shown to be products of mitochondrial translation in ascomycetes[1,4,38] and HeLa[41].

Table III. A Comparison of the Polypeptides Synthesised
    by Ascomycete, HeLa and Higher Plant Mitochondria

Protein Complex	Polypeptide Subunit	
	Yeast	Neurospora
Cytochrome oxidase	42,000[40] 34,500 23,000	41,000[52] 28,500 21,500
Oligomycin-sensitive ATPase	29,000[4] 22,000 12,000 7,500	19,000[53] 11,000
Cytochrome b	-	32,000[39]

Radioactive Polypeptide peak	HeLa[41]	Artichoke
A	42,000	57,500
B	39,000	42,500
C	35,000	34,000
D	31,000	31,500
E	27,500	23,000
F	24,000	19,000
G	22,500	18,000
H	19,500	16,000
I	15,000	10,500
J	11,500	

[Table III. Each value for artichoke represents the average of several estimates made on the basis of the electrophoretic mobility of the individual components relative to that of standard proteins].

The molecular weights of the polypeptides synthesised by
both artichoke tuber and mung bean mitochondria ranged
from 10,500 to 57,500. Bearing in mind the limitations
of the methods, the estimate of the range of polypeptides
synthesised by plant mitochondria are in broad agreement
with those quoted for other organisms.

   Preliminary experiments[45] have shown that more than
95% of the in vitro labelled polypeptides are associated
with the mitochondrial membrane system. This is in agreement with the general consensus that the mitochondrial
translation products are restricted to components of the
inner mitochondrial membrane.

   C. In Vivo Protein Synthesis by Plant Mitochondria

      It was clearly desirable to demonstrate that the
radioactive polypeptide products of in vitro mitochondrial protein synthesis were a true reflection of those
synthesised by the organelle in vivo. To do this, discs
of sterile artichoke tuber tissue were aged for 12 hours
in distilled water, during which time biogenesis of mitochondria is known to occur[51]. The discs were then incubated in 20 µg/ml cycloheximide (a concentration shown
to inhibit cytoplasmic protein synthesis by more than
95%) for 45 minutes, followed by the addition of [^{35}S]-
methionine for a further 6 hours. Mitochondria were
then isolated by the normal method, solubilised with SDS
and the constituent polypeptides analysed by SDS-polyacrylamide gel electrophoresis. A comparison of the
labelled products of in vivo and in vitro protein synthesis by artichoke mitochondria (Fig. 3 A and B) shows
that they are basically very similar. There are minor
differences in the distribution of radioactivity incorporated into the various polypeptides, notably into peaks
C and D. This is not unexpected as the constraints on
the system in vivo and in vitro are almost certainly
different and could easily lead to a difference in the
labelled polypeptide profile. The close similarity between the products of in vivo and in vitro translation
by artichoke mitochondria gives some confidence as to
the validity of using the in vitro approach to the study
of the products of mitochondrial protein synthesis.

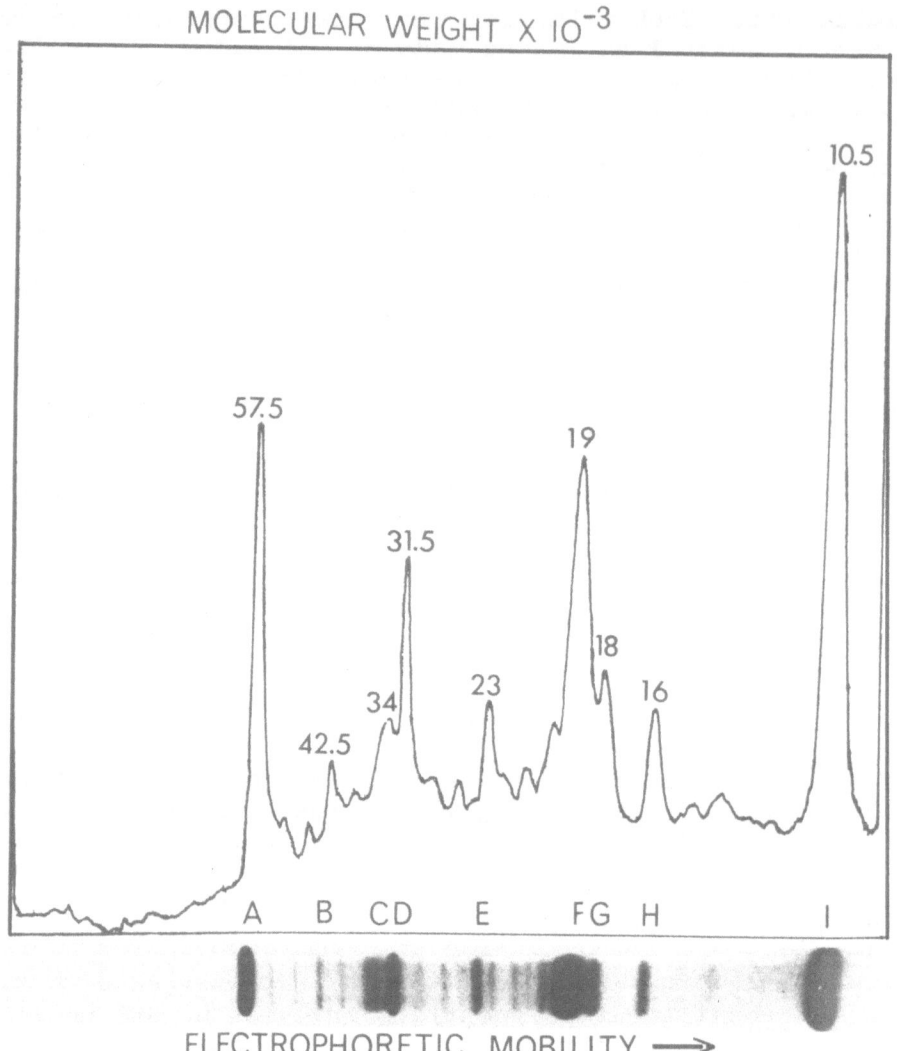

Fig. 4   The Resolution on SDS-polyacrylamide Slab Gels
         of the Polypeptides Synthesised In Vitro by
                Artichoke Tuber Mitochondria

Mitochondrial proteins labelled in vitro as described
in Fig. 3 were fractionated on a 15% polyacrylamide gel
with a 5% stacking gel using the discontinuous buffer
system described by Laemmli   Electrophoresis was for
16 h at 10 mA constant current. Gels were stained with
Coomassive blue and dried to Whatmann 3 MM paper and
exposed to Kodak Blue BandX-ray film. Autoradiographs
were scanned with a Joyce-Loebl microdensitometer.

The preliminary results presented in this section are suggestive, but by no means conclusive, evidence that there has been a conservation during the evolutionary modification of mitochondria, of their ability to synthesise several key polypeptides of the inner mitochondrial membrane. Ultimately it will be important to relate the polypeptide species synthesised by higher plant mitochondria to enzyme complexes of the inner mitochondrial membranes, which are known to contain subunits synthesised on mitochondrial ribosomes.

VI. BIOGENESIS OF MITOCHONDRIA

Relatively little is known about the continuity and replication of mitochondria within the cell. The best evidence indicates that, both in <u>Neurospora</u> <u>crassa</u>[54] and <u>HeLa</u>[2] cells the mitochondria grow by accretion of new materials, inserted randomly into the existing membrane structure and that the mitochondrial population increases by division of pre-existing organelles.

It has been suggested that the biogenesis of mitochondria can be considered to consist of two processes, which under normal conditions are intimately co-ordinated[2]. The first involves the gross formation of mitochondrial membranes, which appears to be solely controlled by the nuclear genome and involves the import of cytoplasmically synthesised protein into the developing organelle. The second process which depends on expression of the mitochondrial genome, is the differentiation of the inner membrane for the function of oxidative phosphorylation. Under certain conditions these two processes can be dissociated, e.g. the work of Linnane[55] and Criddle and Schatz[56] has shown that anaerobically grown yeast or yeast grown on a fermentable carbon source (e.g. glucose), contain very simple mitochondrial structures called promitochondria. These promitochondria lack cytochrome oxidase and other respiratory chain intermediates, as well as having little if any visible ultrastructure. However they do contain mtDNA[30] and possess the ability for protein synthesis. When anaerobically grown cells are aerated or when glucose in the medium is consumed, derepression of the mitochondrial enzyme systems occurs with a parallel increase in internal structure and the development of active organelles capable of oxidative phosphorylation. The application of differential inhibitors of mitochondrial and cytoplasmic protein synthesis to cells undergoing such a transition and also studies on 'pet-ite'mutants[1], have shown that the mitochondrial genome does not play any essential role inthe gross formation and division of mitochondria.

These studies provide evidence that the mitochon-drially synthesised polypeptides are necessary for the correct intercalation of a number of cytoplasmically synthesised proteins into enzyme complexes of the inner mitochondrial membrane. Thus playing a key role in the control of the biogensis of the active mitochondrion.

The foregoing discussion would seem to confirm that the replication and growth of mitochondria is under nu-clear control and is integrated with the overall process of cellular growth. Recent results[57] suggest that reg-ulation of the expression of the mitochondrial genome and acquisition of respiratory competence is directly influenced by physiological stimuli, such as oxygen. In this study, it was demonstrated in isolated yeast mitochondria, that oxygen regulates the synthesis of two of the three mitochondrially synthesised subunits of cytochrome oxidase, directly at the level of trans-lation.

Mitochondrial Biogenesis During Seed Maturation

One example of a dramatic change in plant devel-opment known to be linked to changes in mitochondrial activity, are the events occuring in cotyledons of mat-uring and germinating seeds. In addition, the whole problem of control of dormany and loss of seed viability, may in turn be related to the repression or failure of some aspect of mitochondrial function.

The high metabolic activity found in the cotyledons of many members of the Leguminosae, during the early stages of seed formation, gradually decreases as the seed matures and water is lost. Accompanying this change is a gradual decrease in the phosphorylation efficiency and respiratory control ratios of isolated mitochondria[58], associated with a disorganisation of the fine structure of the mitochondrion[59].

During rehydration and germination of the legume seeds there is a rapid rise in the rate of respiration. Associated with this rise, the respiratory competence of isolated mitochondria, as judged by their coupled oxidative phosphorylation and respiratory control, increases rapidly during the first 20 hours. Electron microscopical studies[59] have shown that accompanying these changes there is a reorganisation of the ultra-structure of the mitochondria, more specifically a reappearance of the inner mitochondrial membrane. Nawa and Asahi[60] found a shift to a higher density of mitochondria fractionated on sucrose gradients, which was associated with an increase in mitochondrial protein and phospholipid content.

Further study will resolve whether these developmental changes in mitochondrial activity, structure and composition are accompanied by considerable synthesis of both new respiratory enzymes and inner mitochondrial membrane, in addition to an increase in the number of mitochondria. This unique aspect of plant development could provide a system for the study of several important aspects of mitochondriogenesis. These include the co-ordination of expression of the nuclear and mitochondrial genomes and the cellular and physiological control of the assembly of the various mitochondrial proteins into a functional organelle. This should help our understanding of the way in which mitochondrial biogenesis and turnover is linked to the main metabolic function of respiration and geared to the development of the plant cell.

VII. THE CODING CAPACITY OF PLANT MITOCHONDRIAL DNA

It has been calculated that the coding capacity of animal mtDNA (mol. wt. 10×10^6) is sufficient to accomodate cistrons for all the mitochondria specific rRNAs and tRNAs and in addition, for all the known mitochondrial translation products. If it is assumed that plant mtDNA contains one cistron for each of the rRNAs and 20 tRNAs and taking into account the estimated number and molecular weight of polypeptides synthesised by plant mitochondria, it is possible to estimate the total required coding capacity (Table IV). The calculation leaves 86% of the potential coding capacity unaccounted for. It is possible that a small proportion of the residual space on the mtDNA could accomodate the cistrons for as yet unidentified translation products or for rRNA and tRNA segments removed during maturation.

Table IV. The Possible Information Content of
Plant Mitochondrial DNA

Mitochondrial Product	Molecular Weight	Molecular Weight of double stranded DNA required for coding
Large rRNA	1.15×10^6	2.3×10^6
Small rRNA	0.78×10^6	1.56×10^6
5S-rRNA	0.035×10^6	0.07×10^6
20 tRNAs	0.56×10^6	1.12×10^6
Mitochondrially Synthesised Polypeptides		
A	5.7×10^4	1.06×10^6
B	4.2×10^4	0.78×10^6
C	3.4×10^4	0.63×10^6
D	3.1×10^4	0.57×10^6
E	2.3×10^4	0.43×10^6
F	1.9×10^4	0.35×10^6
G	1.8×10^4	0.33×10^6
H	1.6×10^4	0.30×10^6
I	1.0×10^4	0.19×10^6
Total		9.69×10^6
Total genome available		70×10^6
∴ % accounted for		14%

Plant mtDNAs have been shown to be circular molecules with a molecular weight of 70×10^6 and with
no evidence of inter- or intra-molecular heterogeneity[8].
The buoyant density of 1.706-1.707 g cm^3 corresponds to
a G + C of 46% in contrast to an average value of 1.679
g cm^3, 18% G + C in the yeast, Saccharomyces cerevisiae[7,
13]. There is thus no evidence for the presence of A +
T-rich 'spacer' sequences in plant mtDNA, which in yeast
may account for 50% of the mitochondrial genome[13]. We
are therefore still left with the problem that a vast
excess of plant mtDNA has no known function.

Acknowledgements: We wish to thank the S.R.C. for
supporting this research.

REFERENCES

1. Schatz, G. and Mason, T. L. (1974). Ann. Rev. Biochem. 43: 51.
2. Attardi, G., Costantino, P., England, J., Lynch, D., Murphy, W., Ojala, D., Posakony, J. and Storrie, B. page 3 in Genetics and Biogenesis of Mitochondria and Chloroplasts. (ed. Birky, C. W., Perlman, P. S. and Byers, T. J.). 1975. Ohio State University Press.
3. Mahler, H. R., Bastos, R. N. Flury, U., Lin, C. C. and Phan, S. H. (1975). see Refn. 2, page 66.
4. Tzagoloff, A., Rubin, M. S. and Sierra, M. F. (1973). Biochim. Biophys. Acta 301: 71.
5. Borst, P. (1972). Ann. Rev. Biochem. 41: 333.
6. Nass, S. (1969). Int. Rev. Cytol. 25: 55.
7. Hollenberg, C. P., Borst, P. and Van Bruggen, E. F. J. (1970). Biochim. Biophys. Acta 209: 1.
8. Kolodner, R. and Tewari, K. K. (1972). Proc. natn. Acad. Sci. U.S.A. 69: 1830.
9. Suyama, Y. and Bonner, W. D. (1966). Pl. Physiol., Lancaster. 41: 383.
10. Wells, R. and Ingle, J. (1970). Pl. Physiol., Lancaster. 46: 178.
11. Leaver, C. J. and Harmey, M. A. (1973). Biochem. Soc. Symp. 38: 175.
12. Ellis, R. J. (1975). Nature 256: 617.
13. Prunell, A. and Bernardi, G. (1974). J. molec. Biol. 86: 825.
14. Eccleshall, T. R. and Criddle, R. S. (1974). p. 31 in The Biogenesis of Mitochondria. (eds. A. M. Kroon and C. Saccone). A. P., New York.
15. Scragg, A. H. (1974). see Refn. 14, page 47.
16. Küntzel, H. and Schäfer, K. P. (1971). Nature New Biol. 231: 265.
17. Kuriyama, Y. and Luck, D. J. L. (1973). J. molec. Biol. 73: 425.
18. Aujame, L. and Freeman, K. B. (1976). Biochem. J. 156: 499.
19. Cooper, C. and Avers, C. (1974). see Refn. 14, page 289.
20. Groot, G. S. P., Flavell, R. A., Van Ommen, G. J. B. and Grivell, L. A. (1974). Nature 252: 167.
21. Eggitt, M. J. and Scragg, A. H. (1975). Biochem. J. 149: 507.
22. Scragg, A. H. (1973). see Refn. 14, page 47.
23. Pring, D. R. (1974). Plant Physiol. 53: 677.
24. Pring, D. R. and Thornbury, D. W. (1975). Biochim. Biophys. Acta 383: 140.
25. Boulter, D. Ellis, R. J. and Yarwood, A. (1972).

Biol. Rev. Cambridge phil. Soc. 47: 113.

26. Barnett, W. E., Brown, D. H. and Epler, J. L. (1967).
    Proc. natn. Acad. Sci. U.S.A. 57: 1175.
27. Leaver, C. J. and Harmey, M. A. (1976). Biochem. J.
    157: 275.
28. Payne, P. I. and Dyer, T. A. (1971). Biochem. J.
    124: 83.
29. Ellis, R. J. (1976). in Perspectives in Experimental
    Biology. Vol. 2.   p. 283. (ed. N. Sunderland). Per-
    gamon Press, Oxford.
30. Schatz, G. and Saltzgaber, J. (1969). Biochim. Bio-
    phys. Acta. 180: 186.
31. Ennis, H. L. and Lubin, M.   (1964). Science 146: 1474.
32. Griffith, D. E. (1975). see Refn. 2, page 117.
33. Ephrussi, B. (1953). in Nucleo-Cytoplasmic Relations
    in Micro-Organisms. Clarendon Press, Oxford.
34. Perlman, P. S. and Mahler, H. R. (1971). Nature New
    Biol. 231: 12.
35. Shapiro, A. L., Viñuela, E. and Maizel, J. V. (1967).
    Biochem. Biophys. Res. Commun. 28: 815.
36. Weber, K. and Osborn, M. (1969). J.Biol. Chem. 244:
    4406.
37. Tzagoloff, A. and Meagher, P. (1971). J. Biol. Chem.
    246: 7328.
38. Sebald, W., Weiss, H. and Jackl, G. (1972). Eur. J.
    Biochem. 30: 413.
39. Weiss, H. (1972). Eur. J. Biochem. 30: 469.
40. Mason, T. L. and Schatz, G. (1973). J. Biol. Chem.
    248: 1355.
41. Constantino, P. and Attardi, G. (1975). J. molec.
    Biol. 96: 291.
42. Moore, A. L., Borck, K. and Baxter, R. (1971).
    Planta 97: 299.
43. Goswami, B. B., Chakrabarti, S., Dube, D. K. and
    Roy, S. C. (1973). Biochem. J. 134: 815.
44. Leaver, C. J. (1975). Ch. 6 in The Chemistry and
    Biochemistry of Plant Proteins. (ed. Harborne, J.
    B. and Van Sumere, C. F.).   A. P., London.
45. Leaver, C. J. and Pope, P. K. (1976). In prep.
46. Douce, R., Christensen, E. L. and Bonner, W. D.
    (1972). Biochem. Biophys. Acta 275: 148.
47. McCoy, G. D. and Doeg, K. A. (1972). Biochem. Bio-
    phys. Res. Commun. 46: 1411.
48. Ibrahim, A. G., Stuchell, R. N. and Beattie, D. S.
    (1973). Eur. J. Biochem. 36: 519.
49. Lamb, A. J., Clark-Walker, G. D. and Linnane, A.
    W. (1968). Biochem. Biophys. Acta 161: 415.
50. Wheldon, L. W. and Lenhinger, A. L. (1966). Bio-
    chemistry 5: 3533.

51. Sakano, K. and Asahi, T. (1971). Pl. Cell Physiol. 12: 417.
52. Sebald, W., Machleidt, W. and Otto, J. (1973). Eur. J. Biochem. 38: 311.
53. Jackl, G. and Sebald, W. (1975). Eur. J. Biochem. 54: 97.
54. Luck, D. J. L. (1965). J. Cell Biol. 24: 461.
55. Linnane, A. W. Haslam, J.M., Lukins, H. B. and Nagley, P. (1972). Ann. Rev. Microbiol. 26: 163.
56. Criddle, R. S. and Schatz, G. (1969). Biochemistry 8: 322.
57. Groot, G. S. P. and Poyton, R. O. (1975). Nature 255: 238.
58. Kollöffel, C. (1970). Planta 91: 321.
59. Bain, J. M. and Mercer, F. V. (1966). Aust. J. Biol. Sci. 19: 49.
60. Nawa, Y. and Asahi, T. (1971). Pl. Physiol. 48: 671.

THE EFFECT OF LIGHT ON RNA AND PROTEIN SYNTHESIS IN
PLANTS

H. MOHR and P. SCHOPFER

BIOLOGICAL INSTITUTE II/UNIVERSITY OF
FREIBURG
7800 Freiburg/Schaenzlestr. 9-11
West Germany

INTRODUCTION

This lecture deals predominantly with the photoregu-
lation of enzyme levels in plants by the photochromic
proteinaceous photoreceptor, phytochrome. While there
are claims that photocontrol of enzymes (enzyme synthesis, degradation, and activation) may be mediated by
light absorption in other pigments as well (e.g. photosynthetic pigments [1]), the emphasis on phytochrome is
justified for two reasons: firstly, phytochrome is the
most important molecule in higher plants for the detection of photosignals from the environment and for making
use of this information to regulate the orderly development of the living system; secondly, detailed knowledge
of the molecular mechanisms through which phytochrome
exerts its control over the development of higher plants
may serve as a useful model for understanding the control
of development in general, i.e. the control of gene expression during development.

The term photomorphogenesis is used to designate the
fact that light controls development (that is, growth,
differentiation and morphogenesis) of a plant independently of photosynthesis. While the specific development
of every living system depends on its genetic information and on its environment, higher plants are particularly sensitive to the environmental factor of light.
Naturally, light does not carry any specific information. However, light can be regarded as an "elective

factor" which deeply influences the manner in which
the genes, present in the particular organism, are used[2].

Figure 1 shows seedlings of the mustard plant illu-
strating the basic phenomena of photomorphogenesis. It
is obvious that photomorphogenesis is just another term
for the normal development of the sporophyte of a higher
plant. Without the light factor, this normal development

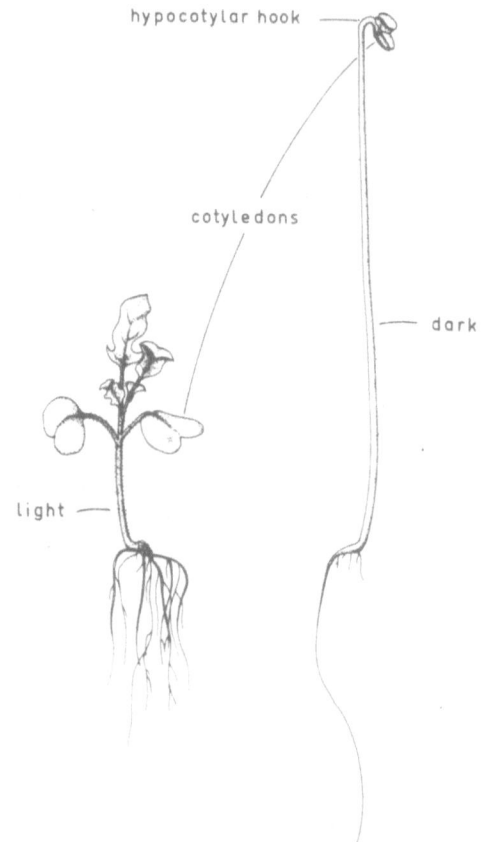

Fig. 1. These two mustard seedlings (Sinapis alba) have
the same chronological age and are virtually identical
genetically. The differences in morphogenesis are due to
light. The drawings emphasize the point that light causes
the cotyledons to develop from storage organs (right) to
photosynthetically active leaves (left). Unlike, for instance, castor bean endosperm, the mustard cotyledon is
not devoid of function after depletion of lipid and protein reserves, since in the light it expands, becomes
green and persists as a photosynthetic organ (after [2]).

cannot occur; rather, the plant will etiolate until it
dies. Thus, when we study the molecular basis of photomorphogenesis, we study in fact the molecular basis of
normal development in higher plants.

We have investigated over the years the development
of the cotyledons of the mustard seedling [2]. These
cotyledons are peculiar organs (Fig. 1). As long as the
seedling develops exclusively in the dark, the cotyledons function as storage organs. Filled with fat and
storage protein they serve the requirements of the ra-
pidly-growing axis system. The cotyledons themselves do
not grow or develop significantly as long as the seedling is kept in complete darkness. However, when the
seedling is illuminated with white light of sufficiently
high irradiance, the cotyledons are transformed rapidly
into photosynthetic organs, very similar in internal
structure and in function to a normal photosynthetically
active leaf. The "mechanism" of the phototransformation
of the cotyledons has been studied in an effort to understand, at the molecular level, the developmental events,
leading to the phenomenon of photomorphogenesis.

THE PHYTOCHROME SYSTEM

The photochromic sensor pigment phytochrome is a
bluish chromoprotein [3]. In the following, we use the
phytochrome model as given in Fig. 2. This model describes quantitatively the properties of the phytochrome
system as it occurs in the cotyledons and in the hypocotylar hook of the mustard seedling during the period
of our experimentation. The Pfr form of the chromoprotein
is the physiologically active form, the effector molecule
of the phytochrome system. The Pr form has no physiological effect. The sum of the amounts of Pr and Pfr is called "total phytochrome", or Ptot. In a dark-grown seedling, only Pr is present. Synthesis de novo of Pr is a
zero order process which does not depend on light. The
physiologically active Pfr originates from Pr through a
first order phototransformation which is photoreversible.
The effector moleculr Pfr is not stable. It disappears
through a first order destruction process which is independent of light. The half-life of Pfr in the mustard
cotyledons is 45 min at 25°C.

Biochemically, phytochrome is a chromoprotein consisting of a protein moiety and a chromophoric group.
The phytochrome chromophore is an open chain tetrapyrrole
(similar to the algal chromophore phycocyanobilin). A
most recent model for the phytochrome chromophore and its

$$P_r' \xrightarrow{{}^{o}k_s} P_r \underset{k_2}{\overset{k_1}{\rightleftharpoons}} P_{fr} \xrightarrow{k_d} P_{fr}'$$

Fig. 2. A model of the phytochrome system as it occurs in mustard seedling cotyledons and hypocotylar hook. The symbols ${}^{o}k_s$, k_1, k_2, k_d represent the rate constants of de novo synthesis of Pr, back and forth phototransformations and Pfr destruction respectively. Only k_1 and k_2 are light dependent (after [4]).

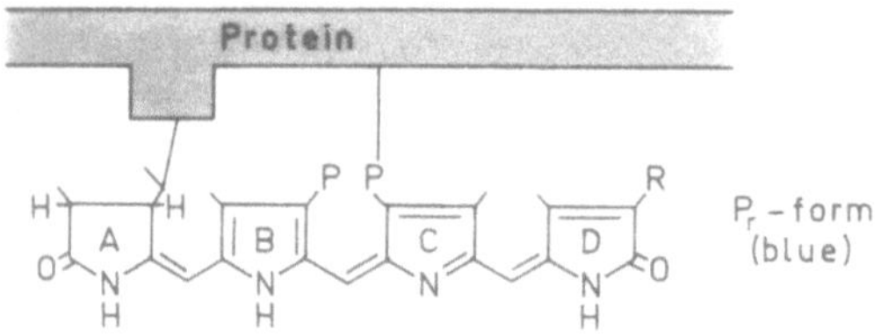

1a, $R = CH{=}CH_2$ $P = CH_2CH_2COOH$ (P_r)
1b, $R = C_2H_5$ (Phycocyanobilin)

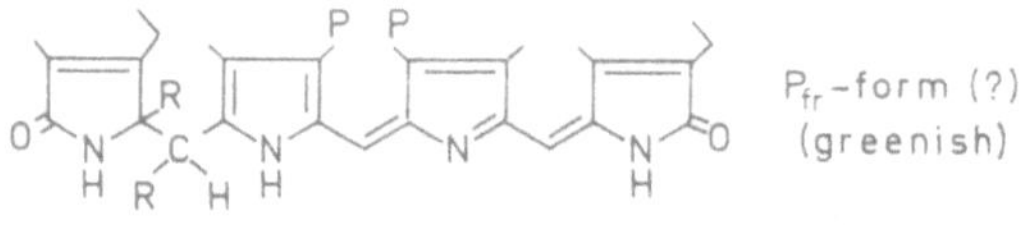

2a, R = H
2b, R = OH

Fig. 3. The most detailed structure proposed so far for the phytochrome chromophore and for the coupling between the chromophore and the protein moiety. The analysis was based on phytochrome from etiolated oat seedlings. The question of which amino acids connect the chromophore with the polypeptide chain is still under debate. The exact structure of Pfr is not yet clear. There is a high similarity with mesobiliviolin (2a) and mesobilipurpurin (2b) but no identity (after [5]).

binding to the protein moiety is shown in Fig. 3. The
model attempts, moreover, to describe the changes of the
chromophore during the transition from Pr to Pfr and vice
versa. Irrespective of details, which are still a matter
of debate and experimentation, one may describe the
photochemical transformations of the phytochrome system
as a combination of a rapid change in the chromophore
and a slower change (relaxation) in the conformation of
the protein part. At physiological temperatures (e.g.
25°C) a protein conformation exists for each of the two
chromophore species giving a stable complex of the chromoprotein.

Biophysically, an important feature of the phytochrome system is that the absorption spectra of Pr (peak
at 660 nm) and Pfr (peak at 730 nm) overlap throughout
the visible spectrum. This overlap is the reason that
photostationary states are a characteristic of the phytochrome system, under conditions of saturating irradiations
[6]. The photostationary state of the phytochrome system
as defined by $\varphi\lambda$ = [Pfr]λ/[Ptot] is rapidly established
in the red and far-red range of the spectrum even at mo-
derate quantum flux density; that is, only "light pulses"
are required virtually to establish photostationary states
in the phytochrome system.

On the basis of this information, the operational
criteria for the involvement of phytochrome in a particular photoresponse can be defined as follows: a photoresponse can be induced by a standard red light pulse
(φred$\approx$0.8); the induction by red light can be fully reversed by immediately following with a far-red light
pulse (φ standard far-red$\approx$0.025; φ 756 nm < 0.001).
In the case of phytochrome involvement, the extent of the
response following the irradiation sequence red plus far-
red will be identical to the extent of the response following a brief treatment with far-red light alone.

Another important property of the phytochrome system
is that it develops a steady state under continuous light.
Based on the model in Fig. 2 any change of total phytochrome can be described by the equation

$$\frac{d\ [Ptot]}{dt} = {}^{o}k_s - {}^{1}k_d\ [Pfr]$$

Under steady state conditions, i.e. "no change of Ptot",
the rate of Pr synthesis is equal to the rate of Pfr
destruction, ${}^{o}k_s = {}^{1}k_d[Pfr]$. Therefore

$$[Pfr] \text{ steady state} = {}^{o}k_s/{}^{1}k_d$$

In words: The steady state concentration of the effector molecule Pfr is only a function of the rate constants for Pr synthesis and Pfr destruction, which are both light-independent. This means that the steady state concentration of the effector molecule Pfr does not depend on the wavelength of the light incident on the system. The only prerequisite is that the incident light is absorbed by both phytochrome forms to an extent sufficient to establish a steady state.

This property of the phytochrome system offers the opportunity to run the steady state of the phytochrome system at a wavelength which does not cause significant chlorophyll synthesis or photosynthesis. We have been using a high irradiance (3.5 Wm^{-2}) standard far-red light source which is equivalent, as far as the phytochrome system is concerned, to the wavelength 718 nm. When we irradiate a mustard seedling grown under rigorously standardized conditions in the dark at $25^{o}C$ at 36 h after sowing, the photo steady state of the phytochrome system in the mustard cotyledons will rapidly be established and maintained over many hours.

CONTROL OF ENZYME LEVELS BY PHYTOCHROME: PHENOMENOLOGY [2]

We assume that development is primarily the consequence of an orderly sequence of changes in the enzyme complement of an organism. Therefore, the investigator of photomorphogenesis will primarily try to explore those phytochrome-mediated responses in which changes in enzyme levels have a well-defined causal role in well-defined developmental steps. We have been engaged in a number of studies on phytochrome-mediated enzyme induction and enzyme repression in the attached mustard cotyledons. With regard to terminology, we emphasize that originally the terms "induction" and "repression" were used operationall in molecular biology to designate the appearance or lack of appearance of an enzyme. We have continued using the terms "induction" and "repression" as originally defined, i.e. without a priori implications about the actual control mechanism. Thus "enzyme induction by phytochrome" means an increase of enzyme level caused by phytochrome, whereas "enzyme repression by phytochrome" means that an increase of enzyme level is arrested by phytochrome.

We do not want to reconsider the abundant but already classical phenomenology of phytochrome-mediated

enzyme induction and repression (see [2]). We are only
repeating the principal findings. Figure 4a shows the
classical example of enzyme induction by Pfr in the
case of a relatively short-lived enzyme, phenylalanine
ammonia-lyase, abbreviated PAL. Figure 4b shows the repression of a probably long-lived enzyme, lipoxygenase,
abbreviated LOG, and Fig. 4c shows an example of those

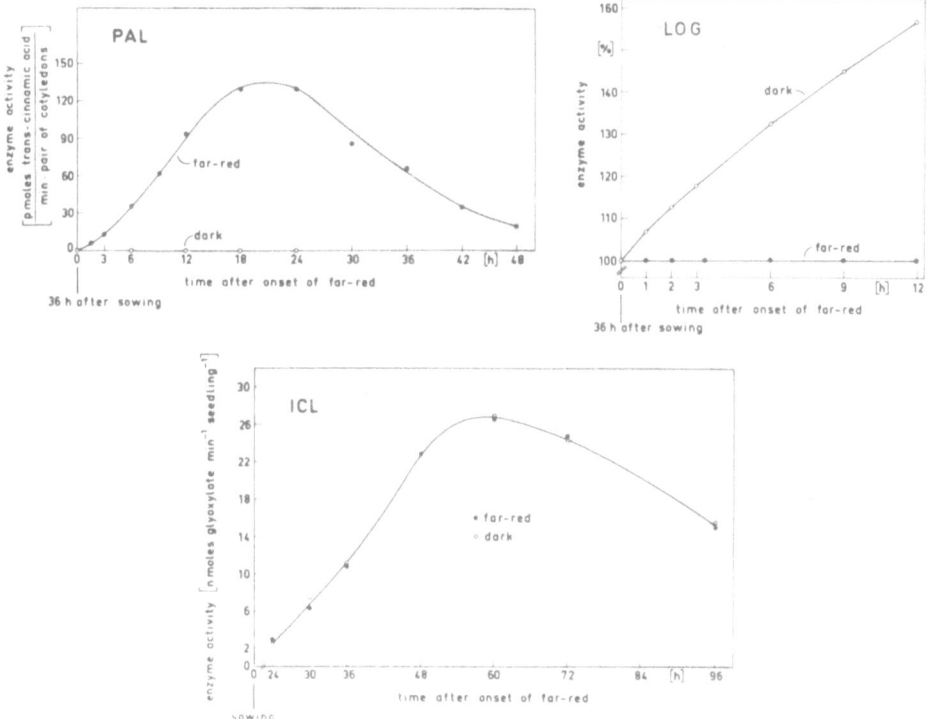

Fig. 4. (a) The influence of continuous far-red light on
the time course of the enzyme phenylalanine ammonia-lyase
(PAL) in the cotyledons of the mustard seedling. Onset of
far-red light: 36 h after sowing. In the dark-grown mustard cotyledons, PAL activity cannot be detected by the
assay which was used in this investigation (after [7]).
(b) The increase of the enzyme lipoxygenase (LOG) in the
dark-grown mustard seedling cotyledons is arrested by
continuous standard far-red light. The available information indicates that LOG is a stable enzyme during the
period of experimentation. A lag-phase of the repression
response is not detectable (after [8]).
(c) Time course of the enzyme isocitrate-lyase (ICL) in
the mustard seedling in darkness and under the influence
of continuous far-red light (after [9]).

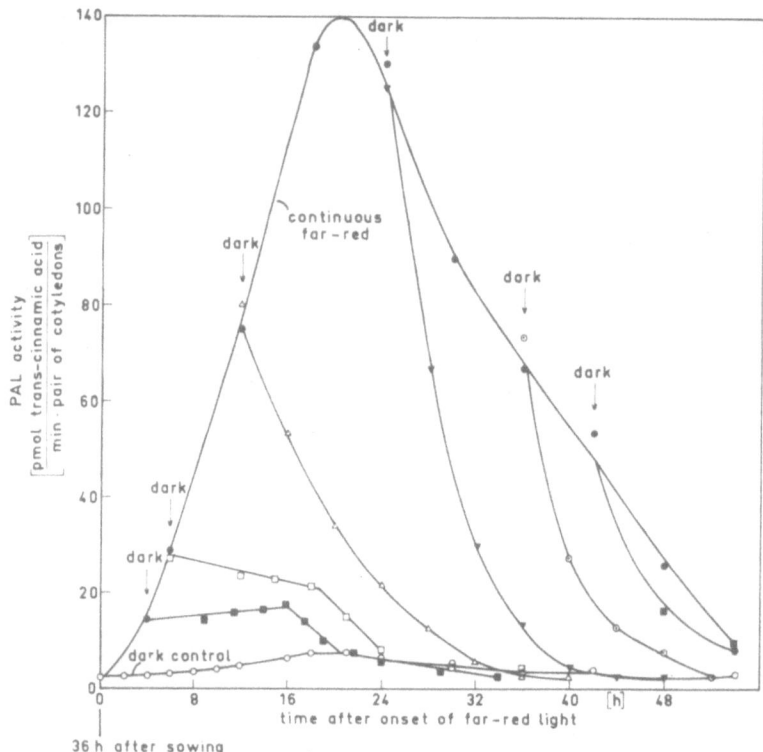

Fig. 5. Time courses (kinetics) of PAL in the cotyledons of the mustard seedling in the dark and under the influence of continuous standard far-red light. Onset of light: 36 h after sowing. In addition, a number of far-red —> dark kinetics are indicated. This term is used to designate those kinetics of the enzyme which are observed after the standard far-red light has been turned off and followed by 5 min 756 nm light (at arrows). The postirradiation with 756 nm light reversed almost all Pfr back to Pr. Thus, the cotyledons are virtually free from the effector molecule Pfr at the beginning of the dark period (after[10]).

enzymes in the mustard cotyledons whose time course of
extractable activity is not changed by phytochrome at
all. The existence of these classes of enzymes documents
the high degree of specificity of phytochrome action in
mustard cotyledons. In the following, we would like to
concentrate on some recent studies which contribute to
our knowledge about the mechanisms of phytochrome control
of enzyme levels.

TWO CASE HISTORIES OF ENZYME INDUCTION BY
PHYTOCHROME IN THE MUSTARD SEEDLING COTYLEDONS

(a) Induction of phenylalanine ammonia-lyase (PAL).
This is an electrophoretically homogeneous key enzyme
of phenylpropanoid biogenesis in plants, which is characterized by a very low dark level (basal level) and
considerable turnover (Fig. 5). Labelling of the enzyme
with deuterium followed by high resolution CsCl density
gradient analysis provides evidence that the low basal
level is due to rapid synthesis and correspondingly rapid degradation of the enzyme (half-life of the order of
4 h) [11]. The time course of deuterium incorporation
during the light-mediated rise in enzyme activity, which
can be derived independently from density shifts and
bandwidth changes of isopycnically banded enzyme (Fig.6),
is only consistent with the concept that phytochrome increases the rate of synthesis de novo [11]. The induced
rise of assayable PAL activity must be attributed to the
fact that synthesis greatly exceeds degradation. In fact,
standard far-red light increases very probably   the
rate of synthesis by 20-fold [10]. With regard to degradation, the elaboration of light —> dark kinetics has
provided additional insights. The basis of these experiments (as far as the phytochrome system is concerned)
is briefly the following: At the moment when the far-
red light is turned off and followed by a 5 min pulse
with 756 nm light, nearly all Pfr molecules are eliminated from the system. Against this background, Fig. 5
shows that the effector Pfr is continuously required in
order to maintain an increase of the PAL level. Secondly,
the light —> dark kinetics show a complex pattern.
Density labelling evidence indicates that the half-life
of PAL is of the order of 4 h under all circumstances.
This means that the conspicuous differences in the light
—> dark kinetics must be attributed to differences in
the rate of PAL synthesis rather than to differences in
the rate constant of enzyme destruction. The biphasic
kinetics which are observed when the light is turned off
after 4 or 6 h have been explained by the existence of
two separate pools of PAL, a small and a large pool.

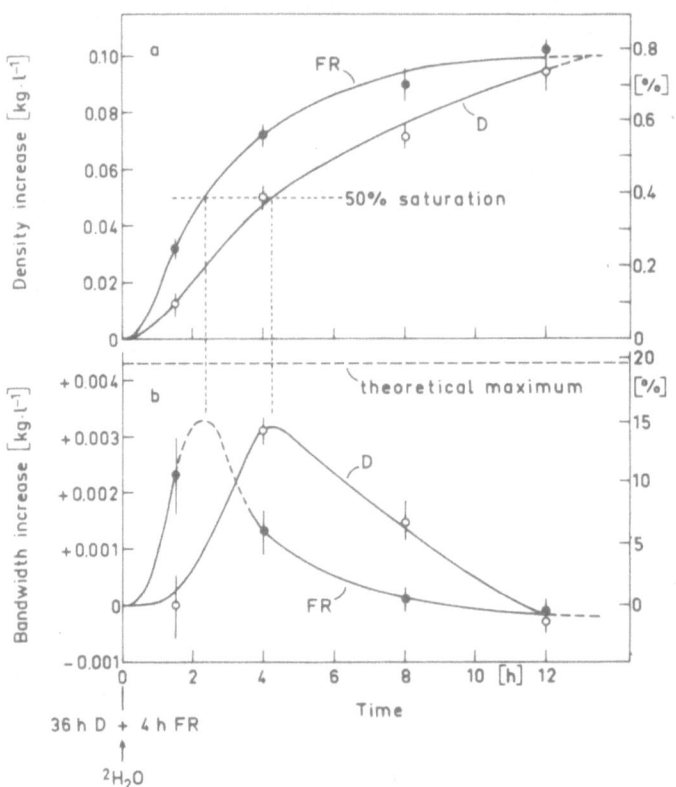

Fig. 6. Time courses of density labelling of PAL with
70% D_2O in far-red light (FR) and darkness (D). a, band
shift (indicating the increase of D in the enzyme pool);
b, bandwidth change (indicating the progression of the
enzyme pool through a state of heterogenously labelled
molecules). The bars indicate standard errors. The dashed
lines indicate half saturation of labelling in light
(ca. 2 h) and darkness (ca. 4 h). The data are calculated
from the position and shape of activity profiles measured
after centrifuging the enzyme to equilibrium on a CsCl
density gradient. It is assumed that maximal band broadening in FR is reached after half saturation of labelling
(after[11]).

With regard to the small pool which dominates the scene
at low amounts of total PAL the rate of synthesis in darkness following a light treatment remains quite high.
With regard to the large pool which dominates the time
course of total PAL after 12 h the rate of synthesis decreases sharply after removal of Pfr from the system by
the 756 nm light pulse. There is evidence that indeed
two differently regulated PAL pools are involved in the
biosynthesis of flavonoids in the mustard cotyledons
[12].Thirdly, the decrease of the PAL level after 20 h
is due to a gradual reduction of the rate of enzyme
synthesis irrespective of the effector Pfr.

   (b) Induction of ribulosebisphosphate carboxylase
(Carboxylase). Induction by phytochrome of enzyme levels,
including Carboxylase in developing plastids is well
known [13]. Figure 7 shows the time course of Carboxylase and NADP-dependent glyceraldehydephosphate dehydrogenase (GPD) in the mustard seedling cotyledons. In contrast to PAL, these enzymes are seemingly stable during
the experimental period, i.e. significant turnover must
not be considered. In the dark-grown mustard seedling,
the level of Carboxylase and GPD in the cotyledons remains low; however both enzymes can be induced by phytochrome, operationally, by continuous far-red light (Fig.
7). The young mustard seedling does not contain Carboxylase nor GPD. Both enzymes cannot be detected before 36 h
after sowing.

   The important finding in our present context is:
For induction of the Calvin cycle enzymes it does not
matter whether Pfr is functioning in the seedling from
24 h after sowing or only from 36 h after sowing onwards.
This observation indicates the action of endogenous regulatory factors which prevent enzyme induction by phytochrome before 42 h after sowing irrespective of light
treatment. Other enzymes in the mustard cotyledons, not
related to photosynthesis, can be induced by phytochrome
much earlier in the course of development. As an example,
PAL becomes inducible by Pfr at 28 h after sowing. There
is obviously a temporal pattern of inducibility of enzymes. The specification of this pattern is not influenced
by phytochrome. We have studied the problem whether the
phytochrome-mediated accumulation of Carboxylase in the
cotyledons of the mustard seedling is related to size,
ultrastructure or organization of the plastid compartment. It turned out that under different light conditions
(e.g. continuous far-red or continuous white light) which
lead to conspicuously different plastids (Fig. 8) the
time course of the Carboxylase level remains precisely

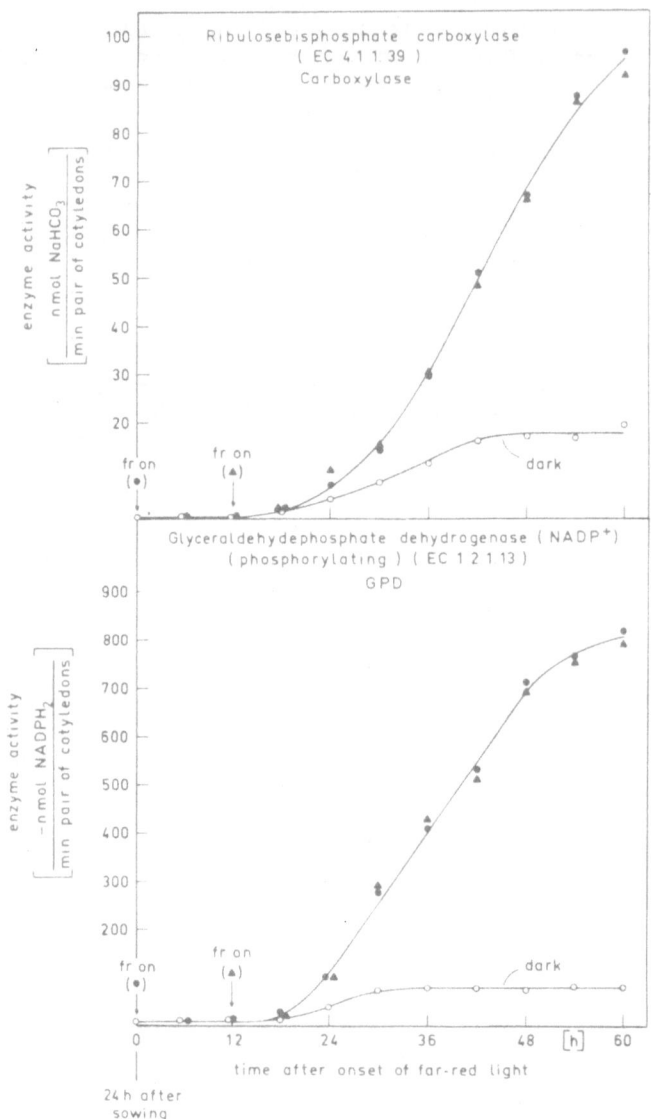

Fig. 7. Time course of the levels of Carboxylase and GPD
in the cotyledons of the mustard seedling in the dark
and under continuous far-red light. Onset of far-red
light (fr) at 24 or 36 h after sowing (after [13]).

the same (Fig. 9). It is concluded that the onset and
the rate of Carboxylase synthesis is not related to the
organizational state of the plastid compartment as dis-
cernible under the electron microscope.

We have further studied the problem whether induction
by phytochrome of PAL and Carboxylase can be traced back
to the same kind of mechanism. The answer to this question is clearly, no. Among several arguments we quickly
analyse the dose [Pfr] and irradiance dependency of both
responses. Figure 10 shows that the induction of Carboxylase is already saturated at a very low level ($\varphi_{756 \text{ nm}} <$
0.1 per cent) and does not depend on irradiance (as an
example, standard red light and 1/10 of standard red
light lead to the same time course). On the other hand
(Fig. 11), PAL induction is a graded response over a
wide range of Pfr doses and depends on irradiance, i.e.

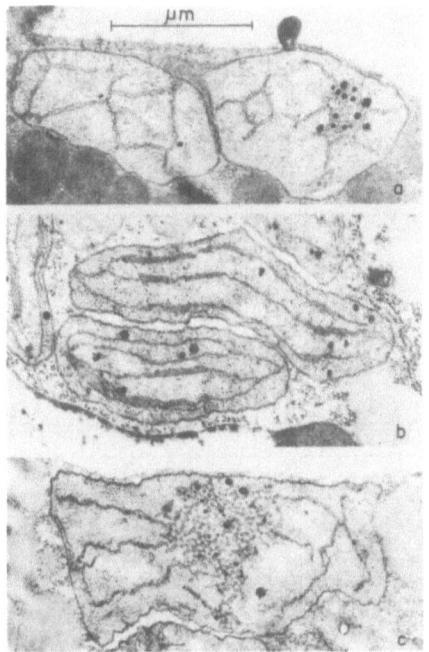

Fig. 8. Representative sections through plastids in
palisade parenchyma cells of cotyledons of mustard
seedlings grown in the dark (a), under continuous white
light (b) or under continuous far-red light (c). Onset
of light at time zero (time of sowing of the seeds).
Fixation was performed 48 h after sowing (after [14]).

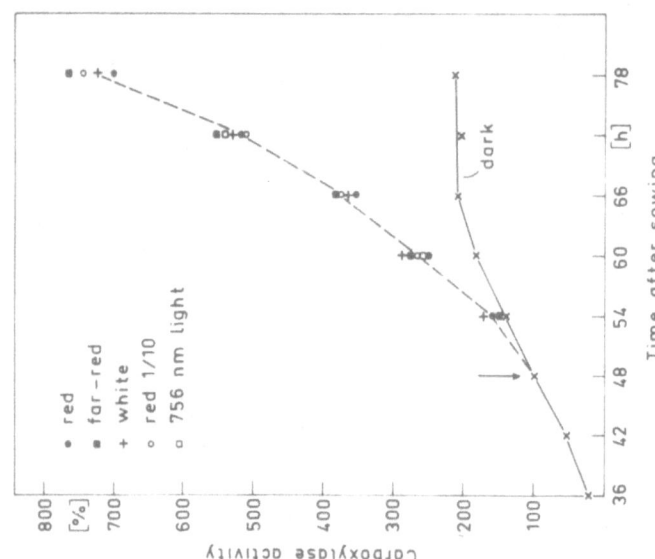

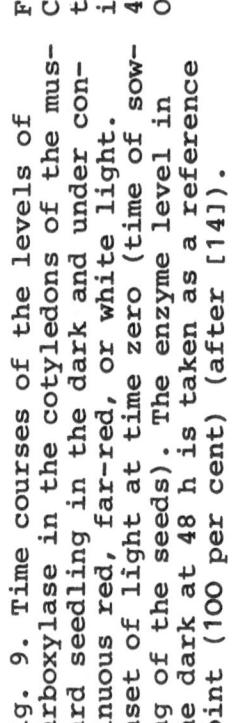

Fig. 9. Time courses of the levels of Carboxylase in the cotyledons of the mustard seedling in the dark and under continuous red, far-red, or white light. Onset of light at time zero (time of sowing of the seeds). The enzyme level in the dark at 48 h is taken as a reference point (100 per cent) (after [14]).

Fig. 10. Time courses of the levels of Carboxylase in the cotyledons of the mustard seedling under different light qualities and irradiances. Onset of light: 48 h after sowing. Recall: $\varphi_{756\,nm} <$ 0.1 per cent (after [15]).

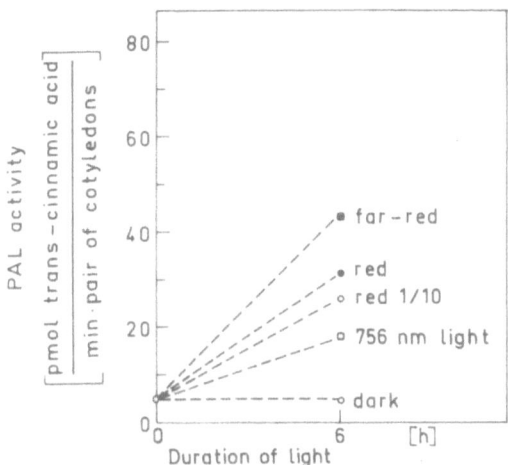

Fig. 11. Increase of PAL levels within 6 h in the cotyledons of the mustard seedling under different light
qualities and irradiances. Onset of light: 36 h after
sowing (after [15]).

the so-called "high irradiance response" (HIR, see [2])
comes into play. The following case history will further
support the notion that there is no common mechanism of
phytochrome action in controlling enzyme levels in the
mustard seedling cotyledons.

A CASE HISTORY OF ENZYME REPRESSION BY PHYTOCHROME[16]

     Figure 12 summarizes a major result of our 6 years
work on control of lipoxygenase levels by phytochrome
in the mustard seedling cotyledons. Increase of lipoxygenase (LOG) activity is controlled by Pfr through a
threshold (all-or-none) mechanism. If the amount of Pfr
exceeds the threshold level, LOG accumulation is fully
and immediately arrested. If the amount of Pfr decreases
below the threshold level, LOG accumulation is immediately resumed at full speed. This pattern of response is
illustrated in Figure 13: LOG accumulates in the dark;
increase of the enzyme is suppressed by Pfr above the
threshold level (red at time zero). As soon as the Pfr
level decreases below the threshold level (far-red at
1.5 h), LOG accumulation is resumed at full speed; as
soon as the Pfr level increases above the threshold level
(red at 5 h), LOG accumulation is again arrested, immediately and totally.

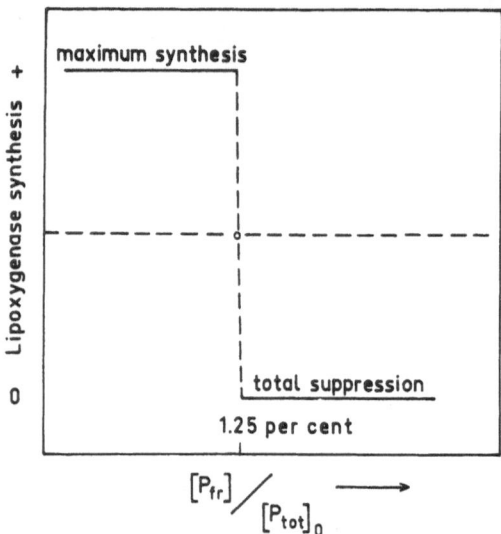

Fig. 12. A scheme to describe the concept of a threshold
regulation of lipoxygenase accumulation in the mustard
seedling cotyledons by Pfr. $[Ptot]_0$ is the total phytochrome at time zero (36 h after sowing). This value is
a constant. The amount of Pfr, [Pfr], is always expressed
as a fraction or percentage of $[Ptot]_0$. Expressed in this
way the threshold level, $[Pfr]_{th}$, is approximately 1.25
per cent (0.0125) (after [16]).

    With regard to the original question ("is there a
common mechanism of phytochrome action in controlling
enzyme levels in the mustard seedling cotyledons?") we
think that there is no escape from the conclusion that
Pfr-mediated control of PAL synthesis (no detectable
threshold, graded response) and Pfr-mediated threshold
control of LOG formation differ in principle even at the
level of the "primary reaction" of phytochrome [18]. It
seems that induction of Carboxylase is also a threshold
phenomenon, a <u>positive</u> threshold response. If this is
the case, the threshold level is certainly much lower
than in LOG regulation: $[Pfr]_{threshold, LOG}$ = 1.25 per
cent (Fig. 12); $[Pfr]_{threshold, Carboxylase}$ < 0.1 per
cent (Fig. 10).

PHYTOCHROME-MEDIATED CHANGES ON THE RNA LEVEL [19, 20]

    So far the molecular analysis has only been success-
ful in the case of ribosomal RNA. Phytochrome, operationally "continuous far-red light", stimulates the accumulation of cytoplasmic and plastid rRNA in the cotyledons

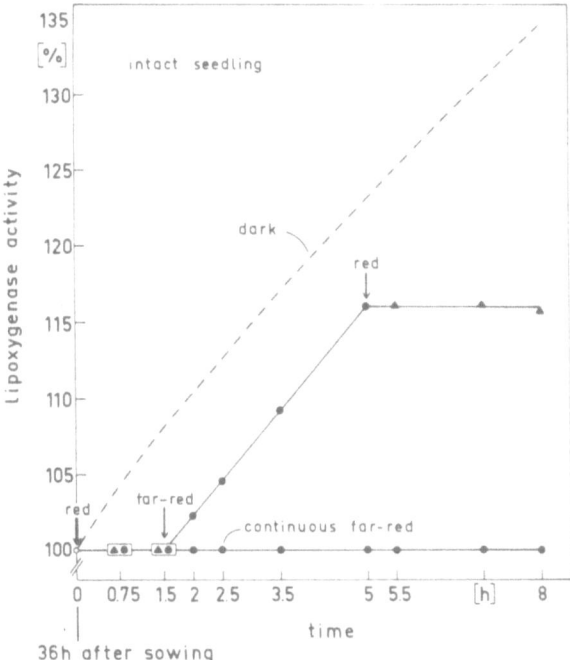

Fig. 13. Time courses of lipoxygenase levels in the mustard seedling cotyledons in darkness (---), under continuous far-red light (●) and under the irradiation
sequence 1.5 h red light (▲) —> 3.5 h far-red light (●)
—> 3 h red light (▲). While red light establishes a
Pfr level above the threshold level under all circumstances, far-red light (given after 1.5 h red light)
establishes a Pfr level below the threshold level. Far-
red light without the red light pretreatment establishes
a Pfr level above the threshold level (after [17]).

of the mustard seedling (Fig. 14). In 36-h-old seedlings
placed under continuous far-red light, the cytoplasmic
and plastid rRNA species will increase after a lag phase
of more than 6 h and reach a maximum level about 24 -
36 h later. These data have been interpreted in terms
of a phytochrome-mediated stimulation of the nuclear and
plastid rDNA transcription rate. Figure 15 gives evidence in support of this conclusion. Short-term labelling
of the high molecular weight nuclear precursor rRNA (pre-
rRNA), which is a direct product of rDNA transcription
also in plants, revealed a stimulating effect of Pfr.
After a 15 min pulse with ^{3}H-uridine the label located
in the pre-rRNA region of the polyacrylamide gel is considerably increased if the seedlings have been irradiated
with far-red light for 12 h prior to labelling. At the

termination of the pulse a small part of the radio-
activity has already reached the 1.3×10^6 and 0.7×10^6 dalton split products, indicating a rapid turnover
of the pre-rRNA. Longer labelling periods lead to a
strong increase in radioactivity of the 1.3×10^6 and
0.7×10^6 dalton rRNA peaks and to an equilization of
labelling in the pre-rRNA peaks obtained from dark-
grown and far-red treated seedlings. This kind of data
supports the conclusion that phytochrome is able to
strongly activate the transcription of RNA cistrons and
the processing of the pre-rRNA. Thus, we may conclude
that phytochrome can modulate, in principle, gene transcription rates. However, because of the non-informational
character of rRNA, these data are clearly not sufficient
to explain the high specificity of phytochrome action on
the enzyme level (see Fig. 6). Rather, the phytochrome-
mediated increase in rRNA synthesis must be considered
as a means to increase the total capacity of protein
synthesis during cotyledon development in the light
which reaches a peak at about 60 h after sowing. So far,
it has not been possible to demonstrate phytochrome

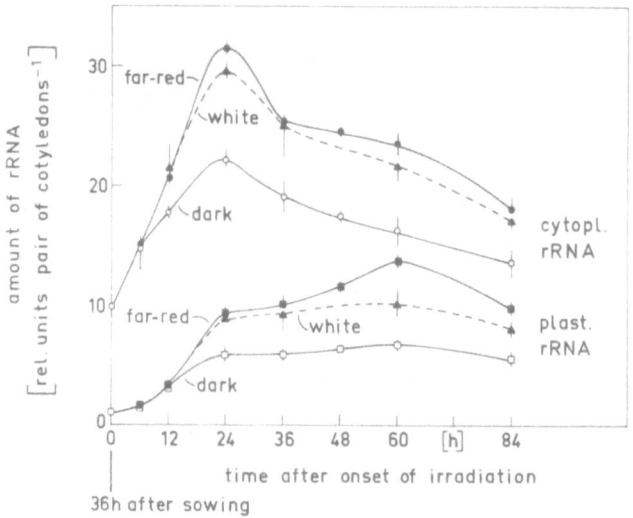

Fig. 14. Time courses of the levels of cytoplasmic and
plastid rRNA in the cotyledons of 36-h-old mustard seedlings kept in darkness (O) or irradiated with continuous
far-red (●) or white (▲) light (cytopl. rRNA: 1.3×10^6
plus 0.7×10^6 dalton RNA; plast. rRNA: 1.1×10^6 plus
0.56×10^6 dalton RNA) (after [19]).

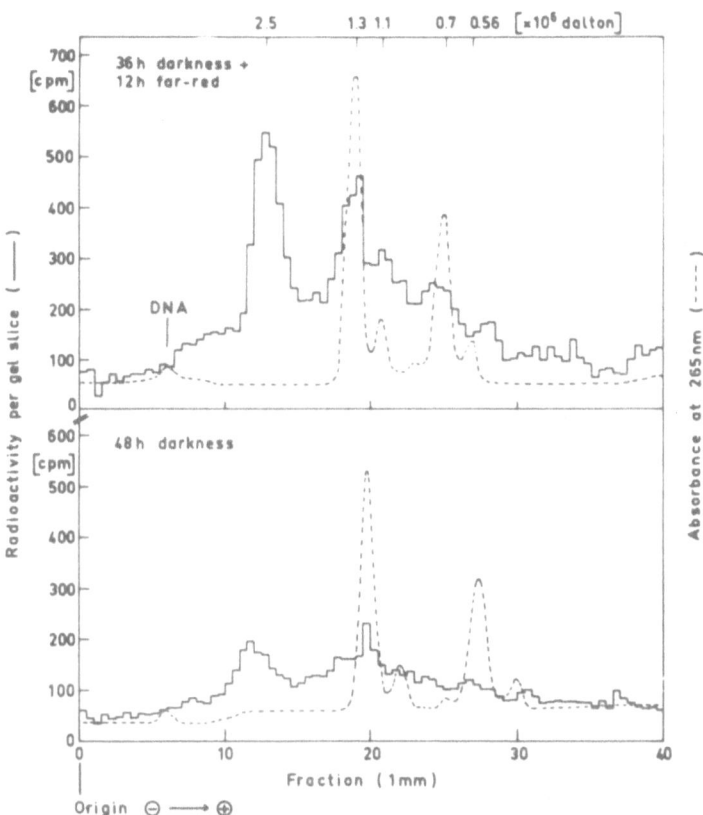

Fig. 15. Stimulation of nuclear precursor-rRNA labelling by a preirradiation with far-red light in the cotyledons of 48-h-old mustard seedlings. Seedlings were grown aseptically in darkness for 36 h and subsequently irradiated for 12 h with standard far-red light or kept in darkness. Cotyledons of 40 seedling were isolated and incubated for 15 min with 2 ml distilled water containing 200 µCi 5-^{3}H uridine. The nucleic acids were extracted and subjected to 2.4% polyacrylamide gel electrophoresis. The gels were scanned for UV absorbance and sectioned in 1 mm slices, the radioactivity of which was measured by liquid scintillation counting. Uptake of label into the cotyledons was not significantly influenced by the light treatment (after [20]).

effects on the synthesis of mRNA by direct means. A considerable amount of inhibitor data are consistent with
this notion, but more direct evidence is lacking.

SUMMARY AND OUTLOOK

Phytochrome-mediated photomorphogenesis - normal
development - is obviously related to differential gene
expression. This can clearly be shown on the enzyme level.
Phytochrome acts by permitting the realization of a pre-
existing specified pattern of inducibility and repressi-
bility of enzymes. The specification of this pattern is
not influenced by phytochrome. The tacitly assumed pre-
mise in much of the phytochrome work that there is a
single primary reaction is not justified. Instead, a
plurality of reactions in which phytochrome directly
participates must be envisaged. Phytochrome controls the
rate of nuclear and plastid rDNA transcription. On the
mRNA level the effect(s) of phytochrome is(are) not yet
clear. There are reports about phytochrome-mediated increases in polyribosome levels [21, 22] and suggestions
that light regulates the level of specific initiation
factors [23]. However, a breakthrough on the mRNA level
can only be expected through assays of the rate of synthesis of <u>specific</u> mRNA species. This kind of approach
appears feasible [24].

REFERENCES

1 Zucker, M. (1972) Ann. Rev. Plant Physiol. 23, 133-156

2 Mohr, H. (1972) Lectures on Photomorphogenesis.
    Springer, Berlin-Heidelberg-New York

3 Mitrakos, K. and Shropshire, W. (eds.) (1972) Phyto-
    chrome. Academic Press, New York

4 Oelze-Karow, H., Schäfer, E. and Mohr, H. (1976)
    Photochem. Photobiol. 23, 55-59

5 Grombein, S., Rüdiger, W. and Zimmermann, H. (1975)
    Hoppe-Seyler's Z. physiol. Chem. 356, 1709-1714

6 Hanke, J., Hartmann, K.M. and Mohr, H. (1969) Planta
    86, 235-249

7 Dittes, L., Rissland, I. and Mohr, H. (1971) Z.
  Naturforsch. 26b, 1175-1180

8 Oelze-Karow, H., Schopfer, P. and Mohr, H. (1970)
  PNAS (USA) 65, 51-57

9 Karow, H. and Mohr, H. (1967) Planta 72, 170-186

10 Tong, W.-F. (1975) Dissertation, University of
   Freiburg i. Brsg.

11 Tong, W.-F. and Schopfer, P. (1975) PNAS (USA), in
   press

12 Wellmann, E. (1974) Ber.Deut.Bot.Ges. 87, 275-279

13 Brüning, K., Drumm, H. and Mohr, H. (1975) Biochem.
   Physiol. Pflanzen 168, 141-156

14 Frosch, S., Bergfeld, R. and Mohr, H. (1976) Planta,
   submitted for publication

15 Frosch, S., Drumm, H. and Mohr, H. (1976), unpublished data

16 Mohr, H. and Oelze-Karow, H. (1976) In: Light and
   Plant Development (H. Smith, ed.) Butterworths,
   London

17 Oelze-Karow, H. and Mohr, H. (1974) Photochem. Photo-
   biol. 20, 127-131

18 Oelze-Karow, H. and Mohr, H. (1976) Photochem. Photo-
   biol. 23, 79-85

19 Thien, W. and Schopfer, P. (1975) Plant Physiol. 56,
   660-664

20 Thien, W. and Schopfer, P. (1975) Planta 124, 215-217

21 Malcolm, A.A. and Russell, D.W. (1974) In: Mechanism
   of Regulation of Plant Growth (Bieleski, R.L.,
   Ferguson, A.R. and Cresswell, M.M., eds.) Bulletin
   12, The Royal Society of New Zealand, Wellington

22 Smith, H. (1976) Eur. J. Biochem. 65, 161-170

23 Travis, R.L., Key, J.L. and Ross, C.W. (1974) Plant
      Physiol. 53, 28-31

24 Higgins, T.J.V., Zwar, J.A. and Jacobson, J.V. (1976)
      Nature 260, 166-169

Research supported by Deutsche Forschungsgemeinschaft
(SFB 46). We thank Mrs. B. Hoffmann for the excellent
preparation of the typescript.

PROTEIN DEPOSITION IN DEVELOPING CEREAL

AND LEGUME SEEDS

Donald Boulter

Department of Botany, Science Laboratories
University of Durham
Durham DH1 3LE, England

INTRODUCTION

The seed is the sexually produced offspring of flowering plants and is an organ of dispersal. Seed development is a preparation for survival during dispersal and for subsequent successful germination. Compared with lower plants, the seed is a more complex propagule than the spore and considerable development of the fertilised ovule, nourished by the maternal plant, takes place before its release. Cell division gives rise to vegetative tissues and then this apparent normal development pattern changes to one geared to ensure a successful future for the offspring as a separate individual. This development includes the deposition of storage reserves, to be used subsequently on germination, then a drastic reduction in metabolic activity and the formation of a protective seed coat, both required for survival during dispersal and the period, sometimes prolonged, prior to germination.

This paper is concerned with only one aspect of this development, i.e. the deposition of protein reserves. It should be borne in mind, however, that some of the enzymes required for the mobilisation of these reserves (Chrispeels and Boulter, 1976) and possibly some messenger RNA molecules required for protein synthesis during germination (Dure, 1975) are also produced during seed development and stored in a suitable form.

In eukaryotes, unlike prokaryotes, transcription and translation of information are intracellularly separated processes and the possibility exists, therefore, for the prolonged storage of transcribed information (mRNA's) before its translation into protein structure.   In animal systems, e.g. the fertilised egg, information is stored in the cytoplasm but development continues actively once set in train.   By contrast, the developing seed has a metabolically active stage after fertilisation of the ovule, followed by a metabolically inactive stage, the mature seed, which, nevertheless, must have the developmental potential to become metabolically active again on germination.   It may prove, therefore, to be an especially useful system for investigation of post-transcriptional storage of information.

Lastly, no discussion of the deposition of proteins in developing seeds would be complete without reference to the importance of legume and cereal grains as a source of protein for man and his animals, and of the progress made to improve the essential amino acid content of these grains which are limiting in sulpho-amino acids for legumes and in lysine for cereals.

The object of this paper is to give a general account of protein deposition in developing seeds and so set the scene for later speakers.   The seed syntheses several hundreds of proteins but only a few are deposited in the seed to be used later.   Various aspects of the possible means of regulation and control of the process, as well as with its general biochemistry, will be covered by other speakers.   There are several excellent reviews of seed protein synthesis, see for example Dure (1975), Millerd (1975), Yarwood (1976), and no attempt has, therefore, been made to cover the extensive literature fully.

DESCRIPTION OF PROTEIN DEPOSITION

General

Since the mature seed is a complex organ consisting of several different tissues of both maternal and zygotic origin, it is necessary to couple the description of the

molecular changes of the process of protein deposition
with a brief description of these tissues.   After the
double and triple fertilization events which give rise
to the zygote and endosperm respectively, initial seed
development involves principally the maternal tissues
(integuments and sometimes the nucellular tissue) and
the triploid endosperm.   The latter particularly
expands in size and cell number, whereas the embryo
grows slowly initially and then more rapidly at the
expense of the endosperm.   In legumes the embryo may
totally absorb the endosperm/nucellus, reserves being
stored in the large cotyledons.   In cereals the
endosperm largely remains, possibly the biologically
more efficient way, and constitutes the storage tissues;
in both the outer integument survives and forms the
seed coat.   Thus, some isomers of particular isoenzymes
in seeds are restricted to the testa and are of maternal
origin (unpublished observations of P. Gates and D.
Boulter).   Eventually, the vascular connection between
the seed and ovary is severed and the seed dries out.
During the course of development nutrients are moving
at different rates to different parts of the seed and
also between the different tissues of the seed itself.
In the first phase characterised by rapid cell division
and maternal and endosperm tissue growth, there is a
build-up of precursors of the storage compounds;   it is
only in the second phase of development, marked by the
cessation of cell division, that deposition of the actual
storage reserves, starch and/or oil, storage protein and
minerals occurs in the storage tissues, i.e. endosperm
of cereals and the cotyledons of legumes.

At the molecular and fine-structural level

Seed proteins are normally extracted and characterized using a modified form (Byers et al., 1976) of the
Osborne solubility method (Osborne, 1924).   This divides
proteins into albumins, water-soluble;   globulins, salt-
soluble;   prolamins, alcohol-soluble;   and glutelins,
acid/alkali-soluble.   When legume seeds are extracted
using these solvents most of the protein is globulin,
whereas with cereals most of the protein usually belongs
to two solubility classes, prolamin and glutelin,
although oats are an exception.   It is the protein of

these same groups which is utilised on germination,
globulins in legumes and prolamins and glutelins in
cereals (Folkes and Yemm, 1956; Boulter and Barber,
1963), thus establishing that reserve or storage
proteins are quantitatively the major part of the
protein of seeds.   Storage proteins, therefore, are
produced in large amounts in the developing seed to be
used later in germination. They have no enzymic function
and are seed, although not cotyledon or endosperm,
specific. Some proteins may occur in relatively large
amounts in seeds and be used as storage proteins in
that they are broken down on germination, even though
they may primarily have other functions, e.g. urease
in jackbeans. The 'true' storage proteins of jackbeans are seed specific, whereas as urease occurs in
other parts of the plant (Granick, 1937).

Each of the storage protein solubility fractions
consists of more than one protein component. The exact
number is not known for certain, but in legumes there
are probably between five and ten globulin storage
proteins (Derbyshire, Wright and Boulter, 1976), and
in cereals at least several proteins in each of the
solubility classes, although the prolamin fraction of
maize may be an exception. Characterization of these
polymeric proteins is an urgent task, and technically
quite difficult due to the association/dissociation
behaviour of their constituent subunits. It is also not
clear to what extent a particular type of storage
protein occurs in other taxa. For example, legumin,
first described in _Pisum_, has been shown to exist in
various other legumes and dicotyledons (see Derbyshire,
Wright and Boulter, 1976), and is presumed to be an
homologous protein.

Examination of electron micrographs of sections of
developing seeds show that most of the protein is
deposited in membrane-bound organelles called aleurone
grains or, more properly, protein bodies (see Fig. 1
and Briarty _et al_., 1969; Khoo and Wolf, 1970).
These occur predominantly in the storage tissues,
although they are also found in the more metabolically-
active embryo and embryo axis of cereals and legumes,
respectively. That the protein of protein bodies is

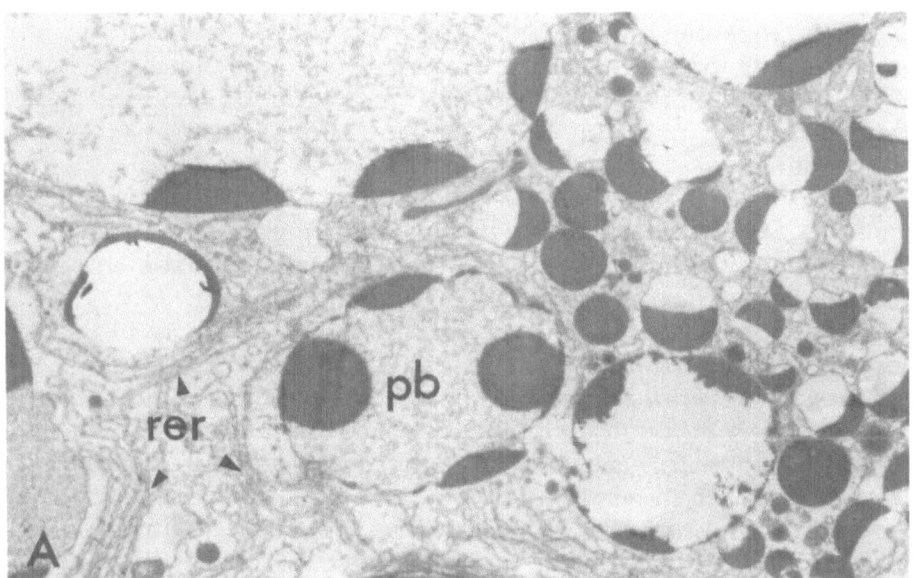

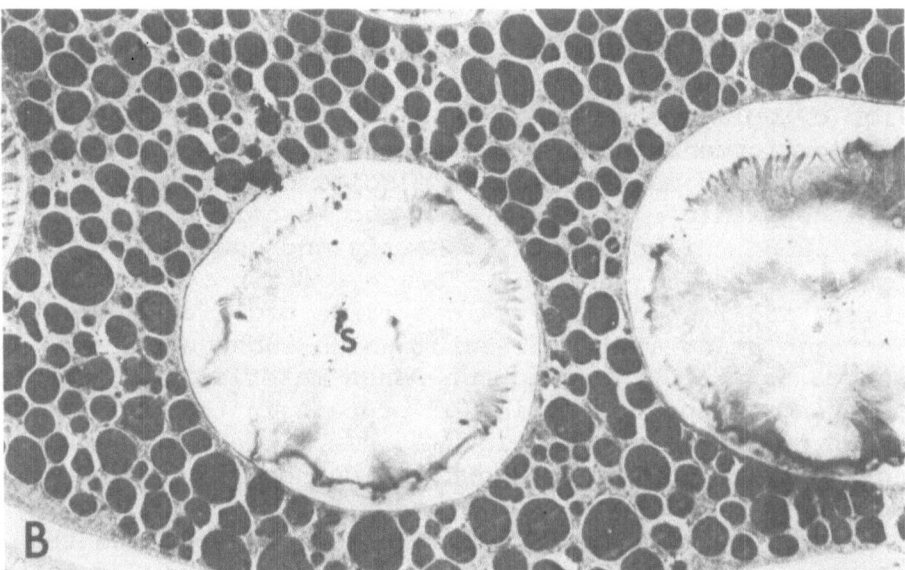

Figure 1.   A. Electron micrograph of part of a cotyledon
cell of a developing Vicia faba seed showing proliferation of the rough endoplasmic reticulum (rer) and
deposition of protein into vesicles and protein bodies
(pb):   X 12,000.
              B. Electron micrograph of part of a cotyledon
cell of mature Vicia faba seed showing protein bodies and
starch grains (s):   X 2,000.

indeed mainly storage protein has been established
directly in some cases from analyses of extracted
protein body preparations, and Graham and Gunning
(1970) have located the storage proteins of Vicia faba
in situ in the protein body by EM fluorescent anti-
body techniques.

Not all storage proteins occur in protein bodies.
In maize, barley and sorghum, the protein bodies appear
to contain mainly the prolamin fraction, zein etc., and
glutelin occurs outside the protein bodies as matrix
protein. This matrix protein is distinct from the
structural metabolic proteins of the membranes, ribosomes etc., although extracted with the latter in the
glutelin protein; it has no function, so far as is
known, other than as a storage protein. In rice, on
the other hand, glutelin is the major storage protein
and both it and prolamin occur in protein bodies
(Harris and Juliano, 1976).

A variety of studies suggests that protein bodies
originate from two sources, either by subdivision of
large cytoplasmic vacuoles seen at an earlier stage of
seed development, or by enlargement of newly formed
vesicles which may develop from Golgi cisterni or from
the endoplasmic reticulum (Bain and Mercer, 1966;
Opik, 1968; Millerd, 1975; Harris and Boulter, 1976).
Just prior to the onset of storage protein synthesis in
cereals and legumes, there is a large proliferation of
the rough endoplasmic reticulum which subsequently
becomes dilated with electron-dense material, and the
beginning of the deposition of protein in the protein
bodies can be seen (see Fig. 1). In Vicia faba Payne
and Boulter (1969) have shown the de novo synthesis of
the rRNA of these membrane-bound ribosomes. Bailey,
C. J. and Boulter, D. (unpublished) have separated the
membrane-bound and free polysomes of developing seeds
of Vicia faba, and have shown that most of the radioactivity incorporated into protein is associated with
the former; since most of the protein being synthesized
is storage protein, this would suggest that the storage
proteins are made on membrane-bound ribosomes. This has
not been unequivocally demonstrated in vitro, however,
since product characterization studies in vitro are not

that advanced. Although storage proteins are found in
the protein bodies there is little evidence, in spite
of earlier reports to the contrary, that protein bodies
can synthesize protein (Millerd, 1975). Recently, however, Burr and Burr (1976) have suggested that zein synthesis in maize occurs on ribosomes attached to the outside of protein bodies.

Bailey, Cobb and Boulter (1970) using EM autoradiography showed an apparent movement of newly synthesized
protein from the endoplasmic reticulum to the protein
bodies of <u>Vicia</u> <u>faba</u> and the evidence suggests that
storage protein is not released into the cytoplasm but
immediately enters the lumen of the endoplasmic
reticulum to which the synthesizing ribosomes are
attached by the large subunit. In animal cells it has
been proposed by Blobel and Dobberstein(1976) that
binding sites in the membrane of the endoplasmic
reticulum react with sites on the large subunit of the
ribosome. Their signal hypothesis postulates that
mRNAs for proteins which will be secreted into the lumen
of the endoplasmic reticulum have signal codons; cell
ribosomes can translate all types of mRNAs but messengers
carrying signal codons, after ribosomal attachment, lead
to the association of these ribosomes with the membrane
by ribosomal binding sites. The protein signal sequence,
about twenty-five amino acids long, is removed subsequently by an endopeptidase located in the endoplasmic
reticulum. A similar mechanism could exist in developing
seeds, although storage proteins are not secreted from
the cell as in the above case.

Most storage proteins are glycoproteins and probably
their constituent carbohydrate moieties are added during
movement along the rough and possibly the smooth endoplasmic reticulum. If this movement is a form of
transport then it is possible that eventually the protein
is packaged into endoplasmic reticulum or Golgi vesicles,
depending perhaps on the type of carbohydrate units
present, since the glycosylating enzymes are different in
smooth and rough endoplasmic reticulum membranes
respectively. Eventually, by collaslescing these
vesicles could give rise to protein bodies. This
process would ensure that the proteins were not exposed

to the aqueous cytosol. Originally it was thought that
storage proteins belonged to the class of secretory
proteins and that secretory proteins were made only on
ribosomes attached to the endoplasmic reticulum;
glycosylation was thought to be required for the passage
of these proteins through membranes. However, it is now
clear that secretory proteins can be made on free polysomes and an alternative explanation for the fact that
storage proteins are usually glycosylated is required.
It is possible that there is some physiological
significance in the fact that lectins which can react
with glycoproteins are of widespread occurrence in
legume seeds.   For example, Concanavalin A (Con A) of
jackbeans recognises and interacts with the major
storage protein, glycoprotein II, of _Phaseolus_ _vulgaris_
(Derbyshire, E. and Boulter, D., unpublished observations), and the question arises as to whether or not an
homologous Con A type protein exists in _P._ _vulgaris_
and, if so, whether this type of interaction is involved
generally in the 'packaging up' of storage proteins?

Most storage proteins are polymeric, but whether
they are transported as subunits or as assembled proteins has not been established. It is possible that
_in_ _vivo_, changes of pH may be involved in the polymerization of the subunits, since Sun _et_ _al._ (1974) have
shown that the major storage protein of _Phaseolus_
vulgaris exists as a protomer at neutral or mildly
alkaline pH values, such as might be encountered in the
cytoplasm, but is reversibly associated to form a
tetramer at pH 4.5, which is approximately the pH value
of the protein bodies.

When purified storage proteins such as the glycinin
(legumin) of soyabean or the legumin of _Vicia_ are
subjected to analytical scanning isoelectric focusing
in urea-dithiothreitol medium, considerable heterogeneity
of subunits has been observed. This heterogeneity
suggests that legumin may exhibit polymorphism, i.e.
that there are several types of legumin comprised of
slightly different subunits (Derbyshire, Wright and
Boulter, 1976).   Thus, it has already been established
that the peanut protein, arachin, exists in nature
in three variant forms (Tombs, 1965; Tombs and Lowe,

1967).

The structure of a storage protein is subject to
considerable constraints. Firstly, it must be compatible
with movement into the endoplasmic reticulum and subsequently into the protein body. Interacting sites must
be available for its polymerization into a form suitable
for storage and also for its re-mobilisation by a cascade
process of proteolytic enzymes during germination. On
hydrolysis its amino acid composition must match the
enzymic machinery of the germinating seed, so that the
necessary transformations can occur to supply the
nitrogen compounds for the latter. Seeds can store
nitrogen in three different ways; in the form of proteins
which may be deposited inside or outside protein bodies
and as unusual amino acids which, for example, in
legumes can represent a considerable proportion of the
dry weight of the seed (Bell, 1976). The significance
of these different methods of nitrogen storage is not
fully understood.

It has now been established that cotyledon cells
become highly polyploid, 16-64C, during seed development
(Dure, 1975) and that after the cessation of cell
division, DNA levels continue to increase well into the
period of rapid storage protein synthesis (Dure, 1975;
Millerd, 1975). Millerd and Whitfield (1973) showed
with Vicia faba that the whole genome was replicated
and not that just some genes were amplified. Millerd
and Spencer (1974) using isolated pea cotyledon nuclei
have shown further that there is a proportional increase
in available template but not in transcriptive activity;
they suggested, therefore, that RNA polymerase was
limiting. However, the seed contains several types of
RNA polymerases and as this aspect was not investigated,
preferential transcription of storage protein genes
cannot be ruled out at present. That differential gene
replication can take place in legume cotyledons is shown
by the results of Cullis and Davies (1975). They
describe a variety of pea, JL813, which had an increased
proportion of rDNA in the cotyledons in comparison with
meristematic root-tip cells and hence can replicate
differentially the rDNA of its genome in polyploid cells.
The alternative explanation originally suggested by

Smith (1973),   that the 'extra' DNA is a store of
nucleotides to be used during germination, has now been
shown not to be so.

   It has already been mentioned that each plant
contains more than one type of storage protein.   In
Pisum, Vicia, Vigna and Phaseolus, it has been shown
that there is a sequential sequence of at least some
of the storage proteins.   A well-documented case is that
of Pisum , in which vicilin is synthesized prior to
legumin (Millerd, 1975).   Millerd (1975) and co-workers
have also shown that detached immature pea cotyledons
will continue to synthesize considerable amounts of
chlorophyll, starch, DNA and RNA in culture over a
period of several days.   If detached prior to the
appearance of vicilin, vicilin was synthesized but a
similar switching on of legumin synthesis was not
realised.   If, however, cotyledons were excised after
the commencement of legumin synthesis, they continued to
synthesize this protein.   In contrast to these results,
when young peapods containing immature seeds with no
storage protein present were detached and cultured, the
seeds developed and synthesized both vicilin and legumin,
emphasizing the important role of the maternal tissues
in seed development.   The important regulatory role of
the maternal tissues in cotton seed development is also
very clear from Dure's work (Dure, 1975;   see also this
volume).   Davies (1973) has shown that in some pea lines
storage protein is maternally inherited and has suggested
that maternal and paternal loci may be differentially
activated.   This suggestion deserves further investigation.   Unfortunately, interpretation of genetic analysis
experiments is hampered by our incomplete knowledge of
the subunit structure of the storage proteins involved.

Nutritional aspects

   Storage proteins occur in large amounts and may be
synthesized at different times, rates and for different
durations in the developing seed.   Furthermore, they
differ in their amino acid compositions.   They are,
therefore, prime targets for manipulation in breeding
programmes, either conventional or mutational, for

improved protein content and quality, and for improvement
via fertiliser regimes. I now give an example for each
of these approaches.

   Cereal and legume grains, which supply most of the
world's food, are primarily nutritionally inadequate in
lysine and sulpho-amino acids respectively, as well as
secondarily in other amino acids (see Harvey, 1970;
Whitehouse, 1973). In 1964 Mertz et al. (1964)
reported the effect of the gene, opaque 2, on the amino
acid composition of proteins of the seeds of maize.
This work stimulated further research, since lysine and
tryptophan, co-limiting amino acids in maize, were
doubled in the opaque 2 endosperm compared with normal.
This was shown to be due to a decrease in the prolamin
fraction (zein), from approximately 37% in normal maize
to 26% in opaque 2, with a concomittant increase in the
glutelin fraction from 29% to 39%. This correlation
between the changed proportion of storage proteins and
the amino acid composition of the meal, is explained by
the fact that prolamin is relatively more deficient in
lysine than is glutelin (see Table 1 for amino acid
composition of barley solubility classes which are
typical of most cereals). More recently, high-lysine
barleys such as hi-proly and the Risø mutant 1508 and
high-lysine sorghums, have been isolated and in every
case the improved amino acid composition of the meal
depends upon a changed proportion of the major proteins,
usually storage proteins, in the mutants. Thus, the
higher lysine of the hi-proly variety is due mainly to
increased proportions of four albumin/globulin proteins
which are richer in lysine than the prolamin and
glutelin fractions (see Table 2). The Risø mutant
1508 has 46% of its total protein in the form of
albumins and globulins as compared with 27% in the
parent variety and a prolamin content of 9% as compared
to 29%; the glutelin fraction remains constant in both
varieties at 39% (Table 2; Rhodes and Jenkins, 1976).

   Although similar high-quality protein mutants have
not yet been described for legumes, it would seem
possible that they will in the future, since the
different storage proteins of legumes, e.g. legumin and
vicilin, differ in their contents of sulpho-amino acids

TABLE   1
Levels of some important amino acids
in Osborne fractions of barley seeds
(N as % of protein N)

	Albumin	Globulin	Prolamin	Glutelin
Amide N	5.9	5.1	23.0	10.3
Glutamate	8.7	6.8	23.0	11.6
Proline	4.2	2.7	15.0	6.6
Lysine	7.9	6.3	0.8	0.9

Data from Folkes and Yemm (1956)

(Boulter et al., 1973).   Furthermore, it is known that
important proteins of legume seeds have a high level of
sulpho-amino acids compared with the main storage protein
for example, trypsin inhibitor compared with glycoprotein
II of Phaseolus vulgaris (Derbyshire, Wright and Boulter,
1976).

Since in all these examples the improvement has
been brought about by a changed proportion of the major
proteins, often the storage proteins, an understanding
of the factors which control the onset, rate and duration
of synthesis of these proteins, is of considerable
importance (Boulter, 1976).

Apart from possible improvements brought about by
plant breeding programmes, there is also the possibility
of improvement of protein quality by better crop manage-
ment, particularly with respect to fertiliser regimes.
It has now been shown in lupin (Blagrove and Gillespie,
1975;   Blagrove et al., 1976) and cowpea (Fox et al.,
1976) that sulphur fertilisation can increase the
sulpho-amino acid content of the seed meals.   Table 3
gives the sulpho-amino acid content of the protein of
cowpea meals from seeds grown under different sulphur
fertiliser regimes, and shows that there is an increase
up to 3-5 parts per million sulphur depending on the
variety.   Fig. 2 shows the effect of fertilisation on
the protein profiles of the meals.   It can be seen from
the cellulose acetate strips that the proportion of the
proteins has changed and that certain low molecular
weight proteins increase in seeds from plants grown with

TABLE 2

Principal protein fractions and their lysine content in cereal grains

Protein fraction (i) per cent of total protein; (ii) lysine per cent of protein fraction.

Cereal grain	Albumins (water soluble)		Globulins (salt soluble)		Prolamins (alcohol soluble)		Glutelins (alkali soluble)	
	(i)	(ii)	(i)	(ii)	(i)	(ii)	(i)	(ii)
Maize (normal)	4	3.8	2	6.1	50–55	0.2	30–45	3.4
Maize (opaque-2)	15	4.1	5	5.2	25	0.1	55	4.7
Barley (normal)	3–4	7.9	10–20	6.3	35–45	0.8	35–45	4.8
Barley (Risø 1508)	(	46	7.5	)	9	2.9	39	5.9
Rice	5	4.9	10	2.6	5	0.5	80	3.5
Oats	1	–	80	–	10–15	–	5	–

As modified from Rhodes and Jenkins (1976).

TABLE 3

Levels of sulpho-amino acids in cowpea grown
under different S fertilisation

Treatment: Levels of sulphate-S (ppm) in soil soln.	g Met/100 g	g CySH/100 g (air-dry basis)	g SMC*/100 g
0	0.272 ± 0.015	0.174 ± 0.017	0.040 ± 0.009
5	0.310 ± 0.002	0.230 ± 0.010	0.226 ± 0.002
15	0.287 ± 0.003	0.207 ± 0.015	0.226 ± 0.006

*S-methyl-cysteine

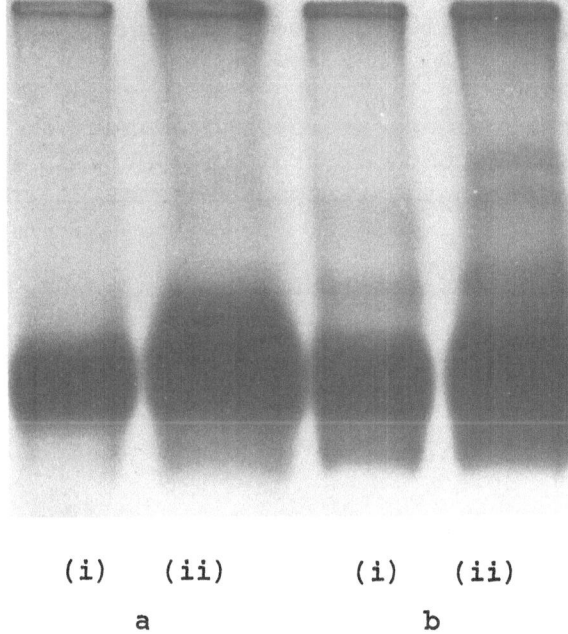

    (i)   (ii)   (i)   (ii)

        a   b

Figure 2.   Protein profiles of cowpea var. Sitao Pole
supplied with (a) 0 ppm S; and (b) 5 ppm S. Proteins
were separated by electrophoresis on cellulose acetate
membranes and stained with Coomassie Blue: equal
volumes of the extracts were loaded and separated on
the same membrane; (i) 0.25 µl; (ii) 0.75 µl.

5 ppm S. The sulpho-amino acid composition of the major
storage protein of cowpeas is extremely low whereas that
of some of the lower molecular weight fractions are
considerably higher in cysteine content (Carasco, J. F.
and Boulter, D., unpublished).   Further work is now
needed to relate these changes in sulpho-amino acid
content and sub-unit profiles to specific seed proteins.
The agricultural importance of S-fertilisation may not
be great, however, since the levels of protein sulpho-
amino acids can only be increased by growing plants with
up to about 8 parts per million, i.e. S-levels which are
normally found in the soil in many regions. However,
some cultivars respond more than others, so further work
may identify highly responsive cultivars. The question

still remains as to the mechanism whereby the changed
proportion of proteins has been brought about.

These examples indicate that in these instances
at least, improvements are due to changes in the
developmental sequence of protein synthesis rather
than to changes in the structural genes of the proteins
themselves.

REFERENCES

Bailey, C. J., Cobb, A. and Boulter, D. (1970)
     Planta (Berl.) 95, 103-118.

Bain, J. M. and Mercer, F. V. (1966) Aust. J. Biol.
     Sci. 19, 49-67.

Bell, E. A. (1976) FEBS Letters 64, 29-35.

Blagrove, R. J. and Gillespie, J. M. (1975) Aust. J.
     Plant Physiol. 2, 13-27.

Blagrove, R. J., Gillespie, J. M. and Randall, P. J.
     (1976) Aust. J. Plant Physiol. 3, 173-184.

Blobel, G. and Dobberstein, B. (1976) J. Cell Biol.
     67, 835-851.

Boulter, D. (1976) In: Proceedings of the Workshop
     on Genetic Improvement of Seed Proteins,
     Washington, 1974.   Nat. Acad. of Sci.,
     Washington.   In Press.

Boulter, D. and Barber, J. T. (1963) New Phytol. 62,
     301-316.

Boulter, D., Evans, I. M. and Derbyshire, E. (1973)
     Qual. Plant. - Pl. Fds. Hum. Nutr. 23, 239-250.

Briarty, L., Coult, D. A. and Boulter, D. (1969)
     J. Ex. Bot. 20, 358-372.

Burr, R. and Burr, F. A. (1976) Proc. Nat. Acad. Sci.
     USA 73, 515-519.

Byers, M., Kirkman, M. A. and Miflin, B. J. (1976)
    In: Proceedings of the 24th Easter School in
    Agricultural Science on 'Plant Proteins',
    Nottingham, 1976. Butterworths. In Press.

Chrispeels, M. J. and Boulter, D. (1975) Plant Physiol.
    55, 1031-1037.

Cullis, C. A. and Davies, D. R. (1975) Genetics 81,
    485-492.

Davies, D. R. (1973) Nature (New Biol.) 245, 30-32.

Derbyshire, E., Wright, D. J. and Boulter, D. (1976)
    Phytochemistry 15, 3-24.

Dure, L. S. (1975) Ann. Rev. Plant Physiol. 26, 259-278.

Folkes, B. F. and Yemm, E. W. (1956) Biochem. J.
    62, 4-11.

Fox, R. L., Kang, B. T. and Nangju, D. (1976)
    Submitted to Agronomy Journal.

Graham, T. A. and Gunning, B. E. S. (1970) Nature
    (Lond.) 228, 81-82.

Granick, S. (1937) Plant Physiol. (Lancaster) 12,
    601-623.

Harris, N. and Boulter, D. (1976) Annals Bot.
    In Press.

Harris, N. and Juliano, B. O. (1976) J. Ex. Bot.
    In Press.

Harvey, D. (1970) Tables of the amino acids in foods
    and feedingstuffs. Tech. Commun. No. 19,
    second edn. Commonwealth Agricultural Bureaux,
    Farnham Royal, Bucks, England.

Khoo, V. and Wolf, M. J. (1970) Am. J. Bot. 57,
    1042-1050.

Mertz, E. T., Vernon, O. A., Bates, L. S. and Nelson,
    O. E. (1964)   Science 148, 1741-1742.

Millerd, A. (1975)   Ann. Rev. Plant Physiol. 26,
    53-72.

Millerd, A. and Whitfield, P. R. (1973)   Plant Physiol.
    51, 1005-1010.

Millerd, A. and Spencer, D. (1974)   Aust. J. Plant
    Physiol. 1, 331-341.

Opik, H. (1968)   J. Ex. Bot. 19, 64-76.

Osborne, T. B. (1924)   The Vegetable Proteins, 2nd
    edition. Longmans, Green & Co., N.Y.

Payne, P. I. and Boulter, D. (1969)   Planta (Berl.)
    84, 263-271.

Rhodes, A. P. and Jenkins, G. (1976)   In:
    Proceedings of the 24th Easter School in
    Agricultural Science on 'Plant Proteins',
    Nottingham, 1976. Butterworths. In Press.

Smith, D. L. (1973)   Annals Bot. 37, 795-804.

Sun, S. M., McLeester, R. C., Bliss, F. A. and Hall,
    T. C. (1974)   J. Biol. Chem. 249, 2118-2121.

Tombs, M. P. (1965)   Biochem. J. 96, 119-133.

Tombs, M. P. and Lowe, M. (1967)   Biochem. J. 105,
    181-187.

Whitehouse, R. N. H. (1973)   In: The Biological
    Efficiency of Protein Production, Ed. Jones,
    J. G. W.   Cambridge University Press, p. 83.

Yarwood, A. (1976)   In: Proceedings of the 24th
    Easter School in Agricultural Science on 'Plant
    Proteins', Nottingham, 1976. Butterworths.
    In Press.

REGULATION OF TRANSLATION OF STORED mRNA OF DICOT SEEDS

LEON S. DURE III and BARRY HARRIS

Department of Biochemistry, University of Georgia

Athens, GA, U.S.A.

For the past two years we have been trying to determine the mechanisms by which the translation of the stored mRNA of cotton seeds is prevented until the commencement of germination. We have collected data that suggests that the stored mRNA is not completely processed until germination begins. More specifically, it does not appear to become polyadenylated prior to seed germination. Before presenting these data, we would like to summarize briefly the evidence for the existence of stored mRNA in cotton seeds.

1. BRIEF RESUME OF EARLIER OBSERVATIONS

Several years ago we collected evidence that most of the protein synthesis necessary to begin germination in cotton cotyledons is directly by mRNA that is transcribed in embryogenesis [1,2]. Further, we were able to show that some of the enzymes translated from this mRNA were unique to germination; that is, were not enzymes found in embryogenesis. One of these enzymes (a carboxypeptidase C) was purified to homogeneity in order to show by isotope labeling that its appearance in germination is the result of de novo synthesis as opposed to zymogen activation, etc. [3,4]. Thus, it appears that some of the mRNA transcribed in embryogenesis but not translated until germination is not simply mRNA for omni-present enzymes that appears to be carried over from embryogenesis into germination, but mRNA for proteins necessary for the subsequent developmental event. In this light the conserved mRNA represents a developmentally significant phenomenon; one that can be viewed as part of the preprogrammed equipping of the seed for successful germination.

279

Subsequently, we were able to implicate the plant growth regulator, abscisic acid (ABA) in the inhibition of the premature translation of this body of mRNA in embryogenesis (5,6). This compound which has been implicated in many inhibitory processes in higher plants (see ref. 7 for review) first becomes demonstratable in ovule tissues at the same time that the existence of the "germination mRNA" becomes demonstratable in embryonic cotyledons. Thus, ABA can be viewed as preventing vivipary in these seeds. (Vivipary in higher plants refers to the premature germination of immature seeds while still in the immature fruit. It constitutes a genetic lethal and is found occasionally in most all angiosperm families.) Oddly, it was found that the inhibition of the translation of the germination mRNA by ABA requires continued RNA synthesis (5,6). That is, if RNA synthesis is blocked in immature, excised embryos by actinomycin D, the inhibition by ABA of germination enzyme synthesis (as followed by carboxypeptidase activity) is overcome. The immature embryos germinate precociously and carboxypeptidase C activity appears. This curious observation concerning the mode of action of ABA has recently been substantiated in the barley aleurone system (8).

Cotton embryos that have not yet reached the point in development when ABA appears and the germination mRNA becomes demonstratable will also germinate precociously when removed from the ovular tissue. This germination is totally sensitive to actinomycin D as is the appearance of the germination enzymes. The precocious germination of these embryos obviously entails a derepression of the cistrons comprising the germination mRNA. Furthermore, cell division in the cotyledons of these very young embryos is stopped when they are removed from the ovular tissue for germination. Cell division stops in vivo in this tissue at the same time the germination mRNA and ABA appears. Furthermore, the vascular connection between the incipient seed and the mother plant atrophies naturally at this point. From this it would seem that the vascular flow from the mother plant is what keeps the embryos developing as embryos; that is, maintaining cell division and keeping the cistrons for the germination mRNA repressed. When this vascular flow is stopped normally by the atrophy of the connecting tissue (the finiculus) or mechanically by the excision of very young embryos, embryonic growth (cell division) stops and the germination cistrons are derepressed. In vivo the advent of ABA prevents the newly synthesized germination mRNA from being translated during the remaining 20 days that the embryos remain in the cotton boll (maturation phase of embryogenesis).

2. MORE RECENT FINDINGS

a. Evidence from Isotope Incorporation That Stored mRNA
Is Polyadenylated in Germination

One possible mechanism by which the translation of the germination mRNA is delayed in cotton cotyledons would be a delay in mRNA
processing from the initial transcription product to functional mRNA
until germination (until the death or disappearance of the ovule
tissues producing ABA). An indication that this may be the case was
the effect of 3'd adenosine (3'd Ado, cordycepin) on visible germination and on protein synthesis and the advent of carboxypeptidase
activity during germination. In these experiments the effects of
3'd Ado and 3'd cytidine (3'd Cyd) on carboxypeptidase activity was
determined in cotyledons that were exposed to these nucleoside
analogues for different time periods during the first 36 hours of
germination. It was assumed at this time that these analogues would
be triphosphorylated in vivo, and that the 3'd nucleoside triphosphates would truncate RNA synthesis by becoming incorporated into
growing RNA chains as has been shown to take place in a number of
tissues (9,10) and shown to occur in cell-free RNA synthesis by
purified RNA polymerases I and II (11,12).

3'd Cyd, analogous to actinomycin D, failed to inhibit visible
cotyledon germination, gross protein synthesis or the advent and
accumulation of carboxypeptidase activity, whereas 3'd Ado inhibited
all of these parameters (Table 1) provided it was supplied the tissue
prior to the 32 hour of germination (13). In fact, the difference
in the degree of inhibition of carboxypeptidase activity obtained
at the different time periods suggested that the 3'd Ado-sensitive
process takes place between the 6th-30th hour of germination (13).

Since these nucleoside analogues bring about an inhibition of
RNA synthesis by totally different means than does actinomycin,
and probably does so with less potential side effects, these data
were considered very exciting at the time. Both analogues were
assumed to stop all RNA synthesis at the high concentrations used,
but only 3'd Ado was assumed lead to an inhibition of mRNA polyadenylation -- and only 3'd Ado prevents visible cotyledon germination, protein synthesis and the advent of carboxypeptidase activity.
We believed that we had put the basis for the processing of stored
cotyledonary mRNA during germination on a better foundation by
lessening our reliance on data obtained with actinomycin.

To establish that the nucleoside analogues through their
putative 3'd nucleotide triphosphates do indeed abort the synthesis
of all classes of RNA, their effect on the synthesis of rRNA, tRNA
and mRNApoly(A) during early germination was studied by following

isotope incorporation into these species (14). The use of poly(U)-
Sepharose and oligo dT-cellulose affinity chromatography made it
possible for the first time to quantitatively follow the effect of
3'd Cyd on mRNApoly(A) synthesis. Since 3'd Ado presumably leads
to the truncation of polyadenylation, its effect on mRNA synthesis
could not be followed, since the sequestration and concentration on
the affinity columns of mRNApoly(A) synthesized during the labeling
period cannot occur in the absence of the poly(A) region of the
chain.

Table 1.   RNA and Protein Synthesis in Inhibitor-treated Cotyledons

RNA Synthesis 10-22 hrs of Germination	Per Cent of Control		
	Act D (20 µg/ml)	3'D Cyd (2.5 mM)	3'd Ado (2.5 mM)
Total RNA	12	15	4
25, 18, 5S RNA	9	9	4
tRNA	18	22	5
mRNApoly(A)	38	100	...
mRNA	29	100	...
poly(A)	70	100	...
Protein synthesis 24th-30th hr of germination	130	130	20
Carboxypeptidase activity at 72 hr of germination	95	95	5

In the case of rRNA (including 5S RNA) and tRNA, both nucleoside analogues caused essentially complete inhibition of isotope
incorporation in germinating cotyledons (Table 1). However, to
our surprise 3'd Cyd had no effect on the synthesis of mRNApoly(A).
By analogy, 3'd ATP derived from 3'd Ado probably is also ineffec-
tive in truncating mRNA synthesis, since it seems unlikely that the
RNA polymerase can sense the lack of a 3' OH group on one nucleotide
triphosphate but not on another. However, this presumed synthesis
cannot be demonstrated by these methods if the poly(A) moiety is
missing as mentioned above. These data indicate a difference in
substrate recognition between RNA polymerases I and III that transcribe rRNA, 5S RNA and tRNA and RNA polymerase II that is presumed
to transcribe the structural genes. It has been shown in Hela cells

that the synthesis of hnRNA, the presumed initial transcription product of RNA polymerase II, is also not affected by the nucleoside analogues (9,10). Yet 3'd ATP has been shown to cause chain termination in cell-free transcription by purified RNA polymerase II. This curious inconsistency has not been resolved to our knowledge.

The net effect of the latter experiments was that we were once again in the position of justaposing the effects of actinomycin D and 3'd Ado on visible germination, protein synthesis and carboxy-peptidase activity. These results taken at face value indicate that although the synthesis of the germination proteins in cotyle-dons during the first 3 days of germination may not require mRNA synthesis, it does require polyadenylation.

The ability to sequester mRNApoly(A) on the affinity columns made it feasible to directly test the possibility that poly(A) is added onto preformed RNA during the early stages of germination. This idea received some additional reinforcement when we measured the effect of actinomycin on mRNApoly(A) synthesis using the affin-ity columns. Dosages of actinomycin that inhibit essentially all rRNA and tRNA synthesis were found to stop about 70% of isotope incorporation into mRNA but only 30% of isotope incorporation into poly(A) (Table 1). If only newly synthesized mRNA is polyadeny-lated during early germination, isotope (^{32}P) incorporation into poly(A) should be inhibited to the same extent as is its incorpora-tion into mRNA; that is, only the mRNA that is synthesized during the labeling period should get polyadenylated. In actuality much more gets polyadenylated.

In view of the above suggestive evidence that preformed mRNA becomes polyadenylated during early germination, we designed exper-iments that allowed us to measure the incorporation of ^{32}P and ^{3}H adenosine into mRNApoly(A) and into the mRNA and poly(A) portions of these molecules individually. The rationale of these experi-ments was that if the polyadenylation of some/all of the preformed mRNA takes place during the labeling period, the mRNA portion of the mRNApoly(A) fraction will be underlabeled relative to the poly(A) portion. That is, any preformed mRNA that is polyadeny-lated during the labeling period will contain isotopes only in the poly(A) component. When the apparent average chain length of the mRNA portion is then calculated from its isotope content relative to the isotope content of the poly(A) portion, and the value obtained is compared with the true average chain length obtained from gel electrophoresis under fully denaturing conditions (in 99% formamide), the calculated apparent chain length should be smaller than the true length. Furthermore, the difference between these values should be an indication of the ratio of preformed to newly synthesized mRNA that is polyadenylated during the labeling period.

The ability to sequester mRNApoly(A) coupled with other conventional techniques makes it possible to ascertain the following facts (15,16).

1. The average chain length of the poly(A) portion of mRNApoly(A)
2. The base composition of mRNApoly(A)
3. The base composition of the mRNA portion of mRNApoly(A)
4. The % ^{32}P of mRNApoly(A) that is contained in the poly(A) or mRNA portion exclusively
5. The % ^{3}H adenosine of mRNApoly(A) that is contained in the poly(A) or mRNA portion exclusively

From these 5 bits of information it is possible to determine the apparent average mRNA chain length based on isotope incorporation by 3 different means (15,16).

If these data are obtained from cotyledons germinated in the presence of actinomycin D, any apparent underlabeling of the mRNA portion of mRNApoly(A) fraction should be accentuated, since 70% of the synthesis of new mRNA will have been inhibited which, in turn, should confine most of the isotope incorporation to the poly(A) portion of the preformed mRNA.

Conversely, if these same measurements of the distribution of radioactivity between mRNA and poly(A) are made at a time in germination after the putative stored mRNA has presumably been processed into functional polysomal mRNA, the distribution of radioactivity should show no underlabeling of mRNA, and the calculated mRNA average chain length should be consonant with that found by gel electrophoresis.

Initially we established that the average poly(A) chain length found in mRNApoly(A) isolated from any period in germination is 100-110 AMP residues. Messenger RNA synthesized during the first day of germination was found to contain 22% AMP. In order to illustrate how the five items of information listed above can be used to determine mRNA chain length, Table 2 shows how ^{32}P and ^{3}H adenosine would be distributed between mRNA and poly(A) synthesized during the labeling period if the mRNA average chain length were 1500 nucleotides, using the observed 100 nucleotide poly(A) chain length and the 22% AMP mRNA base composition as known values. In such a case, mRNApoly(A) should contain 27% AMP and 6.3% of the ^{32}P should be contained in the poly(A) moiety as well as 23% of the ^{3}H adenosine. These theoretical values assume that RNA and poly(A) synthesis draw on nucleotide triphosphate pools of the same specific radioactivity.

Table 2 also shows the experimentally determined values obtained for the distribution of the isotopes between mRNA and poly(A) from cotyledons germinated about 1 day (the period of greatest sensitivity to 3'd Ado) ± actinomycin D, and from cotyledons labeled much later in germination when carboxypeptidase synthesis is no longer effected by 3'd Ado. By this latter time in germination presumably the stored mRNA has been totally processed. From the elevated level of AMP in mRNApoly(A) and from the large amount of ^{32}P and ^{3}H adenosine contained in the poly(A) alone, it is apparent that the mRNA found in the one-day germinated cotyledons appears highly underlabeled. This underlabeling is accentuated in the mRNA from actinomycin treated cotyledons. When the apparent average mRNA chain lengths are calculated from these data, absurdly short chain lengths result. However, when the same measurements and calculations are carried out on mRNApoly(A) from cotyledons labeled later in germination, the distribution of isotopes between mRNA and poly(A) is such that the calculated mRNA chain length is about 1500 nucleotide residues.

Each of these labeling experiments was repeated at least 4 times, and the data in Table 2 is the average of all the values obtained. An obvious explanation for the apparent short length of the mRNA obtained from the one-day germinated cotyledons would be that the mRNA has been hydrolytically shortened by ribonucleases during purification. Figure 1 shows the range of sizes of mRNA poly(A) molecules isolated from cotyledons labeled during the 12th-20th hour of germination as revealed by electrophoresis in 99% formamide. Both the mass average and number average size is over 1500 nucleotides indicating that the molecules have not been hydrolytically shortened by nucleases during their extraction and purification. Taken at face value the calculated mRNA chain length of about 500 nucleotides suggests that two thirds of the mRNA that is polyadenylated during the first day of germination exists in the dry seed, and, hence, does not become radioactive. The fact that later in germination the calculated mRNA size approaches that given by gel electrophoresis under fully denaturing conditions strengthens this interpretation.

The results of these experiments further substantiate the idea that some/all of the stored cotyledon mRNA is not processed until germination -- and that this failure to process the germination mRNA in embryogenesis is an aspect of the mechanism by which its premature translation in embryogenesis is prevented. However, such experiments are not definitive. Factors that are as yet unperceived could be responsible for the observed distribution of radioactivity between mRNA and poly(A) during the various labeling periods.

Table 2. Calculation of Apparent mRNA Chain Length Based on Distribution of Radioactivity

	Expected Values for mRNA of 1500 Nucleotides[1]	Experimental Data		
		10-22 hrs Germination	10-22 hrs Germination in Actinomycin D	32-40 hrs Germination
Average poly(A) length	100	100	100	100
% AMP in mRNA	22	22	22	28.4
% AMP in mRNApoly(A)	27	38.3	43.1	33.4
% 32P of mRNApoly(A) in poly(A)	6.3	18.8	35.7	7.5
% 3H adenosine of mRNApoly(A) in poly(A)	23	45.4	55	20
		Calculated mRNA Chain Length		
From % AMP in mRNApoly(A)	1500	420	300	1520
From distribution of 32P	1500	475	200	1360
From distribution of 3H	1500	600	410	1510
Average of data	1500	500	300	1460

[1]Values expected should mRNA average 1500 nucleotides using experimentally derived value of 100 for poly(A) length and 22% as the percent AMP in mRNA.

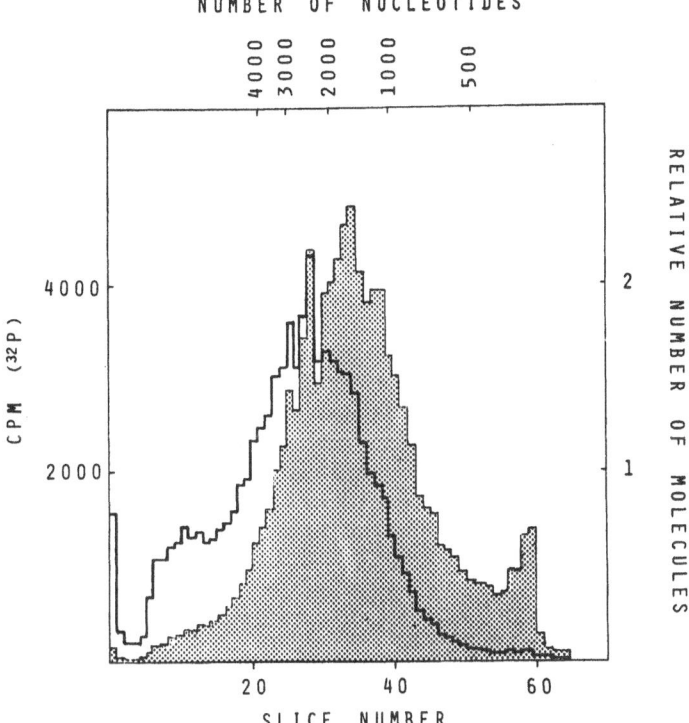

Fig. 1. SDS gel electropherogram in 99% formamide of mRNApoly(A) fraction.

b. The Isolation of Optically Measurable Amounts of mRNApoly(A)

We have recently attempted to reinforce our hitherto indirect evidence for the polyadenylation of preexisting mRNA in early germination by measuring the actual amounts of mRNApoly(A) contained in the cotyledons of dry seeds and in cotyledons germinated 24 hours ± actinomycin and 3'd Ado spectrophotometrically. Such a procedure requires that a very large amount of starting material be used, since the amount of total RNA that is mRNApoly(A) is generally very small in all tissues that have been examined critically. Of paramount importance to these measurements is the recycling of the initial mRNApoly(A) fraction from the oligo dT cellulose column back through the column under fully denaturing conditions to rid the fraction of contaminating rRNA (17). This rRNA is not retained by the column in high salt solutions except in the presence of mRNApoly(A), which indicates that it is RNA:RNA interactions that cause it to contaminate the mRNApoly(A) fraction.

Table 3 presents the data obtained from these optical measurements. There is no discernable change in the total amount of high molecular weight RNA in cotyledons during the first day of germination as shown in Table 3 by the amount of material precipitable with 2 M LiCl. However, the amount of RNA that is mRNApoly(A) does increase. The increase is less in actinomycin treated cotyledons. However, the increase in both actinomycin treated and untreated may be larger than perceived, since the residual mRNApoly(A) present in dry seed cotyledons appears to have been de-adenylated or destroyed during the first day of germination as shown by its total absence in cotyledons germinated 1 day in 3'd Ado. That is, it appears that none of the mRNApoly(A) found after 1 day of germination exists in a polyadenylated form in the dry seed. (Conversely, it could be argued that 3'd Ado brings about a rapid and unnatural loss of mRNApoly(A) during the first day of germination in addition to simply preventing additional polyadenylation from occurring.)

Table 3. Amount of mRNApoly(A) in Cotyledons as Determined by
         Spectrophotometry

Cotyledon Source	A_{260} Units/100 Cotyledon Pairs (+15%)	
	LiCl Precipitated RNA	mRNApoly(A) Fraction
Dry seed	500	4 (0.8%)
24 hr germinated	500	8 (1.6%)
24 hr germinated + Act. D	500	6 (1.2%)
24 hr germinated + 3'd Ado	500	0

Assuming that all the mRNApoly(A) found after 1 day of germination has been polyadenylated during the 1 day period, the data show that in actinomycin the mRNApoly(A) fraction has increased to about 75% that of untreated cotyledons. Since actinomycin inhibits only 70% of the synthesis of new mRNA during this period, some of the increase found in actinomycin is new mRNA and some of it preexisting. The difference in the amount of mRNApoly(A) between actinomycin treated and untreated cotyledons is 2 A_{260} units which should represent the 70% of the mRNA not formed in actinomycin. This suggests that the total amount of newly synthesized mRNA is about 2.9 A_{260} units, which indicates that about 5 A_{260} units of mRNApoly(A) are derived from stored mRNA.

3. SUMMARY OF EVIDENCE FOR THE POLYADENYLATION DURING
    GERMINATION OF STORED mRNA

We have attempted to measure the polyadenylation of pre-existing mRNA during early germination by three methods, the results of which are presented in Table 4. The first of these methods is based simply on the fact that actinomycin has a much greater inhibitory effect on the incorporation of radioisotopes into mRNA than into poly(A). This fact alone suggests that RNA not made radioactive during the labeling period (preexisting RNA) is polyadenylated. Algebraic computation based on the difference in inhibition of synthesis of the two moieties of mRNApoly(A) by actinomycin indicate that a little over one half of the RNA poly-adenylated in early germination preexists in the dry seed (stored mRNA).

The second method is based on the analysis of the distribution of isotopes between poly(A) and mRNA in uninhibited cotyledons. Once the average length of the poly(A) chains and the percent AMP contained in mRNA are determined, an apparent mRNA chain length can be computed from the distribution of isotopes by three different computations. These computations do not yield identical values, but they suggest that the calculated average chain length is about 500 nucleotide residues. The same measurements made on mRNApoly(A) from older cotyledons predict average chain lengths of about 1500 nucleotides. Gel electrophoresis of mRNApoly(A) under fully dena-turing conditions indicate a true mass average chain length well over 1500 nucleotides. Thus, these data indicate that only a third of the mRNA polyadenylated during early germination was made radioactive (newly synthesized).

The third method is based on the isolation of amounts of mRNApoly(A) that are measurable spectrally. The amount of mRNA-poly(A) found in actinomycin treated cotyledons indicate again that about one third of the mRNApoly(A) found in 1 day germinated cotyle-dons is newly synthesized and the remaining two thirds originate from stored mRNA.

In spite of the fact that all of these data seem to point to the same conclusion, namely that some/all of the stored mRNA of cotton cotyledons is polyadenylated in germination, one must be cautious in accepting this conclusion. There is some evidence that in higher animal cells, mRNApoly(A) gets further adenylated in the cytosol (10-15 residues) after the initial and more extensive adenylation has taken place in the nucleus (18). If this process also occurs in plants, and if the ATP pool sustaining this adenyla-tion reaches a higher specific radioactivity than does the nuclear pool during early germination, our results from the first two methods of analysis might simply indicate that only the cytosol adenylation of stored mRNA awaits germination.

Table 4. Summary of Calculations of the Relative Amount of Stored and Newly
          Synthesized RNA Polyadenylated During the 1st Day of Germination

Basis	Stored mRNA	New mRNA
1. Differential inhibition by actinomycin D of isotope incorporation into mRNA and poly(A)	57.5%	42.5%
2. Low level of isotope incorporation into mRNA relative to poly(A)	67.0%	33.0%
3. Increase in mRNApoly(A) in actinomycin D relative to control as measured optically	64.0%	36.0%

Our third method of analysis suffers from the fact that it is very difficult to prove that the mRNApoly(A) fraction is totally mRNApoly(A), even after a second passage over the affinity column. The optical measurement of the amount of poly(A) recoverable from the mRNApoly(A) fraction and a knowledge of its average chain length should allow one to determine if most of the molecules in this fraction have poly(A) chains. In the case of cotton cotyledons this entails the recovery of 1/1000th of the starting RNA as poly(A) in amounts that would allow for its unmistakable identification as poly(A) spectrally.

If indeed stored mRNA in cotton seeds fails to get polyadenylated when it is first synthesized, the next question becomes what is the role of ABA in this delay?

REFERENCES

1. Waters, L.C. and Dure, L.S. III (1966) J. Mol. Biol., $\underline{19}$, 1.
2. Ihle, J.N. and Dure, L.S. III (1969) Biochem. Biophys. Res. Comm., $\underline{36}$, 705.
3. Ihle, J.N. and Dure, L.S. III (1972a) J. Biol. Chem., $\underline{247}$, 5048.
4. Ihle, J.N. and Dure, L.S. III (1972b) J. Biol. Chem., $\underline{247}$, 5034.
5. Ihle, J.N. and Dure, L.S. III (1970) Biochem. Biophys. Res. Comm., $\underline{38}$, 995.
6. Ihle, J.N. and Dure, L.S. III (1972c) J. Biol. Chem., $\underline{247}$, 5041.
7. Milborrow, B.V. (1974) Ann. Rev. Plant Physiol., $\underline{25}$, 259.
8. Ho, D.T. and Varner, J.E. (1976) Plant Physiol., $\underline{57}$, 175.
9. Siev, M., Weinberg, R. and Penman, S. (1969) J. Cell. Biol., $\underline{41}$, 520.
10. Abelson, H.T. and Penman, S. (1972) Biochim. Biophys. Acta, $\underline{277}$, 129.
11. Blatti, S.P., Ingles, C.J., Lindall, T.J., Morris, P.W., Weaver, R.F., Wienberg, F. and Rutter, W.J. (1970) Cold Springs Harbor Symp. Quant. Biol., $\underline{35}$, 649.
12. Horowitz, B., Goldfinger, B. and Marmur, J. (1974) Fed. Proc., Fed. Amer. Soc. Exp. Biol., $\underline{33}$, 1418.
13. Walbot, V., Capdevila, A. and Dure, L.S. III (1974) Biochem. Biophys. Res. Comm., $\underline{60}$, 103.
14. Harris, B. and Dure, L.S. III (1974) Biochemistry, $\underline{13}$, 5463.
15. Walbot, V., Harris, B. and Dure, L.S. III (1975) in Developmental Biology of Reproduction (C. Markert, Ed.), pp. 165-187, Academic Press, New York.
16. Harris, B. and Dure, L.S. III, Submitted for publication.
17. Bantle, J.A. and Hahn, W.E. (1976) Cell, $\underline{8}$, 139-150.
18. Darnell, J.E. (1976) in Progress in Nucleic Acid Research and Molecular Biology, Vol. 19 (W. Cohn & E. Volkin, Eds.), In press.

HORMONAL CONTROL OF PROTEIN SYNTHESIS

J. E. Varner

Biology Department, Washington University

St. Louis, MO 63130

There are many instances in which a change in tissue concentration of a hormone or the exogenous addition of a hormone specifically changes the level of activity of one or more enzymes. In no case is it known how the tissue controls the level of enzyme activity in response to the changed hormone concentration. It is instructive to review the best known of these responses as we try to develop concepts that will allow us to understand them.

I shall describe only the cereal grain aleurone layer responses to gibberellins, abscisic acid and ethylene.

Several other systems show great promise for the study of hormone control of protein synthesis. These include the soybean hypocotyl responses to auxin and cytokinins (Guilfoyle, et al., 1975; Teissere, et al., 1975; Vanderhoef and Stahl, 1975); the pea seedling response to auxin and ethylene (Byrne, et al., 1975; Verma, et al., 1975); the Rhoeo leaf section responses to auxins and abscisic acid (DeLeo and Sacher, 1970; Sacher and Davies, 1974); the oat internode responses to gibberellins (Kaufman, 1965; Adams, et al., 1975); and the Agrostema response to cytokinins (Borriss, 1967; Dilworth and Kende, 1974).

Heterotrophic growth of cereal seedlings depends upon mobilization of the proteins, carbohydrates and minerals stored in the endosperm (Fig. 1). This mobilization is brought about by the living cells of the outer layers of the endosperm - the aleurone layer. The agents of this mobilization are such hydrolases as α-amylase, protease, β-glucanase, ribonuclease, α-glucosidase, limit dextrinase, and acid phosphatase.

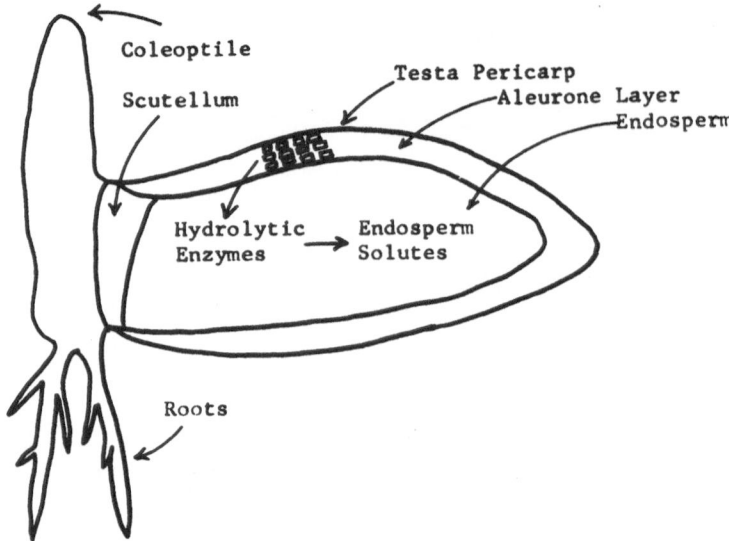

Fig. 1.   Schematic drawing of barley seedling.
         (Jones and Armstrong, 1971)

        The striking increases that occur in α-amylase (Yomo and
Varner, 1971) and protease in embryoless half-seeds and in isolated aleurone layers treated with gibberellic acid (GA₃) are due
to <u>de novo</u> synthesis. The smaller increases in ribonuclease and
in β-glucanase are also due to synthesis that is only partly dependent on added GA₃ (Bennett and Chrispeels, 1972). Although the
small increase in acid phosphatase is apparently due to synthesis,
the principal observable change in phosphatase is its release from
the wall into the surrounding medium (Ashford and Jacobsen, 1974).
This release is part of the aleurone cell's response to GA₃ and
results from the extensive cell wall degradation that follows the
secretion of pentosanases (Briggs, 1963; Taiz and Honigman, 1976).
The principal polysaccharides of the wall are arabinoxylans (McNeil,
et al., 1975). The kinds of proteins being synthesized and secreted
by the aleurone layer is controlled by the growing embryo through
the mediation of gibberellins synthesized in the embryo. The response of the aleurone layers to GA₃ can be modified by the addition of abscisic acid (ABA) (Yomo and Varner, 1971). In addition,
the response of GA₃-treated layers to ABA can be modified by the
addition of ethylene (Fig. 2; Jacobsen, 1973). It is not certain
that the germinating seedling uses ABA and ethylene to modulate
the aleurone layer's responses to GA₃. One might, however, argue

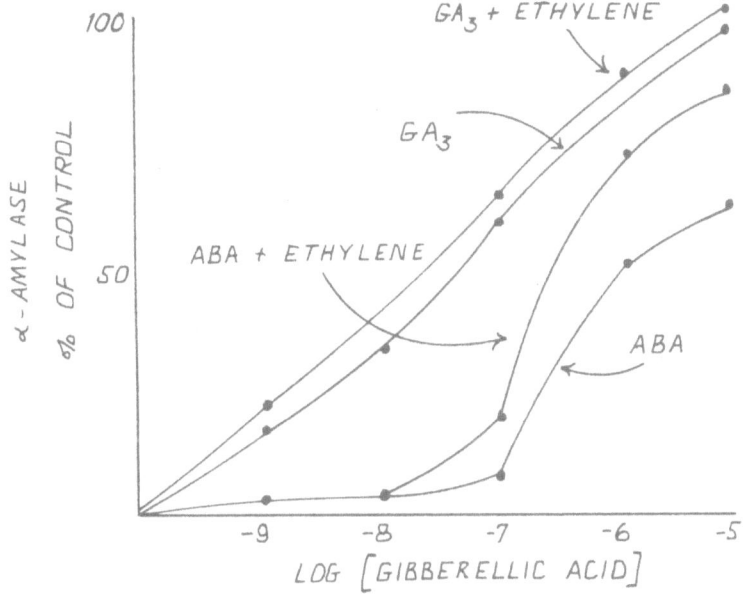

Fig. 2. Reversal of ABA inhibition of amylase synthesis by increasing amounts of GA in the presence and absence of ethylene. All incubations were for 24 hours. The values given are for released α-amylase only and all are relative to the α-amylase production in the presence of 10^{-4} M GA_3 which has been given the value 100. [ABA] = 2 x 10^{-7} M, [ethylene] = 10 µl/liter. (Jacobsen, 1973)

that since the tissue has the ability to respond to ABA and ethylene such a modulation probably occurs in the course of normal germination.

The barley aleurone system was characterized in review by photographs, light micrographs, electron micrographs, tables, and figures in 1971 (Yomo and Varner). The role of the aleurone layer in malting was reviewed in 1973 (Palmer); and there was a general review of the physiological role of gibberellins in 1973 (Jones, 1973b).

When isolated aleurone layers are treated with gibberellins, the rapid synthesis of α-amylase and protease beginning at about eight hours is preceded by the proliferation of rough endoplasmic reticulum as shown by electron micrographs (see Yomo and Varner,

(1971). A GA$_3$-dependent increase in the incorporation of labeled choline and of labeled phosphate into phospholipids reflects this membrane and proliferation although a cautious interpretation of the observation is recommended (Firn and Kende, 1974). An earlier effect, the GA$_3$-dependent increase in phosphorylcholine cytidyl transferase and in phosphorylcholine glyceride transferase as measured in cell-free extracts is detectable at two hours (Fig. 3; Johnson and Kende, 1971) and even earlier as we shall see.

The principal approach to finding out how the aleurone cells respond to gibberellins has been to describe the most easily observable responses, e.g. the massive secretion and release of hydrolases, the extensive cell wall degradation, the release of reducing sugars, the vacuolation of the cells and the degradation of the reserve proteins. These accumulated changes are the expression of a redirection of the cell's metabolism that occurs following treatment with GA$_3$. Such a redirection is - current common sense tells us - dependent on the presence of a GA$_3$ receptor.

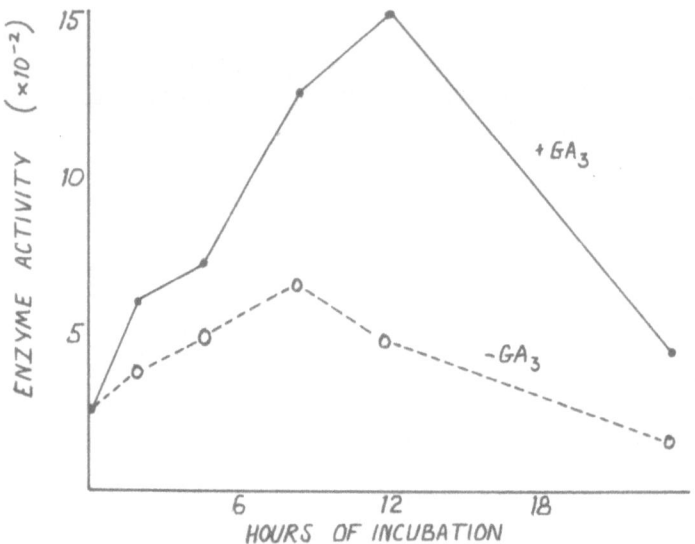

Fig. 3. Effect of GA on phosphorylcholine-cytidyl transferase activity. Fifty half-seeds were incubated. Aleurone layers were isolated and the enzyme assayed in pellets prepared from cell-free extracts. Each set of points represents an average of 6, 9, 5, 3, and 2 experiments respectively. (Johnson and Kende, 1971)

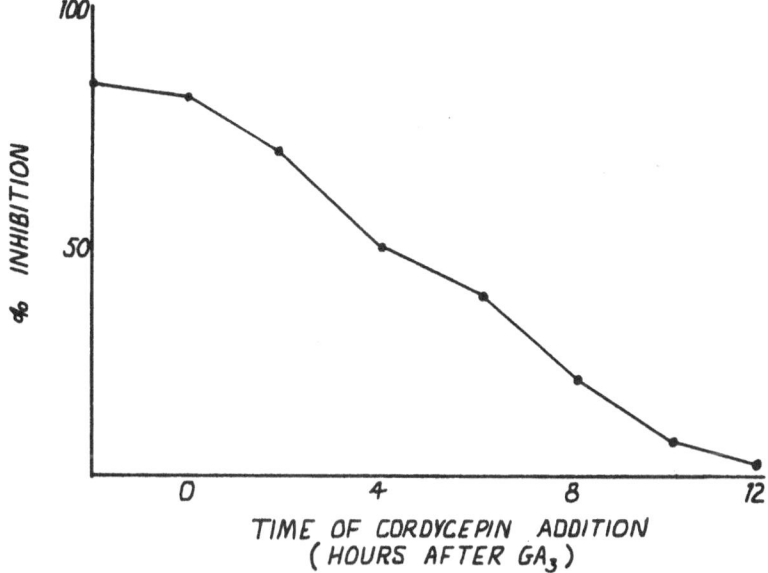

Fig. 4. Effect of cordycepin on α-amylase activity. Cordycepin was added at different times after GA3 as indicated, and aleurone layers were further incubated until 24 hr after addition of GA3 when the activity of α-amylase was assayed. (Ho and Varner, 1974)

We review first the late consequences of GA3-treatment and work our way back in response time toward the earliest event - the formation of a gibberellin-receptor complex, an event not yet observed.

Endosperm modification begins only after the release of hydrolases from the aleurone layers into the starchy endosperm. Such release requires that the aleurone cell walls be bathed in a medium of a certain minimum ionic strength (Varner and Mense, 1972). The aleurone cells are the source of these ions. The ions are secreted from the aleurone cells in response to GA3 (Jones, 1973a) and also during normal germination (Clutterbuck and Briggs, 1974). Enzyme release from the aleurone cells parallels, and requires degradation of the cell walls (Ashford and Jacobsen, 1974). Release of the hydrolases from the cell walls is, of course, preceded by secretion of the hydrolases from their site of synthesis into the periplasmic space. This secretion requires energy (Varner and Mense, 1972).

Aleurone layers, in midcourse production of α-amylase, secrete most of the newly synthesized protein within 60 minutes of the time

of synthesis. At the end of a 10 minute labeling period with
tritiated amino acids, the radioactivity is associated chiefly with
the rough endoplasmic reticulum. From autoradiographs there is no
evidence that the labeled protein is ever packaged in a vesicle as
a part of the secretion process (Chen and Jones, 1974a, 1974b),
although it has been reported that $\propto$-amylase and protease are found
in lysosomal-like vesicles in cell homogenates (Gibson and Paleg,
1972; Firn, 1975; Gibson and Paleg, 1975).

The GA_3-dependent synthesis of hydrolases is prevented by RNA
synthesis inhibitors added at the same time or shortly after the
addition of GA_3 (Yomo and Varner, 1971). However, the synthesis
of $\propto$-amylase is no longer susceptible to an RNA synthesis inhibitor added 10-12 hours or later after the addition of GA_3 (Fig.
4). Thus all kinds of RNA, including mRNA, required for $\propto$-amylase
synthesis are stable in vivo for at least 14 hours. Because ABA
added 12 hours after GA_3 inhibits $\propto$-amylase (Fig. 5), it appears
that ABA can control $\propto$-amylase synthesis at the level of translation. Density labeling with ^{13}C-amino acids (Table I) rules out
the possibility that the $\propto$-amylase activity appearing during 12 to
24 hours after GA_3-treatment was synthesized as a prozymogen during
the 0-12 hour period. The control may well be indirect because
the addition of cordycepin immediately relieves the ABA inhibition
of $\propto$-amylase synthesis (Fig. 5). This observation should be of

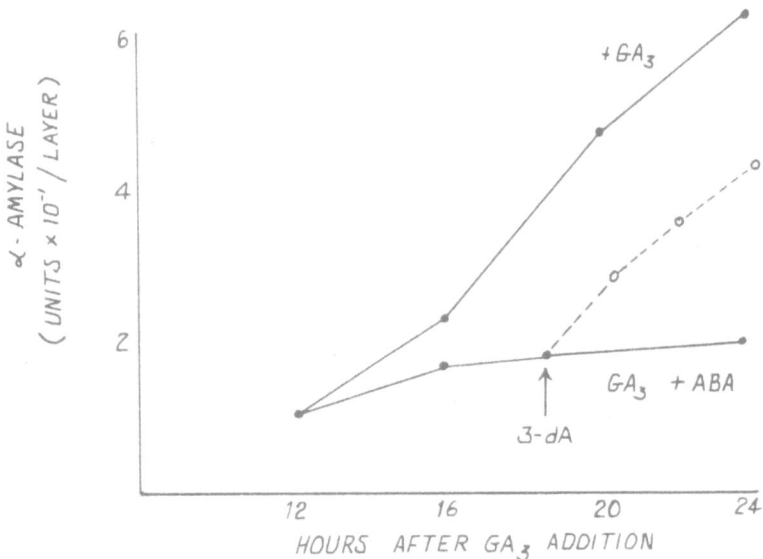

Fig. 5. Effect of midcourse addition of ABA and cordycepin on the
synthesis of α-amylase. (Ho and Varner, 1976)

Table 1. De novo synthesis of α-amylase as demonstrated by ^{13}C-amino acid density labeling.

Treatment		Density (g/ml)		% de novo synthesis
		α-amylase	radioactivity peak	
0 - 24 hr:				
^{12}C		1.300	1.308	
^{13}C		1.317	1.326	
		$\Delta = 0.017$	$\Delta = 0.018$	95%
0 - 12 hr:	12 - 24 hr:			
^{12}C	^{13}C	1.310	1.318	
		$\Delta = 0.010$	$\Delta = 0.010$	100%
^{13}C	^{12}C	1.302	1.311	

The aleurone layers were incubated with either ^{13}C-amino acids
(at a concentration of 10 mM), radioactive leucine, buffer and
GA_3 as indicated. After incubation α-amylase was prepared from
both layers and incubation medium and the buoyant density of
proteins was measured by equilibrium centrifugation in a CsCl
gradient. Radioactivity peak in the gradient represents the distribution of newly synthesized proteins. (Ho and Varner, unpublished data)

great help in sorting out possible sites of action of ABA even
though there is yet no direct evidence that ABA is involved in the
control of α-amylase synthesis in the growing seedling. Is it
possible that ethylene as it relieves the ABA inhibition of α-
amylase synthesis (Jacobsen, 1973) also exerts its control at the
level of α-amylase translation?

During the initial phase of the rapid accumulation of α-amy-
lase and protease, there is a GA_3-dependent increase in the rate of
synthesis of poly (A) RNA (Fig. 6) which can be presumed to be mRNA
because it contains poly (A) and because it has an appropriate
molecular weight. It has been shown, by translation of the mRNA
in a cell-free system, that α-amylase mRNA synthesis parallels the

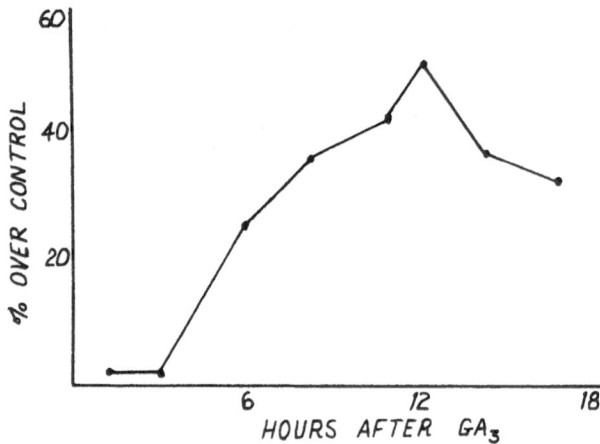

Fig. 6.   Effect of GA₃ on poly(A)-RNA synthesis in barley aleurone
layers.   (Ho and Varner, 1974)

GA$_3$-enhanced synthesis of poly (A) RNA (Fig. 7; Higgins, et al.,
1976).   This direct assay of mRNA should allow a resolution of the
possibility that some or all of the ⋌ -amylase mRNA is present
before the addition of GA$_3$ (Carlson, 1972).

Coincident with the increased rate of synthesis of poly (A)
RNA, there is an increased rate of incorporation of ^{32}P$_i$ into
phospholipids (Koehler and Varner, 1973).   The rate of labeling of
phospholipids is proportional to the concentration of GA$_3$ added
and at a constant concentration of GA$_3$ is inhibited proportionately
to increasing concentrations of ABA.   Therefore the response of
the cells to ABA seems not to lie solely at the level of translation of ⋌ -amylase.

Changes in the kinds of proteins being synthesized by GA$_3$-
treated cells as compared with control aleurone cells are readily
detected within two to four hours after the addition of GA$_3$.   During
midcourse ⋌ -amylase production about 40% of the total protein being
synthesized is ⋌-amylase (Varner, Flint and Mitra, 1975; and D.
F. Flint, unpublished data) and ⋌-amylase constitutes a major
fraction of the total protein secreted or released (Jacobsen and
Knox, 1974; Ho and Varner, 1976).   Aleurone layers of barley contain large amounts of sucrose (30-40 ug/mg fresh weight).   Treatment of the layers with GA3 causes the release of sucrose and other
sugars from the cell starting at two hours after the hormone (Fig.

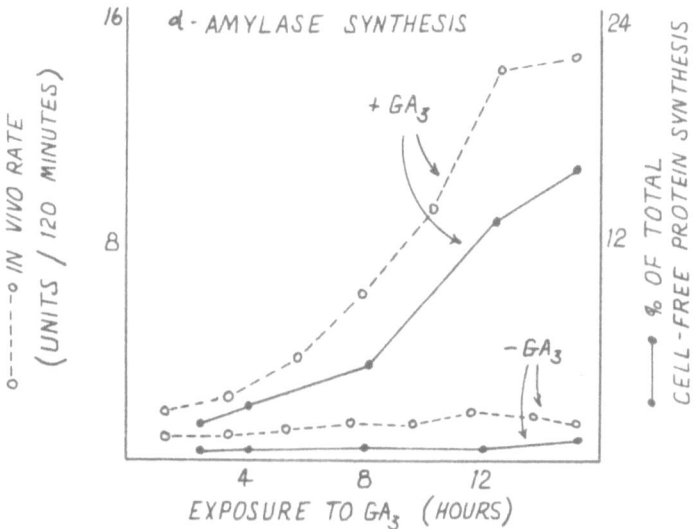

Fig. 7. The time course of GA₃-enhanced level of translatable mRNA for α-amylase and the rate of <u>in vivo</u> α-amylase synthesis in GA₃-treated aleurone layers. (Higgins et al., 1976)

8). Sucrose release probably does not require protein synthesis because it is not inhibited by high concentrations of osmotica such as polyethylene glycol and mannitol. Sucrose released from the aleurone tissue during germination is presumably used by the growing embryo.

The GA₃-dependent increase of phosphorylcholine-cytidyl and phosphorylcholine-glyceride transferases is readily detectable within two hours of GA₃ treatment (Fig. 3). In wheat aleurone tissue the turnover rate of nucleotides, especially CTP is enhanced in 15 to 90 minutes after GA₃ (Collins, et al., 1972). All these events seem to be a prerequisite of the later membrane proliferation because phosphatidylcholine (lecithin) is the major phospholipid in barley aleurone membranes (Koehler and Varner, 1973).

1. Choline + ATP $\xrightarrow{\text{choline kinase}}$ P-choline + ADP

2. P-choline + CTP $\xrightarrow{\text{phosphorylcholine-cytidyl transferase}}$ CDP-choline + PPi

3. CDP-choline + 1,2-diglyceride $\xrightarrow{\text{phosphorylcholine-glyceride transferase}}$ lecithin + CMP

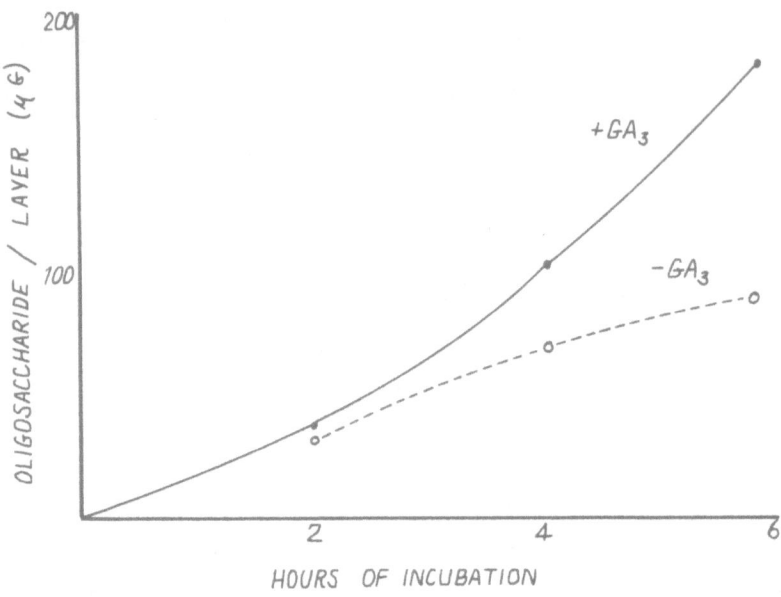

Fig. 8. Time course of the effect of GA₃ on the release of acid-
hydrolyzable oligosaccharide from isolated aleurone layers.
(Chrispeels et al., 1973)

The GA_3 enhanced membrane proliferation does not result in a net
synthesis of phospholipids (Koehler and Varner, 1973; Firn and
Kende, 1974). Because aleurone tissue has much lipid stored in
spherosomes, membrane proliferation is probably the consequence
of a hormone enhanced phospholipid turnover, i.e. from storage
phospholipid to membrane phospholipid.

      Apparently the increase in phosphorylcholine-glyceride trans-
ferase activity is due to some kind of activation because amino
acid analogues, or cordycepin, do not prevent the increase (Ben-Tal
and Varner, 1974). If activation accounts for the early increase
of phosphorylcholine-glyceride transferase activity following GA_3
treatment, one might expect to see some evidence for such activation in vitro by mixing homogenate of hormone-treated tissue with
homogenate of control tissue. Such an effect is observed within
minutes of GA_3 addition (Ben-Tal and Varner, 1974; Ben-Tal, 1974)
(Table II). Phosphorylcholine-glyceride transferase activity is
inhibited by Ca^{++}. Therefore, the activity change of this enzyme could result from changes in concentrations of Ca^{++} or the
concentrations of Ca^{++} complexing agents, such as citrate or

Table II. Evidence for the activation of phosphorylcholine
glyceride transferase in cell-free preparations.

Hormone Treatment of Aleurone Layers during Incubation	Incubation Time	Homogenate No.	Relative Activity	
			Observed	Expected
Control	4 hr	1	50	
GA$_3$	4 hr	2	72	
10^{-5} M ABA	4 hr	3	50	
GA$_3$ and 10^{-5} M ABA	4 hr	4	47	
		1 and 2	158	122
		2 and 3	162	122
		1 and 4	96	97
		3 and 4	102	97
Control	8 hr	1	50	
GA$_3$	1 hr	2	84	
10^{-5} M ABA	8 hr	3	57	
GA$_3$ and 10^{-5} M ABA	1 hr	4	61	
		1 and 2	202	134
		2 and 3	199	141
		1 and 4	122	111
		3 and 4	118	118

Means of 3 different experiments. The expected values for lecithin formed are the sums of the separate activities of cell-free
preparations involved in a given mixture. The amount of lecithin
formed in the control tissue after 4 hr in the first part, and
after 8 hr incubation in the second part was taken as 50% activity.
From Ben-Tal, 1974.

phytate. Since aleurone tissue contains much phytate, a redistri-
bution of phytate or of calcium or an early release of calcium to
the medium due to a change in membrane permeability could account
for the activation of phosphorylcholine-glyceride transferase.
The activity of membrane bound enzymes, such as phosphorylcholine-
glyceride transferase, can be altered by surfactants, such as
lysolecithin which is one of the products of the hydrolysis of
storage phospholipids. The activation of phosphorylcholine-di-
glyceride transferase (as measured in cell-free extracts) begins
within a very short time after hormone treatment; therefore, the
mechanism of this activation process should provide useful clues
about the primary action of the hormone.

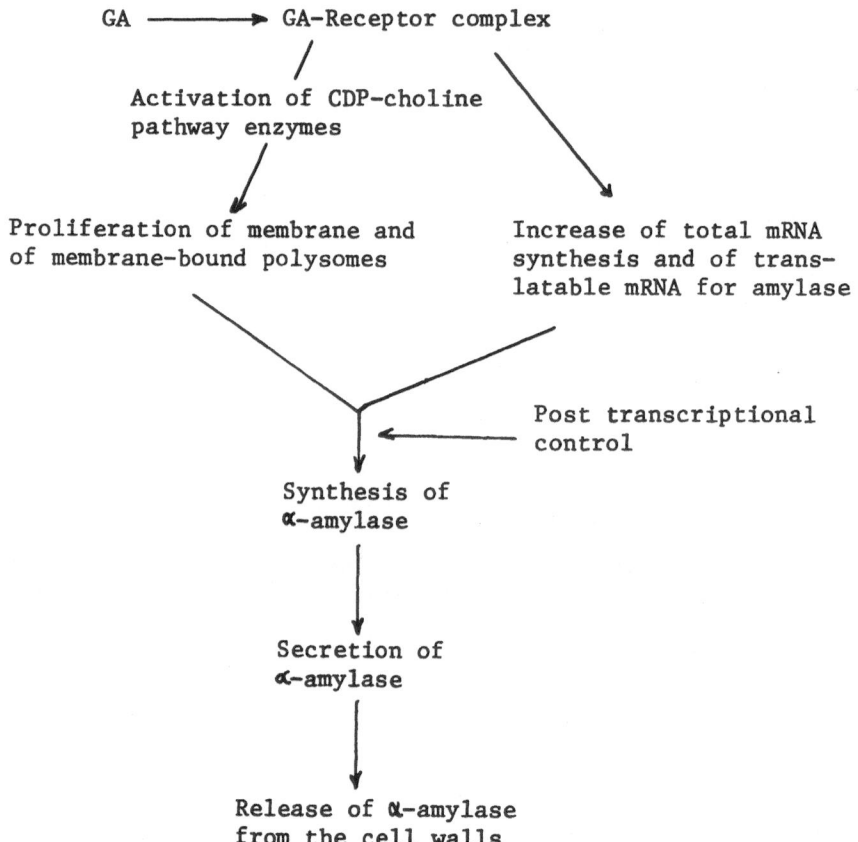

Fig. 9.   Summary of the GA$_3$ effects in barley aleurone layers.

Another early response of aleurone cells is the GA_3-dependent susceptibility to the toxic effects of 1,10 phenanthrolene (Goodwin and Carr, 1970).

Searches have been made for the primary binding site of gibberellins in barley aleurone tissue but no helpful information is yet available. Although aleurone cells do metabolize gibberellins (Musgrave, et al., 1972) and ABA influences this metabolism (Nadeau, 1972), the overall rate of metabolism is insufficient to account for the rapid effects of ABA.

Exogenous cyclic AMP does not mimic the effects of GA_3 in aleurone tissue (Clutterbuck and Briggs, 1974). Careful investigation has ruled out the possibility that cyclic AMP is synthesized in response to GA_3 in barley aleurone layers (Keates, 1973). So far no convincing demonstration of adenyl cyclase has been obtained in plant tissues (see Lin, 1974) and although nucleotide phosphodiesterase is present in various plants, it has much less activity toward 3',5'-cyclic AMP than toward 2',3'-cyclic AMP which appears to be the product of RNA catabolism (Lin and Varner, 1972; Amrhein, 1974). Therefore, the occurrence of cyclic AMP in plant tissue is in doubt and it seems unlikely that cyclic AMP mediates the action of GA_3 in aleurone layers.

In summary (Fig. 9), it appears that the enhanced synthesis of α-amylase depends on the faster rate of formation (transcription and/or processing) of its specific mRNA. The availability of a cell-free assay for the α-amylase specific mRNA (Higgins, et al., 1976) should allow further elucidation of the roles of abscisic acid and of ethylene in the control of α-amylase synthesis.

References

Adams, P. A., Montague, M. J., Tepfer, M., Rayle, D. L., Ikuma, H. and Kaufman, P. B. 1975. Plant Physiol. 56, 757-760.
Amrhein, N. 1974. Planta 118, 241-258.
Ashford, A. E. and Jacobsen, J. V. 1974. Planta 120, 81-105.
Bennett, P. A. and Chrispeels, M. J. 1972. Plant Physiol. 49, 445-447.
Ben-Tal, Y. 1974. Ph.D. Thesis. Michigan State Univ.
Ben-Tal, Y. and Varner, J. E. 1974. Plant Physiol. 54, 813-816.
Borriss, H. 1967. Wiss. Z. Univ. Rostock. Math. Naturwiss. Reihe 16, 629-639.
Briggs, D. E. 1963. J. Inst. Brew. 69, 13-19.
Byrne, H., Christou, N. V., Verma, D. P. S. and Maclachlan, G. A. 1975. J. Biol. Chem. 250, 1012-1018.
Chen, R. and Jones, R. L. 1974a. Planta 119, 193-206.
Chen, R. and Jones, R. L. 1974b. Planta 119, 207-220.

Chrispeels, M. J., Tenner, A. J. and Johnson, K. D. 1973. _Planta_
    113, 35–46.
Clutterbuck, V. J. and Briggs, D. E. 1974. _Phytochem_. 13, 45–54.
Collins, G. G., Jenner, C. F. and Paleg, L. G. 1972. _Plant_
    Physiol. 49, 404–410.
DeLeo, P. and Sacher, J. A. 1970. _Plant Physiol_. 46, 806–811.
Dilworth, M. F. and Kende, H. 1974. _Plant Physiol_. 54, 826–828.
Firn, R. D. 1975. _Planta_ 125, 227–233.
Firn, R. D. and Kende, H. 1974. _Plant Physiol_. 54, 911–915.
Gibson, R. A. and Paleg, L. G. 1972. _Biochem_. J. 128, 367–375.
Gibson, R. A. and Paleg, L. G. 1975. _Aust_. _J_. _Plant Physiol_. 2,
    41–49.
Goodwin, P. B. and Carr, D. J. 1970. _Cytobios_. 7–8, 165–174.
Guilfoyle, T. J., Lin, C. Y., Chen, Y. M., Nagao, and Key, J. L.
    1975. _Proc_. _Nat_. _Acad_. _Sci_. 72, 69–72.
Higgins, T. J. V., Zwar, J. A. and Jacobsen, J. V. 1976. _Nature_
    260, 166–168.
Ho, D. T.-H. and Varner, J. E. 1974. _Proc_. _Nat_. _Acad_. _Sci_. _U.S.A_.
    71, 4783–4786.
Ho, D. T.-H. and Varner, J. E. 1976. _Plant Physiol_. 57, 175–178.
Jacobsen, J. V. 1973. _Plant Physiol_. 51, 198–202.
Jacobsen, J. V. and Knox, R. B. 1974. _Planta_ 115, 193–206.
Johnson, K. D. and Kende, H. 1971. _Proc_. _Nat_. _Acad_. _Sci_. _U.S.A_.
    68, 2674–2677.
Jones, R. L. 1973a. _Plant Physiol_. 52, 303–308.
Jones, R. L. 1973b. _Ann_. _Rev_. _Plant Physiol_. 24, 571–598.
Jones, R. L. and Armstrong, J. E. 1971. _Plant Physiol_. 48, 137–
    142.
Kauffman, P. B. 1965. _Physiol_. _Plant_. 18, 703–724.
Keates, R. A. B. 1973. _Nature_ 224, 355–357.
Koehler, D. E. and Varner, J. E. 1973. _Plant Physiol_. 52, 208–214.
Lin, P. P.-C. 1974. _Adv_. _Cyclic Nucleotide Res_. 4, 439–461.
Lin, P. P.-C. and Varner, J. E. 1972. _Biochim_. _Biophys_. _Acta_ 276,
    454–474.
McNeil, M., Albersheim, P., Taiz, L. and Jones, R. L. 1975. _Plant_
    Physiol. 55, 64–68.
Musgrave, A., Kays, S. E. and Kende, H. 1972. _Planta_ 102, 1–10.
Nadeau, R., Rappaport, L. and Stolp, C. F. 1972. _Planta_ 107, 315–
    324.
Palmer, G. H. 1973. _J_. _Inst_. _Brew_. 79, 513–518.
Sacher, J. A. and Davies, D. D. 1974. _Plant and Cell Physiol_. 15,
    157–161.
Taiz, L. and Honigman, W. A. 1976. _Plant Physiol_. In press.
Teissere, M., Penon, P., Van Huystee, R. B., Azou, Y. and Ricard,
    J. 1975. _Biochim_. _Biophys_. _Acta_ 402, 391–402.
Vanderhoef, L. N. and Stahl, C. A. 1975. _Proc_. _Nat_. _Acad_. _Sci_.
    72, 1822–1825.
Varner, J. E., Flint, D. and Mitra, R. 1975. Workshop on Genetic
    Improvement of Seed Proteins. National Academy of Science/
    National Research Council. In press.

Varner, J. E. and Mense, R. M. 1972. *Plant Physiol*. 49, 187-189.

Verma, D. P. S., Maclachlan, G. A., Byrne, H. and Ewings, D. 1975. *J. Biol. Chem*. 250, 1019-1026.

Yomo, H. and Varner, J. E. 1971. *In* "Current Topics in Developmental Biology," ed. A. A. Moscona and A. Monray. 6, 111-114. New York Academic.

THE CONTROL OF PLANT GROWTH BY PROTEIN KINASES

A. Trewavas and B. R. Stratton

Department of Botany, University of Edinburgh

Mayfield Road, Edinburgh

A substantial majority of animal hormones have their biological effects mediated by the adenyl cyclase system (1). Until Kuo and Greengard (2) produced their unifying theory it was not clear how cyclic AMP initiated the subsequent molecular events associated with hormone action. Their theory (2) proposed that the effects of cyclic AMP were mediated by cyclic AMP-dependent protein kinases in the responsive tissue. Since it is the catalytic function of protein kinases to phosphorylate proteins, this theory places our understanding of hormone action firmly on knowing which cellular proteins are phosphorylated. The list of known phosphorylated protein is very extensive and covers proteins in all of the major cellular groups (3). Evidence that phosphorylation (or dephosphorylation) modifies the biological activity of the phosphorylated protein has however only been critically demonstrated in a few cases (3). Despite the apparent absence of cyclic AMP in plants, this field of research offers such considerable promise for dissecting out the regulatory systems of plants that its study can hardly be avoided.

Our own research interests in this field are to uncover the possible role of protein phosphorylation in regulating plant growth and differentiation. Our initial studies have investigated three areas, ribosomal protein phosphorylation, histone I phosphorylation and nuclear acid protein phosphorylation, and we will deal with each of these in turn.

RIBOSOMAL PROTEIN PHOSPHORYLATION

Sterile cultures of <u>Lemna</u> <u>minor</u> have been incubated in $^{32}P_i$ and the ribosomal proteins have been examined for radioactivity (4). In relatively short labelling periods a radioactive protein was detected which ran as a single component in both urea/acetic acid and sodium lauryl sulphate gel electrophoresis but as a multiple component containing 4-5 derivatives using isoelectric focussing (Trewavas, unpublished data). Other phosphory- lated proteins appear on much longer labelling periods. Acid hydrolysis of this single protein revealed that the phosphate was attached to serine phosphate. After labelling <u>Lemna</u> to equilibrium with $^{32}P_i$, calculations indicated the presence of 0.6 to 0.75 atoms of protein phosphorus/ribosome. The phosphorylated protein is found is both polysomes and derived monosomes and appears to be located in the small ribosomal sub-unit (figure 1).

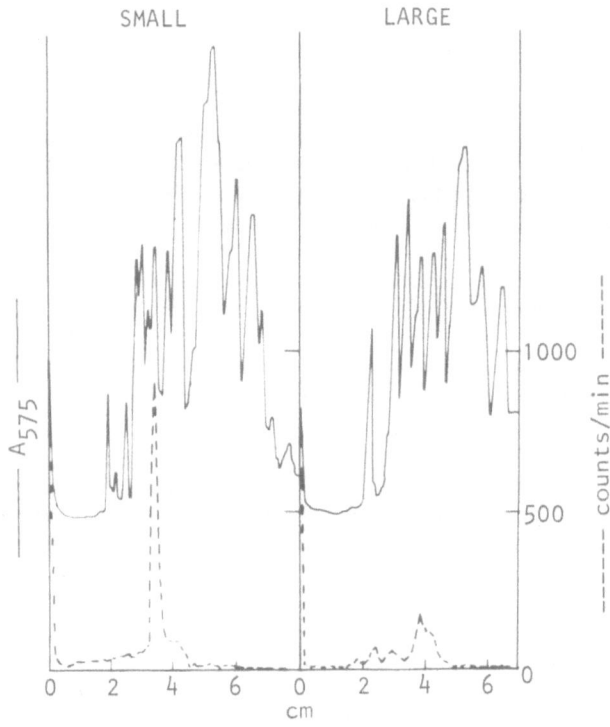

Figure 1. Acid-urea gel electrophoresis of ^{32}P labelled proteins from the large and small sub-units of ribosomes of <u>Lemna</u>.

The apparent molecular weight of this protein is about 42,000. Addition of growth inhibiting concentrations of abscisic acid did not alter the apparent degree of labelling of this protein in five hours but after 24 hours the total protein phosphorus was reduced from 0.75 atoms/ribosome to around 0.36 atoms. There is considerable similarity between ribosomal protein phosphorylation in Lemna and that in rat liver which has already been commented upon (5). The single liver ribosomal protein (S6) which is phosphorylated has multiple derivatives and a molecular weight of about 36-38,000. This protein however is basic in contrast to that in Lemna which is probably acidic.

Protein kinase has been found associated with ribosomes from pea stem and Lemna (6). This enzyme catalyses the phosphorylation of ribosomal proteins 'in vitro' and may be involved in the formation of phosphoproteins 'in vivo'. Protein kinase sediments with ribosomes during sucrose density gradient centrifugation through buffers containing 0.02 M KCl. About two-thirds of the enzymatic activity dissociates from the ribosomes in 0.3 M KCl but the remaining protein kinase is firmly bound even in 0.7 M KCl. The solubilised protein kinase activity separates as a single enzymatic specie on DEAE cellulose columns (figure 2).

Cyclic AMP does not modify the activity of these protein kinases 'in vitro' and only small effects of benzyladenine can be detected out of several growth substances tested. Treatment of Lemna with nitrogen or abscisic acid (to dissociated polysomes) leads to only slight losses of ribosome associated protein kinase (6).

The function of ribosomal protein phosphorylation is still very much a mystery. There is abundant evidence in animals which indicates little or no direct relationship between protein synthesis and ribosomal protein phosphorylation (3) and our own, whilst on a more limited basis, would tend to confirm this. Gressner and Wool (5) have suggested that the level of ribosomal protein phosphorylation is related to the rate of ribosome synthesis. Our data shows that abscisic acid reduces the level of ribosomal protein phosphate (4) and it also reduces ribosome synthesis (12) in tentative agreement with this hypothesis.

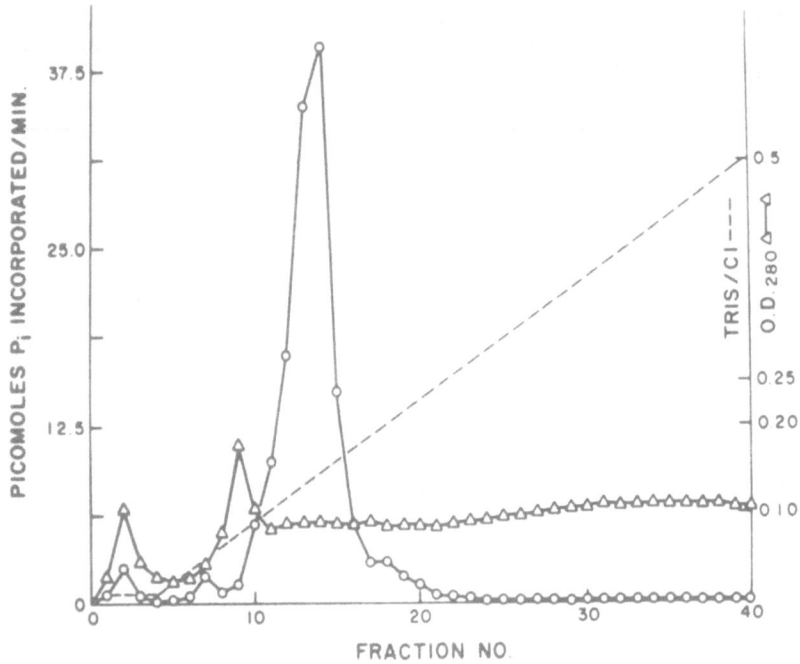

Figure 2.   DEAE cellulose chromatography of the protein
kinase solubilised from pea ribosomes with 0.3 M KCl.
Details in reference 6.

THE PHOSPORYLATION OF HISTONE I IN ARTICHOKE

The experiments of Bradbury and coworkers (7) have
demonstrated a critical relationship betseen chromosome
condensation and histone I phosphorylation in Physarum.
Following the original observations of Lake (8), protein
kinase with high specificity for histone I was found to
increase some 20-40, fold in late G2 of the cell cycle
and then to decrease just prior to M. The phosphorylation of histone I was found to follow a comparable although slightly delayed time course. Microscopic observation indicated a direct association between chromatin
condensation and the level of histone I phosphorylation.
Addition of exogenous histone I kinase can actually alter
the time of mitosis in Physarum implicating the production of histone I kinase as one of the timing steps in
the division cycle.

In our own studies we have confined ourselves to demonstrating that histone I is phosphorylated in at least one higher plant and that the kinetics of histone I phosphorylation in the cell cycle is in general agreement with the pattern of events in Physarum. For these studies we have used the partially synchronous artichoke explant system developed at Edinburgh (9). Dormant tissue of artichoke was induced to divide by removing sterile explants into auxin and mineral salts medium. After labelling with $32P_i$, the histones were extracted and separated by gel electrophoresis (20). Radioactivity was found to coincide with the histone I band and this radioactivity was phosphatase labile. Labelled histone I was eluted from gels, acid hydrolysed and the products separated by paper electrophoresis (4). Radioactivity in the approximate proportions 6:2:1 was located in P_i, serine phosphate and threonine phosphate. The detection of threonine phosphate in histone I in the artichoke is in agreement with data from animal cells which shows that phosphorylation of threonine and serine residues of histone I takes place in dividing cells. In non-dividing cells phosphorylation only takes place on serine residues (10). The large amount of inorganic phosphate was unexpected and whilst approximately half can be accounted for as resulting from the degradation of serine and threonine phosphates during acid hydrolysis, the remainder suggests that other amino acid phosphate linkages cannot be excluded (11).

To try and demonstrate increased phosphorylation of histone I during the G2/M period of the cell cycle, initial attempts were made using the enzymatic method described by Chalkley and associates (13). This method which uses phosphatase treatment of histone I was found to suffer from severe difficulties and we have had to resort to labelling studies with $32P_i$. Detection of enhanced phosphorylation by the incorporation of $32P_i$ also suffers two drawbacks. Even in non-dividing cells a certain level (anywhere from 1-10%) of histone I is phosphorylated. Histone synthesis is confined to the S phase of the cell cycle. Maintenance of a resting level of phosphorylation must ensure incorporation of $32P_i$ into histone I at this time thus obscuring a relatively specific change in G2/M. Secondly, the artichoke cells used for these experiments are only partially synchronous and whilst cultures can, for example, be obtained rich in cells in G2/M, these will still contain cells which are in S and so incorporating phosphate into newly-synthesised histone. To overcome

TABLE 1

Variation of Histone I phosphorylation through the cell cycle of artichoke

Cell cycle phase	Early G1	Late G1	S	G2/M
^{32}P attached to histone I*	1,224	961	6,513	5,112
^3H lysine incorporated into histone I*	29,520	88,010	110,120	30,140
Ratio $^{32}P/^3H$	0.04	0.011	0.059	0.17
^3H lysine incorporated into acidic proteins	16,270	28,980	64,320	56,200

*units = counts/min./unit area of histone I on gel adjusted for isotope uptake

Sterile explants of artichoke were prepared and incubated in auxin mineral salts medium. The phases of the first cell cycle were determined using thymidine labelling of DNA and cell counts. Labelling was conducted for four hours with ^{32}Pi and ^3H lysine, the chromatin was prepared and histones extracted and separated by gel electrophoresis (20). After staining and scanning the gel was sliced, the slices dissolved and counted. Acidic proteins were measured as the acid-insoluble residue after solution of the histones.

these difficulties, explants have been doubly-labelled
with ^{3}H lysine and ^{32}P$_i$. The incorporation of ^{3}H lysine
can be used as an assessment of histone synthesis. If
a comparable variation in histone I phosphorylation
occurs in artichoke as it does in Physarum we would predict that the ^{32}P/^{3}H ratio of histone I will be higher
in the G2/M period than in S. The results of such an
experiment are shown in table 1 and are in agreement
with the prediction. The ^{32}P/^{3}H ratio is threefold
higher in G2/M cells compared to S and earlier phases of
the cell cycle.

These results indicate then that histone I phosphorylation may be associated with the process of chromatin condensation in artichoke as it is in Physarum.
As a possible control point for cell division in plants,
histone I kinase offers exciting possibilities. In two
systems known to us, dormant buds of chickpea (14) and
pericycle cells in radish (15), the cells are resting
mainly in G2. Addition of cytokinin in the first case
and auxins in the second initiates division within 1-2
hours suggesting that such cells are resting in late
G2. The possibility that growth substances may control
histone I kinase levels is one that can now be seriously
entertained.

CHROMOSOMAL ACIDIC PROTEIN PHOSPHORYLATION

Sterile embryos of barley and cultures of Lemna
have been labelled with ^{32}P$_i$ and the chromatin proteins
prepared and separated by acid-urea and SLS gel electrophoresis (16, 17). Under these condtions chromatin proteins become labelled with 80-90% of the label in the
acidic protein fraction. Most of this label is recoverable after acid hydrolysis as serine phosphate. By
labelling Lemna to isotopic equilibrium with ^{32}P$_i$ it
was found that the ratio of DNA phosphate to chromatin
protein phosphate was about 100, a figure comparable to
that found in animal systems. Nuclei have been isolated
from Lemna and barley embryos and found to possess endogenous kinase activity. In vitro labelling can be accomplished using γ^{32}P-ATP but not with α^{32}P-ATP.

As judged by 'in vivo' labelling the phosphorylation profile of Lemna chromatin proteins is altered by
abscisic acid and this is shown in figure 3.

Although substantial changes can be seen in at

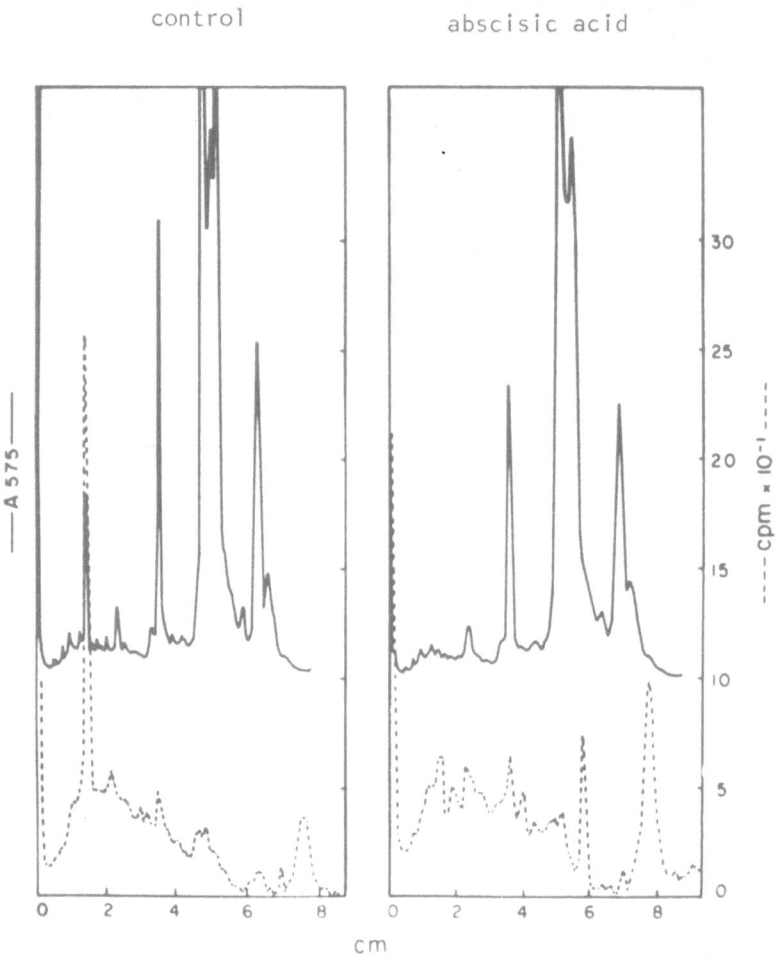

Figure 3. Protein profiles and phosphorylation patterns
of <u>Lemna</u> chromatin proteins labelled in vivo with $^{32}P_i$
in the presence or absence of abscisic acid. <u>Lemna</u> was
labelled for 8 days and abscisic acid added for the last
24 hours. The chromatin proteins were prepared and separated by gel electrophoresis (20). After staining and
scanning the gels were sliced and counted.

least three components the complexity of this one dimensional profile probably obscures the extent and significance of the alterations. To investigate protein
phosphorylation changes more fully we have taken advantage of the recent advent of two dimensional systems.

These methods require however more protein for separation and thus we have been limited in the first instance to plant systems which provide suitable yields of
nuclei. Of these probably the most useful is the germinating embryo. Nuclei can be obtained from barley
embryos in high yield and can be purified by isopycnic
banding on Ludox. Furthermore we had already obtained
preliminary data, using one-dimensional electrophoresis,
which indicated that the chromosomal protein phosphate
pattern was altered during the first 24 hours germination. Our first 2D work indicated the considerable
promise of this method since it revealed not only substantial qualitative changes in phosphorylated nuclear
proteins but also in the unphosphorylated ones as well
(19). This work did suffer from two drawbacks however.
Firstly, proteins were labelled by incubating nuclei in
^{32}P-ATP. As will be discussed later, this may not show
direct correspondence to 'in vivo' conditions. Secondly,
we used 3.5 mm thick slabs for the second SLS dimension
of the separation. We were unable to dry these thick
slabs down and autoradiographs could only be prepared
by laying the wet slabs in a polythene bag next to the
X-ray films. This results in a considerable lowering of
the sensitivity and discrimination of autoradiography.
We have now greatly improved these original data by
labelling embryos 'in vivo' with ^{32}P$_i$ and by adopting
the O'Farrell 2D method with its much greater resolution and sensitivity (18).

     Sterile barley embryos have been germinated for 2,
5, 9, 13 and 18 hours and labelling for the last 2-4
hour period with ^{32}P$_i$ or ^{35}S-methionine. The nuclear
proteins have been extracted, separated two-dimensionally
and autoradiographs prepared of the dried gels. These
experiments have enabled us to make the following preliminary conclusions.

     About 60 phosphorylated proteins can be extracted
from barley nuclei and of these just over 40 are invariant through the first 18 hours germination. About
12 phosphorylated proteins disappear during the 5-9
hour period of germination and several new ones appear
at this time. Of the dozen which disappear at least
nine were found to incorporate methionine during the
first five hours germination. Their synthesis then may
be switched off and this may represent an active rather
than a passive loss from the nucleus. The synthesis of
DNA commences in the embryo during this 5-9 hour period
and it is tempting to relate the changes in phosphorylated

proteins to the onset of this important stage of the
cell cycle.

^{35}S   Nearly 200 nuclear proteins can be detected using
^{35}S labelled methionine and 2D gel analysis. The vast
majority of these remain invariant throughout the first
18 hours germination but there are qualitative changes
in about 40 of those during the first nine hours of germination. These presumbaly reflect the changing patterns
of protein synthesis occurring in the cytoplasm. Germination is therefore accompanied by changes in phosphorylated
and unphosphorylated nuclear proteins much of which is
the result of alterations of the synthesis of nuclear
proteins.

We have also been able to compare the types of proteins which are labelled in vivo with ^{32}P$_i$ or in vitro
with ^{32}P-ATP. It is clear that our in vitro labelling
grossly underestimated the complexity of protein phosphorylation in the plant nucleus. This is as already
described almost certainly the result of using a much
less sensitive method for autoradiography. We have
also noticed that whilst many proteins are certainly
labelled in common between the two methods there are
notable exceptions of proteins which become labelled
'in vitro' which do not become labelled 'in vivo'. As
with ribosomal proteins therefore (6) data obtained by
labelling chromatin proteins 'in vitro' must be treated
with caution.

ACKNOWLEDGEMENTS

Part of this work was carried out whilst one of us
(B.S.) was in receipt of a studentship from the S.R.C.
Other unpublished research was supported by a grant
from the S.R.C. to A.T.

REFERENCES

1. Jost, J. P. and Rickenberg, H. V. 1971.
      Ann. Rev. Biochem. 40: 741.

2. Kuo, J. F. and Greengard, P. 1969.
      Proc. Nat. Acad. Sci. U.S. 64: 1349.

3. Trewavas, A. J. 1976.
      Ann. Rev. Plant Physiol. (in press)

4. Trewavas, A. J. 1973.
    Plant Physiol. 51: 760.

5. Gressner, A. M. and Wool, I. G. 1974.
    J. Biol. Chem. 249: 6917.

6. Keates, R. A. B. and Trewavas, A. S. 1974.
    Plant Physiol. 54: 95.

7. Bradbury, E. M., Inglis, R. J., Matthews, H. R.
   and Langan, T. A. 1974.
    Nature 249: 553.

8. Lake, R. G. 1973.
    Nature 242: 145.

9. Yeoman, M. M. and Evans, P. K. 1967.
    Ann. Bot. 31: 323.

10. Langan, T. A. and Hohmann, P. 1974.
    Fed. Proc. 33: 1597.

11. Smith, D. L., Bruegger, B. B., Halpern, R. M. and
    Smith, R. A. 1973.
    Nature 246: 103.

12. Trewavas, A. J. 1970.
    Plant Physiol. 45: 742.

13. Balhorn, R., Chalkley, R. and Granner, D. 1972.
    Biochemistry 11: 1094.

14. Usciati, M., Coddaccioni, M. and Guern, J. 1972.
    J. Exptl. Bot. 23: 1009.

15. Blakeley, L. Personal communication.

16. Van Loon, L. C., Trewavas, A. and Chapman, K. S. R.
    1975.
    Plant Physiol. 55: 288.

17. Chapman, K. S. R., Trewavas, A. and Van Loon, L. C.
    1975.
    Plant Physiol. 55: 293.

18. O'Farrell, P. 1975.
    J. Biol. Chem. 250: 4007.

19. Trewavas, A. J. 1976.
    Phytochem. 15: 363.

20. Panyim, S. and Chackley, R. 1969.
    Arch. Biochem. Biophys. 130: 337.

GENERAL PROPERTIES OF ISOLATED PROTOPLASTS AND UPTAKE OF FOREIGN GENETIC MATERIAL

E. C. Cocking

Department of Botany, University of Nottingham

University Park, Nottingham NG7 2RD, England

GENERAL PROPERTIES

The general properties of isolated protoplasts have been described in several comprehensive reviews (Cocking, 1974a; Cocking 1974b) and it will not be profitable to repeat all these details here.

Plant protoplasts are plant cells from which the cell wall has been removed by suitable enzymatic degradation, leaving the haemostatic unit of the plant cell intact. Because of the absence of the cell wall, uptake of foreign genetic material is greatly facilitated (Cocking, 1976). Fusion is also readily possible beween isolated protoplasts, thus enabling the ready mixing together of the genomes of different species. Not only is inter-plant species fusion possible, but also inter-fungal (Kevei and Peberdy 1976), plant-fungal (Davey and Power, 1975) and plant-animal (Ahkong et al, 1975). A significant feature of the development of isolated plant protoplasts is that, when suitable cultural conditions have been established, they regenerate a new cell wall, and divide to produce small aggregates. Such cell aggregates are sometimes capable of developing into an organised callus mass, producing roots and shoots and subsequently whole flowering plants. Progress in this field of investigation is therefore dependent on the extent to which the genetical consequences of uptake of genetic material into protoplasts, and protoplast fusion, are stabilised and expressed in developing cells and in regenerated plants.

Protoplasts can be isolated from leaf mesophyll tissue, epidermal tissues, petals, roots and root nodules, germinating pollen grains, tetrads, fruits and <u>in vitro</u> cultured plant tissues. Leaf material is the preferred source of protoplasts, since after removal of the lower epidermis by peeling or an enzyme treatment, large numbers of cells are accessible for conversion into protoplasts. This consideration also applies to cultured plant tissues, particularly cell suspensions, but protoplast release is often critically dependent upon the stage in the growth cycle of such tissues. Yeilds are variable, but from leaf tissue an average of 5×10^5 protoplasts per gram of fresh weight can be expected.

Although the isolation of protoplasts presents no major problems, successful culture is still restricted to a relatively few species, and even to certain varieties of a given species.

Protoplasts of most species regenerate a new cell wall after 2-3 days, and following dedifferentiation, those of a few species enter division. Division,if maintained, results in the formation of cell colonies after 3 weeks. At this point in the regeneration process the concentration of the plasmolyticum is progressively reduced so that after 8-12 weeks callus can be transferred to media lacking added plasmolytica. The growth of protoplast-derived callus and its handling for plant regeneration may parallel that already established for callus obtained directly from the plant. The time course of events is dependent largely on the plant species in question, but flowering plants can be reproduced from individual protoplasts after 4-6 months.

Any discussion of the uptake of foreign genetic material by isolated protoplasts will require a survey of the situation with particular reference to the uptake of DNA, the uptake of organelles and micro-organisms and uptake as a consequence of protoplast fusion. Whilst it may be fairly readily possible to be clear about the situation regarding uptake, the extent of integration and subsequent expression of this foreign genetic material is far from clear. Recently I have reviewed these considerations comprehensively (Cocking, 1976) and I will only attempt here to summarise the general situation from this review to which the reader is referred for fuller details.

UPTAKE OF DNA AND VIRUSES

Lurquin and Hotta (1975) have emphasised that this area of investigation remains a controversial one. Some studies have provided evidence for covalent bonding between the absorbed and the host DNA, whilst other studies have challenged the reality of the published facts. There are many technical difficulties one of which is well demonstrated by the recent evidence (Kleinhofs <u>et al</u> 1975) that bacterial contamination was responsible for early claims

of the integration of isolated DNA into plant chromosomes. It
could be stated that it is now more or less accepted that uptake of
foreign DNA into plant cells does take place. It is the subsequent
fate of these informative molecules which still remains a matter
of controversy.

One of the early attractions of employing isolated protoplasts
was the suggestion that less degradation of high molecular weight
DNA would occur, since the DNA could be presented directly to the
plasma membrane. Ohyama et al. (1972) studied the uptake of
E. coli ^{14}C-DNA by isolated Ammi visnaga protoplasts. A fundamental
difficulty in the interpretation of these results, and also of
those working with Petunia protoplasts,(Hoffmann and Hess, 1973;
Hoffmann, 1973) was that although up to 20% of the foreign DNA
taken up was recoverable in the acid-precipitable fraction, it was
not possible to be certain that DNA degradation products were not
being used for de novo DNA synthesis. Indeed, the persistence of
heterologous DNA intracellularly, without degradation, is in itself
unexpected (Lurquin and Hotta, 1975). These studies using isolated
protoplasts did not help to resolve the problem of the need to
distinguish between uptake and integration of high molecular weight
DNA, and its degradation and reuse.

In the early studies involving the use of isolated protoplasts
Holl (1973) compared DNAse activity in tissues, and in protoplasts,
because it seemed that the DNAse activity of the host material
might play a role in determining the ultimate fate of the fed DNA.
Holl observed that DNAse activity was inhibited by DEAE-dextran but
not by poly-L-ornithine. Ryser (1967) has discussed the importance
of interaction at the plasma membrane in the uptake of DNA by
animal cells, and it may well be that comparable interaction
phenomena are occurring with plant protoplasts. Any major
perturbation of the plasma membrane may enable high molecular weight
nucleic acids to penetrate. Sarker et al (1974) noted that alkaline conditions greatly facilitated the infection of isolated
protoplasts by TMV RNA and suggested that comparable alkaline
conditions might facilitate uptake of DNA by isolated protoplasts.
The situation is very complex however. Although, as we have noted,
poly-L-ornithine, which often greatly enhances virus infection of
isolated protoplasts (Takebe, 1975) does not inhibit DNAse activity,
it does apparently stimulate uptake of DNA into plant protoplasts
(Holl, 1973). Two laboratories have claimed successful DNA-
mediated correction of genetic deficiences in plants (Ledoux et al.,
1974; Hess, 1972). It would, however, seem likely that the results
reported by Hess (1972) relating to DNA mediated change of white
flowering seedlings to red flowering seedlings can probably be explained in other ways, because he did not take the existing
differentiation of the shoot apex in his seedling material into
account (for a critical detailed discussion see Bianchi and
Walet-Foederer, 1974). Whether alternative explanations are also

possible in the case of the DNA-mediated genetic correction of thiamineless <u>Arabidopsis thaliana</u> it is not, as yet, clear.

Even in bacteria, transformation is a rare phenomenon (Cocking, 1973) and the mechanism of uptake of the DNA bringing about such transformation is far from being fully understood. In the case of plant protoplasts, our knowledge of the mechanism of uptake of high molecular weight DNA is almost negligible. There is some suggestion that the use of isolated protoplasts may minimise the opportunity for the degradation of high molecular weight naked DNA before it is taken up into the cellular environment. There is apparently no advantage in the use of protoplasts to offset the problem of degradation of DNA by nucleases within the cell.

Isolated protoplasts have not been used, and would not offer any major experimental advantage, in experiments aimed at substantiating the earlier observations of Ledoux of 'hybrid' density DNA peaks. The peaks appear following treatment of eukaryotic cells (e.g. Barley and <u>Arabidopsis</u> germinating seeds, (Ledoux 1972) with heteropycnic bacterial DNA. Kleinhofs et al., (1975) has noted that bacterial contamination can produce such 'hybrid' density peaks. It should also be noted that very significant amounts of foreign DNA may become integrated with the endogenous DNA without resulting in detectable density shifts, outside the limits of error inherent in preparative caesium chloride gradients (Lurquin and Behki, 1975).

Knowledge of the interaction between isolated protoplasts and plant viruses has greatly improved our understanding of the basic process of virus infection of plant cells (Cocking, 1970; Takebe, 1975). This information is highly relevant to any studies which might be planned in which bacteriophages are employed to facilitate the uptake of genetic material by isolated protoplasts. The phage protein coat will provide protection for the DNA from nuclease digestion.

There have been no detailed studies of the interaction between phages and isolated protoplasts. Most workers have assumed that the cell wall does not act as a barrier to the penetration of phages such as λ phage. Comparisons of uptake of phage into isolated protoplasts and into cultured cells would be particularly useful in this respect. Sander (1976) has recently developed the isolated protoplast system for the multiplication of phage, and this system now looks promising, particularly if greater multiplication of the phage can be obtained. Until this work of Sander's there had been only one preliminary report (without adequate experimental details) of the effect of treatment of isolated barley leaf protoplasts with T3 phage (Carlson,1973). Isolated protoplasts will undoubtedly serve in future years as a very useful cell system for the study of the molecular biology of bacteriophage-plant cell interactions.

UPTAKE OF ORGANELLES AND MICRO-ORGANISMS

Studies on uptake of viruses into isolated protoplasts (Cocking, 1970) was later extended to a more detailed study of the uptake of small (approximatelu 0.1μ) polystyrene latex particles. This work showed, quite unambiguously, that such particles are taken up by an endocytotic process. Freeze-etching enabled the process to be clearly visualised in the electron microscope, and provided important data from face view membrane fractures (Willison et al., 1971). Detailed freeze-etch work showed that uptake was initiated by adhesion of the latex sphere to the plasma membrane, resulting in a depression which extended solely over the area of contact of the sphere. The membrane bound particle continued to travel towards the cytoplasm and eventually came to lie in a plasma membrane invagination closely surrounded by the membrane. It was clear, however, that negligible endocytotic uptake took place with particles greater that approximately 0.5μ, and that special stress conditions were necessary for uptake with larger particles. This led Davey et al., (1973) to devise a method of plasmolytic uptake of micro-organisms such as Rhizobia. Potrykus (1973) and Potrykus and Hoffmann (1973) obtained good light microscopic evidence for the uptake of both chloroplasts and nuclei by isolated protoplasts of Petunia, N. glauca and Zea mays. Unfortunately, no electron microscope studies were carried out, nor was it possible for pro-toplasts to survive for more than 18 hr. Davey and Power (1975) showed that protoplasts isolated from suspension-cultured cells of P. tricuspidata took up yeast cells, yeast protoplasts, and blue-green algae cells when treated with polyethylene glycol.

The basic concept of the modification of individual plant cells as a result of uptake of organelles and organisms with the horizon of regeneration of whole "modified" plants is particularly attractive, combining as it does the somatic with the sexual genetics of higher plants. This concept has already been discussed by Cocking (1973), and even at this earlier date, it was emphasised that only a systematic study of each stage in this multi-stage process would enable a basic foundation to be laid. It would seem likely that the survival of transplanted organelles, or micro-organisms, will be dependent on a very fine balance between their division and the division of the recipient plant cell. Clearly then, the greater the extent of the integration into the actual cytoplasm the better. This would allow the regulating interaction between nucleus and cytoplasm to proceed smoothly. Indeed, if such interaction is required between nuclei and chloroplasts, then actual fusion of the cell types involved may be biologically far more meaningful. As we shall see in our further discussion of

genetic uptake, transplantation is readily achieved as a consequence
of the actual fusion of protoplasts or of subprotoplasts.

UPTAKE AS A CONSEQUENCE OF PROTOPLAST FUSION

Uptake of genetic material by higher plant somatic cells, as
a result of fusion of one somatic cell with another, could be regarded as an extrapolation of our earlier discussions of interaction
between isolated protoplasts and genetic material presented as nuclei
acid itself, or as some form of virus, or as an organelle. Indeed,
such uptake, as a result of fusion, is probably the most realistic
way of achieving genetic modification of somatic plant cells.

The somatic hybridisation of plant species can be divided
into four main stages:- The isolation of protoplasts; fusion
of protoplasts; selective culture of fusion products; and their
regeneration into plants.

In order to recover these somatic hybrids it is necessary to
select a few colonies from amongst many thousands or even millions
derived from the parental protoplasts. It is for this reason that
a great deal of effort is now being directed towards the development
of stringent selection systems for the recovery of somatic hybrid
plants following the fusion of protoplasts of different species.
(Power et al., 1976).

REFERENCES

Ahkong, Q.F., Howell, J.L. Lucy, J.A., Safwat, F., Davey, M.R. and
    Cocking, E.C. (1975). Nature, 225, 66.
Bianchi, F. and Walet-Foederer, H.G. (1974). Acta Bot. Nerl.
Carlson, P.S. (1973). Proc. Nat. Acad. Sci. U.S.A. 70, 598.
Cocking, E.C. (1970). Inter.Rev. Cytol. 28, 89
Cocking, E.C. (1973). In "Protoplastes et fusion de cellules
    somatique vegetales" (J. Tempé, ed.) 73-1, 327. Inst. Nat. Rech.
    Agron., Versailles.
Cocking, E.C. In Genetic Manipulations with Plant Material (L. Ledoux
    ed.) Plenum Press, New York (1974a)
Cocking, E.C. In Scienza and Technica 74, 199, (Arnoldo Mondadori),
    Milano (1974b).
Cocking, E.C. (1976). International Review of Cytology,
    (In the Press)
Davey, M.R., Cocking, E.C., and Bush, E. (1973). Nature (London).
    244. 460.
Davey, M.R. and Power, J.B. (1975). Plant Science Letters 5. 269
Hess, D. (1972) Z. Pflanzenphysiol. 66. 155
Hess, D. (1973) Zeit, für Pflanzenphysiologie 68, 432.

Hoffmann, F. (1973). Z. Pflanzenphysiol. 69. 249.

Hoffmann, F. and Hess, D. (1973). Z. Pflanzenphysiol. 69, 81.

Holl, F.B. (1973). In "Protoplastes et fusion de cellules somatiques vegetales". (J. Tempé, ed.) 73-1 509. Inst. Nat. Rech. Agron., Versailles.

Kevei,F. and Peberdy, J.F. (1976). Submitted for publication.

Kleinhofs, A., Eden F.C., Chilton, M.D.and Bendich, A.J. (1975). Proc. Nat. Acad. Sci. U.S.A. 72, 2748.

Ledoux, L., Huart, R. and Jacobs, M. (1974). Nature (London). 249, 17.

Lurquin, P.F. and Behki, R.M. (1975). Mutation Research 29. 35.

Lurquin, P.F. and Hotta, Y. (1975). Plant Science Letters 5, 103.

Ohyama, K., Gamborg, O.L. and Miller, R.A. (1972). Canadian J. of Botany 50 2077.

Potrykus, I. (1973). Zeit. für Pflanzenphysiologie 70. 364.

Potrykus, I. and Hoffmann, F. (1973). Zeit fur Pflanzenphysiologie 69. 287.

Power, J.B., Frearson, E.M., Hayward, C., George, D., Evans, P.K., Berry, S.F. and Cocking E.C. (1976). Nature (in the press).

Ryser, H.J.P. (1967). J. Cell Biol. 32, 737.

Sander, E. (1976). Personal Communication.

Sarker, E., Upadhya, J.D. and Melchers, G. (1974). Molecular and Gen. Genetics 135, 1.

Takebe, I. (1975). Ann. Rev. Phytopathology 13, 105.

Willison, J.H.M., Grout, B.W.W. and Cocking, E.C. (1971). Bioenergetics 2, 371.

AGROBACTERIUM TUMEFACIENS : WHAT SEGMENT OF THE PLASMID

IS RESPONSIBLE FOR THE INDUCTION OF CROWN GALL TUMORS ?

By J.Schell, M.Van Montagu, A.De Picker, D.De
Waele, G.Engler, C.Genetello, J.P.Hernalsteens,
M.Holsters, E.Messens, B.Silva, S.Van den Els-
acker, N. Van Larebeke and I.Zaenen

Laboratory of Genetics, and
Laboratory of Histology and Genetics

University of Ghent, B-9000 Ghent, Belgium

I. CROWN GALL DISEASE

Several causes of plant tumors have been described
(1). Particularly intriguing is the case of the crown
gall tumors. The agent inducing this disease is a Gram-
negative bacterium, *Agrobacterium tumefaciens* belonging
to the family of the *Rhizobiaceae*. Until now it is the
only documented case where an oncogenic agent is of
bacterial origin.

Agronomists and botanists have studied the conditions of crown gall formation and the development of the
tumors in different plant families. They have been
joined by microbiologists for a systematic study of the
bacteria isolated from the galls and for obtaining non-
oncogenic derivatives of A.*tumefaciens*. An account of
this earlier work in crown gall research can be found in
several readily accessible review articles (2, 3).

Molecular biologists may be stimulated to study
crown gall as a model system in oncogenesis, because the
system allows the use of genetic techniques that are far
better developed for bacteria than for eukaryotic cells
and their viruses. Such a genetic analysis of the tumor
inducing organisms will be essential for the identification of the genes responsible for the neoplastic trans-

formation and for understanding the mechanism of action of these genes.

The recent progress in *Agrobacterium* work, which will be discussed here, indicates that an interest in *Agrobacterium* is validated not only because of its oncogenic properties. Indeed there are hints that A.*tumefaciens* possesses genes which can be transferred and expressed in plant cells, and techniques are becoming available to isolate these genes. If this is confirmed the *Agrobacterium* plasmid can become a test DNA for studying eukaryotic gene expression. Furthermore if it is correct that a segment of the plasmid becomes integrated into the plant cell DNA, one can postulate that it must be possible to use this segment as vector for integrating other prokaryotic or eukaryotic DNA into plant cells. These various aspects of crown gall research may encourage "classical" molecular biologists to initiate research in plant molecular biology.

II. THE EXTRA-CHROMOSOMAL DNA OF AGROBACTERIUM

The properties of a large variety of *Agrobacterium* strains, isolated in nature, have been compared (4, 5). It is striking that some strains have very similar characteristics except that one is oncogenic, the other not. Nevertheless, from the taxonomist point of view the former were classified as A.*tumefaciens*, the latter as A. *radiobacter*. As it was not unlikely that this "one property" difference could be coded by extra-chromosomal DNA, we investigated the possible presence of episomes in *Agrobacterium*.

Many strains were found to be lysogenic for an inducible prophage (6) but until now there is no indication that part of a phage genome participates in the process of oncogenesis. A major advance was the finding that all oncogenic *Agrobacterium* strains contained at least one large plasmid (7). The A.*radiobacter* strains, which by "definition" are non-oncogenic, did not contain plasmids. Curing and reintroduction of the plasmid into the *tumefaciens* strains proved that this plasmid is essential for tumor induction (8).

However, upon investigation of several more *radiobacter* strains, we found that some strains contained a large plasmid (Table I). Several properties suggested that these plasmids should be considered as mutant Ti-plasmids. We also analysed A.*rhizogenes* strains, which induce root proliferation and A.*species* bacteria which induce tumors but which do not conform to any previously described *Agrobacterium* strain (9) As seen in Table I, all these pathogenic strains contained a large plasmid. The relatedness between these plasmids and the A.*tumefaciens* plasmids has not yet been examined.

The plasmids are routinely prepared in 30 µg amounts via ethidium bromide-cesium chloride gradients. As characteristic features for these plasmids we determined :

a) the length of the circular molecules as measured after spreading on a water hypophase by the classical Kleinschmidt technique (7)

b) the "restriction pattern" obtained after digestion of each plasmid with several restriction endonucleases (Sma, EcoRI, HindIII, SacII and XhoI) and separation of the DNA fragments by electrophoresis on agarose gels (10)

c) the presence of a set of genetic markers (see section III)

These studies have shown that :

a) the plasmids are particularly large, the sizes ranging between 95 and 156 megadaltons

b) many strains contain more than one plasmid

c) all the genetic information for rendering an *Agrobacterium* strain oncogenic is carried by one of these plasmids. Such a plasmid we call a Ti+-plasmid.

d) a set of genes, closely linked to the determinants for oncogenicity, was indentified, specifying the degradation and possibly the synthesis of unusual amino acids. The most studied of these compounds are octopine (N^2-(D-1-carboxy ethyl)-L-arginine and nopaline (N^2-(1,3-dicarboxy propyl)-L-arginine

e) the Ti+-plasmids, studied until now, belong to two general types. This classification is based both on their "restriction pattern" and on the presence of genes specifying the utilisation of

Table I : Electron microscope evidence for the presence
          of large plasmids in non-*tumefaciens* strains.

Bacterial strains		length of the plasmid in μm
A.*rhizogenes*	TR 101	48,1 ± 2,5
A.*rhizogenes*	TR 7	46,0
A.*rhizogenes*	TR 107	no plasmid detected
A.*rhizogenes*	11325	61,5 ± 1,7 ⎫
		45,1 ± 1,2 ⎭
A.*rhizogenes*	8196	n.m.
A.*rhizogenes*	15834	n.m.
A.*rhizogenes*	223	61,1 ± 1,4
A.*species*	R 1	64,8 ± 1,2
A.*species*	R 13	58,9 ± 1,0
A.*species*	TR-8-3	n.m.
A.*radiobacter*	K 84	n.m.
A.*radiobacter*	K 57	42,0

n.m. : the plasmid was visualised in the EM but
       the length was not measured

octopine and nopaline. They will be referred to
as octopine and nopaline plasmids respectively.

Some aspects of this work will be further documented in the following sections.

III. GENETIC MARKERS LOCATED ON THE Ti-PLASMID

a) Agrocin Sensitivity

The strains A.*radiobacter* K 84 and A.*tumefaciens*
396 produce a low molecular weight antibiotic, called

agrocin by Kerr (11). The chemical nature of this sub-
stance has not yet been elucidated. It is an acidic
compound, with a molecular weight (measured by gel filtration on Biogel P-5) lower then 3.000 dalton. It does
not withstand incubation at a pH higher then 9,5 or lower
then 4,0 and it is soluble in methanol.

Most A.*tumefaciens* strains of the nopaline class
are killed by this agrocin. This sensitivity is Ti-plasmid born (12) and agrocin resistant mutants are either
cured for the plasmid or are plasmid mutants. The genes
responsible for this sensitivity are rather closely linked to the oncogenicity genes. This is deduced from the
fact that several of the Ti-deletion mutants, obtained
after selection for agrocin resistance, are no longer
oncogenic.

b) Phage Exclusion

A broad host range *Agrobacterium* phage AP-1 is excluded, when the infected cell harbours a Ti-plasmid.
The Ti$^+$-cells will be killed by the phage but no infec-
tive progeny is formed. This type of phage exclusion is
similar to the exclusion of the coli phage T7 by the F$^+$
coli strains.

The AP-1 exclusion seems to be a very general phenomenon, as it is shown by all oncogenic Ti-plasmids
tested. In some cases where the exclusion could not be
tested directly (e.g. strain B6 and derivatives) due to
the phage resistance of the host bacteria, it is still
possible to show that the plasmid carries the exclusion
locus by transfer of the plasmid to a phage sensitive
Agrobacterium strain.

It appears that the exclusion locus is very closely
linked to the oncogenicity genes. Screening for exclu-
sion-minus mutants almost always yielded strains which
are no longer oncogenic. Inspection of the restriction
pattern obtained from octopine and nopaline plasmids
(Fig. 1) indicates that a segment (arrow) was in common
to all types of Ti-plasmids. If this segment would belong to a "transposon" than one can hypothesize that
oncogenicity and phage exclusion may be localised on
this "transposon" as both are characteristic for all
Ti-plasmids.

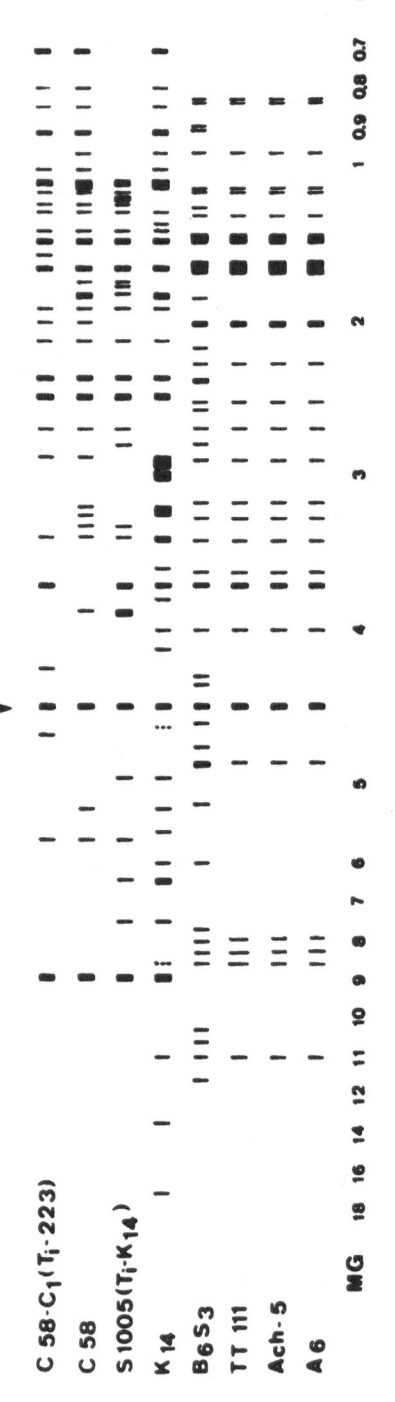

Fig. 1 : Eco RI restriction patterns of DNA from some characteristic Ti-plasmids

Plasmid DNA of the indicated A.tumefaciens strains was purified on CsCl-EtBr gradients, digested with Eco RI endonuclease and subjected to electrophoresis on 0.8 % agarose for 15 hr at 2V.cm⁻¹ in the presence of EtBr. The first four lines show patterns obtained with nopaline plasmids. The C58-C1 (Ti 223) is the strain C58 cured for the Ti-C58 and in which the Ti-plasmid of strain 223 was introduced. S1005 (Ti K 14) is a radiobacter strain harbouring now the Ti-plasmid of K 14. Strain K 14 contains two plasmids. The last four lines show patterns of Ti-octopine. The strain B6S3 harbours two plasmids. The full homology between the octopine plasmids is striking, a fact already observed in our EM heteroduplex analysis.

Another case of phage exclusion was observed with phage S-18. Unfortunately this phage has a rather narrow host range. The K-14 Ti-plasmid, the only case analysed till now, carries a locus for S-18-exclusion that is distinct from the AP-1 exclusion locus. This was shown by isolating separate mutants for each of these markers.

c) The Synthesis of Octopine or Nopaline in Crown Gall Tumors

It has been shown (13) that crown gall tumors contain unusual arginine derivatives, never detected in normal plant tissue. Some publications opposing this view have been withdrawn. The nature of the arginine compound present in the tumors is determined by the bacteria that induced the tumor and not by the plant cell (14).

The specificity for synthesis of a given compound is strictly related to the capacity for degradation of the same compound. A.tumefaciens strains which induce octopine containing tumors can catabolise octopine but not nopaline (15). As the utilisation of octopine or nopaline is a plasmid determined property it is not unlikely that one of the crucial enzymes in the biosynthesis of these compounds is also plasmid encoded. This is suggested by this strict correlation between specificity of degradation and synthesis not only of octopine and nopaline but also of homo-octopine (16). Plasmid transfer experiments further substantiate this hypothesis (17). The plasmid from an octopine degrading strain has been conjugated into an Agrobacterium cured for a nopaline plasmid. The latter strain now induces tumors containing octopine, where the original plasmid induced nopaline tumors.

Nevertheless, isolation of plasmid mutants which abolish or modify the synthesis of octopine/nopaline will be necessary in order to prove that the plasmid carries the structural gene for the dehydrogenase active in the plant cell.

However, we think that it is an important working hypothesis to consider already the possible presence on the plasmid of an octopine/nopaline-synthesis marker. The deletion and insertion mutants of the oncogenic Ti-plasmids are now checked for their ability to induce octopine or nopaline.

d) Degradation of Octopine or Nopaline

A.*tumefaciens* strains can utilise either octopine
or nopaline as a nitrogen or carbon source (13). This
specificity is strictly controlled by the Ti-plasmid as
was shown by plasmid transfer experiments (17). The expression of the genes necessary for the degradation of
octopine or nopaline is under negative control (18).
Constitutive mutants were isolated for both substrates.
A set of mutants no longer able to utilise octopine was
isolated. Even in the cases where the mutation was localised on the plasmid (proven by plasmid transfer) it
has not yet been possible to analyse the biochemical nature of the defect.

Deletion mutants of Ti-plasmid have been isolated
which have lost oncogenicity together with the capacity
to degrade octopine/nopaline, thus demonstrating a linkage between oncogenicity and octopine/nopaline utilisation. However the deletions were always extensive (between 10 and 20 % of the plasmid length) for both the
octopine and the nopaline plasmids, so that a more precise mapping is not yet possible.

e) Oncogenicity

Although tedious, it is possible to screen several
hundred A.*tumefaciens* colonies for oncogenicity. On appropriate test plants a response will be obtained after
ten to fifteen days. It may be particularly beneficial
to screen for oncogenicity-minus mutants after creating
circumstances where an efficient generation of deletion
or insertions could have occurred. In Fig. 1, the arrow
indicates the restriction fragment present in all Ti-plasmids. This fragment is absent in the non-oncogenic dele-
tion mutants isolated from K-14 (Ti onc$\overline{\Delta}$). These indica-
tions make it already worthwhile to chose this segment
for in vitro cloning and to check by complementation with
the deletion mutant if this restriction segment can render the Ti$^+$ phenotype to a *tumefaciens* Ti onc$\overline{\Delta}$.

If one accepts that a segment of the Ti-plasmids becomes integrated into the plant DNA in the course of oncogenic transformation, then this 4,5 10^6 dalton segment is
a serious candidate for such integration. Nick translation (19) of this restriction fragment and hybridisation
(20) to DNA of crown gall cells may prove the correctness
of this hypothesis.

If this segment behaves as a transposon (21), as its presence in all Ti-plasmids suggest, then it could be flanked at each side by DNA sequences (IS or insertion sequence) that are homologous but inverted with respect to each other. As a result of this structure, these DNA fragments, if sufficiently long, will form, after denaturation and self annealing of single strands, a typical "mushroom" like structure with double stranded stems (the inverted repeats) and a single stranded loop (the genes in between the inverted repeats) (22). Indeed we observed, after denaturation and self annealing of the Ti-plasmids in strain K-14, such a mushroom with a stem of 0,3 µm and a loop of 3,3 µm. The common restriction fragment could be part of this single stranded loop. Non-oncogenic mutants arise spontaneously from this plasmid. In each case examined the mushroom structure is no longer present.

However in most Ti-plasmids the inverted repeats could not be observed in this way. It is possible that they were not observed because the segment in between the inverted repeats was too long, resulting in unstable mushroom structures.

f) Insertion of Drug-Resistance Markers in the Ti-Plasmid

Several transposons (Tn) have been already described which are able to insert genes conferring resistance to antibiotics (21). Although it is not yet clear if insertion can occur at any nucleotide position, the possible insertion sites are so frequent that insertion can occur in each gene present in the target DNA.

Inserting transposons in Ti-plasmids facilitates the identification of Ti-genes. The physical mapping of the insertions sites is possible as the transposon will be visualised by heteroduplex hybridisation in the electron microscope, as a mushroom structure (22). The distances between insertions measured, relative to a chosen reference insertion, can then be compared with the changes in the restriction enzymes fingerprinting of each plasmid that carries an insertion. In this way one can order on the Ti-map the restriction fragments in which a Tn integrated.

A gene is inactivated when a transposon is inserted. By screening among the Ti::Tn plasmids for the cases where one of the known genetic markers is "mutated" it becomes possible to localise these genes on the Ti-plasmid. We have had success with this technique by using Tn C, a transposon that confers resistance to trimethoprim and streptomycin (23).

The Ti::Tn C plasmids will also be of fundamental help when in vitro cloning of Ti-fragments is attempted. Once a Tn C insertion has proven that a given restriction segment contains information for either oncogenicity or octopine catabolism/synthesis, one can easily clone this segment by selecting for the drug-markers of Tn C. In the obtained clone it may be possible to restore the expected phenotype by recombination (24) with an analogous, independently isolated clone, as both Tn C's may be expected to be in slightly different positions.

g) Integration of RP4 in the Ti-Plasmids

The presence of similar IS sequences in two different plasmids can lead, via homology recombination to a co-integrated plasmid (25). We have tested whether it was possible to integrate a whole RTF factor into a Ti-plasmid. We chose the wide host range P-like plasmid RP4 (26). This plasmid codes for resistance markers towards kanamycin, carbenicillin and tetracycline and can readily be introduced into all *Agrobacterium* strains. We have obtained a co-integrate between the Ti- and the RP4-plasmid. For this we conjugated a A.*tumefaciens* (Ti$^+$) (RP4) with a cured acceptor strain resistant to erythromycin and rifampicin. The colonies obtained, after selection for acceptors having a Ti-marker (octopine or nopaline utilisation) and RP4-markers, were again used as donor for a second cross to an acceptor resistant to spectinomycin and streptomycin. Among the acceptors of the second cross, approximately 10 % jointly received the Ti-marker and the RP4 marker. Using these colonies for further crosses we showed a 100 % linkage between the RP4 markers and all the Ti-plasmid markers. Hybridisation in the electron microscope and restriction enzyme analysis have proven that the whole RP4 plasmid has been integrated into a full length Ti-plasmid. These coïntegrates were obtained from octopine plasmids as well as from nopaline plasmids. In both cases the integration was in the same RP4 site 12,3 μm, clockwise, from the single Eco RI cut in RP4.

Both the fertility and the host range of the RP4 factor have been conserved. This means that now, the Ti-plasmid as coïntegrate, can readily be transmitted in a natural way to most Gram-negative bacteria.

These Ti-RP4 coïntegrated plasmids, like the Ti::Tn C plasmids are very convenient for EM heteroduplex studies.

For in vitro cloning these coïntegrates offer the additional advantage that the *tra* and *ori* regions of RP4 can be used, after partial digestion with restriction enzymes that do not cut RP4 DNA. Further on deleted and rearranged Ti-RP4 clones may be obtained after digestion with enzymes for which there are a few sites on RP4. In both classes, the transformation can be done in coli, with a consecutive conjugation to *Agrobacterium* as the yield of transformants is still several orders of magnitude higher in coli than in *Agrobacterium*.

4. THE TRANSFER OF Ti-PLASMIDS

The demonstration that the Ti-plasmids are essential for tumor formation was possible once the techniques became available to reintroduce a Ti-plasmid into a Ti-cured *A.tumefaciens*. Until now this curing was only reproducibly possible in a few *A.tumefaciens* strains, such as C-58 (an octopine strain). Each time curing was obtained by growth at 37°C (8). Luckily all *A.radiobacter* strains tested were acceptors for Ti-plasmid. When they were rendered Ti+ they express all markers of the Ti+-plasmid (cfr. Section III) including oncogenicity. The first reintroduction of Ti-plasmids (8) (27) was possible by using a conjugation *in planta* (28). Afterwards Ti-plasmids were transferred with wide host range plasmids. Recently Kerr et al. (29) and we (30) have demonstrated that Ti-plasmids have sex factor activity, i.e. that they are able to promote their own transfer via a conjugation mechanism. The *tra* locus of the Ti-plasmid is inducible and constitutive mutants have been isolated.

The transformation of plasmid DNA is still only possible with a low yield (10^7 transformants/receptor cells in the case of Ti-plasmid) by using a cold shock method (31).

Now that the coïntegrates between Ti-RP4 have become available, the Ti-plasmids can be very efficiently conjugated into any Gram-negative bacteria. This means that we have finally reached a situation in which efficient molecular genetic studies of the Ti-plasmids can begin.

5. CONCLUSIONS

Although it is not yet proven that a segment of the Ti-plasmid is integrated into the plant cell, there is sufficient progress in the molecular biology "technology" and in the variety of Ti-plasmid mutants that can be obtained, that one may reasonably expect a clear answer in the near future. At that moment it will become particularly challenging to see if foreign DNA (a transposon) integrated into this Ti-segment will be transferred to a plant cell upon tumor induction. If some of the drug-markers (mercury or trimethoprim resistance) can be expressed in plant cells, then it will become possible to select for such cases.

It will then become conceivable to try to introduce plant DNA into a teratogenic Ti-plasmid, to see if by this type of engineering new properties can be introduced into plants.

ACKNOWLEDGEMENTS

The authors wish to thank Dr.R. Schilperoort, Dr. A. Kerr and Dr. J. Tempé for many helpful discussions and the communication of results prior to publication.

This work was supported by grants from the "Kankerfonds van de A.S.L.K." and from the "Fonds voor Kollektief Fundamenteel Onderzoek" (no. 10316) to J.S. and M.V.M. M.H. is a "Aangesteld Navorser" of the "Nationaal Fonds voor Wetenschappelijk Onderzoek. A.D.P. and D.D.W. are indepted to the Belgian I.W.O.N.L. for a fellowship.

LITTERATURE

1. Butcher, D.N. (1973). In "Plant and Tissue Culture" Ed. H.E. Street, Blackwell, London, p. 356

2. Drlica, K.A. and C.I. Kado (1975) Bacteriol. Rev. 39, 186

3. Lippincott, J.A. and B.B. Lippincott (1975) Ann. Rev. Microbiology 29, 377

4. Keane, P.J., A. Kerr and P.B. New (1970) Austr. J. Biol. Sci. 23, 585

5. Kersters, K. and J. De Ley (1973) J. Gen. Microbiol. 78, 227

6. Vervliet, G., M. Holsters, H. Teuchy, M. Van Montagu and J. Schell (1975) J. Gen. Virol. 26, 33

7. Zaenen, I., N. Van Larebeke, H. Teuchy, M. Van Montagu and J. Schell (1974) J. Mol. Biol. 86, 109

8. Van Larebeke, N., Ch. Genetello, J. Schell, R.A. Schilperoort, A.K. Hermans, J.P. Hernalsteens and M. Van Montagu (1975) Nature 255, 742

9. Kerr, A. (1969) Austr. J. Biol. Sci. 22, 111

10. Roberts, R. (1976). In "CRC Critical Reviews in Biochemistry". Ed. G. D. Fasman. Vol. 4, issue 4, in press. Chemical Rubber Company, Cleveland, Ohio.

11. Kerr, A. and Htay, K. (1974) Physiol. Plant. Pathol. 4, 37

12. Engler, G., J.P. Hernalsteens, M. Holsters, M. Van Montagu, R. Schilperoort and J. Schell (1975) Molec. Gen. Genet. 138, 345

13. Menagé, A. and G. Morel (1964) Compt. Rend. Acad. Sci., serie D, 259, 4795

14. Goldmann, A., J. Tempé and G. Morel (1968) Compt. Rend. Soc. Biol. 162, 630

15. Petit, A., S. Delhaye, J. Tempé and G. Morel (1970) Physiol. Vég. 8, 205

16. Petit, A. and J. Tempé (1976) Compt. Rend. Acad. Sci., serie D, 282, 69

17. Bomhoff, G., P.M. Klapwijk, H.C.M. Kester, R.A. Schilperoort, J.P. Hernalsteens and J. Schell (1976) Molec. Gen. Genet. 145, 177

18. Petit, A. and J. Tempé (1975) Compt. Rend. Acad.
    Sci., serie D, 281, 467

19. Maniatis, T., A. Jeffrey and D.G. Kleid (1975) Proc.
    Nat. Acad. Sci. 72, 1184

20. Southern, E.M. (1975) J. Mol. Biol. 98, 503

21. Bukhari, A.I., J. Shapiro, S. Adhya. DNA Insertion
    Elements, Plasmids and Episomes. Cold Spring Har-
    bor Laboratory, Cold Spring Harbor, New York, 1976,
    in press.

22. Davidson, E.H., D.E. Graham, B.R. Neufeld, M.E.
    Chamberlin, C.S. Amenson, B.R. Hough and R.J. Britten
    (1973) Cold Spring Harbor Symp. Quant. Biol. 38, 295

23. Barth, P.T., N. Datta, R.W. Hedges, N.J. Grinter
    (1976) J. Bacteriol. 125, 800

24. Hernalsteens, J.P., R. Villarroel-Mandiola, M. Van
    Montagu and J. Schell (1976). In "DNA Insertion
    Elements, Plasmids and Episomes". Eds. A.I. Bukhari
    J. Shapiro, S. Adhya. Cold Spring Harbor Laboratory.
    Cold Spring Harbor, New York, in press

25. Davidson, N., R.C. Deonier, S. Hu, E. Ohtsubo (1975)
    In "Microbiology 1974". Ed. D. Schlessinger. Ame-
    rican Soc. for Microbiology, Washington, D.C., p.56

26. Datta, N., R.W. Hedges, E.J. Shaw, R.B. Sykes and
    N.H. Richmond (1971) J. Bacteriol. 108, 1244

27. Watson, B., T.C. Currier, M.P. Gordon, M.D. Chilton
    and E.W. Nester (1975) J. Bacteriol. 123, 255

28. Kerr, A. (1969) Nature 223, 1175

29. Kerr, A., P. Manigault and J. Tempé (1976) submitted
    to Nature

30. Genetello, C., N. Van Larebeke, M. Holsters, A. De
    Picker, M. Van Montagu and J. Schell (1976) submit-
    ted to Nature

31. Holsters, M., D. De Waele, M. Van Montagu and J.
    Schell. Manuscript in preparation.

NUCLEOTIDE SEQUENCES OF PLANT VIRAL RNAS

J.P. Briand, C. Fritsch, H. Guilley, G. Jonard,
C. Klein, D. Lamy, K. Richards and L. Hirth
Laboratoire de Virologie, Institut de Biologie
Moléculaire et Cellulaire du CNRS,
15, rue Descartes, 67000 Strasbourg, France

The genome of plant viruses is either carried by large RNA molecules containing the whole genetic information (tobacco mosaic virus for example) or by several pieces of nucleic acid each containing a part of the total genetic information (alfalfa mosaic virus, brome mosaic virus). But whatever the type of virus the shortest RNA piece it contains is on the order of 1000 nucleotides. Under these circumstances the sequence work on plant viral RNA's has been aimed at solving relatively limited sequences of strategic importance rather than at trying to sequence the entire molecule. The sequence problems of interest belong to three main categories.

The precise localization of genes along the RNA chain. The localization of the coat protein gene is of particular interest in view of its efficient translation in vivo and in vitro.

The determination of the primary and secondary structure of the 3' terminal region of the RNA, in particular with reference to the amino acid acceptor properties of the 3'OH end of some plant viral RNA's.

The identification and determination of the nucleotide sequences playing a role in the in vitro morphogenesis of helical viruses, particularly of TMV.

343

The present paper will report the recent findings obtained in our laboratory with respect to the above strategic sequences. Most of our studies have been performed on the RNA of TYMV or of TMV and in order to avoid confusion we shall consider independently the results obtained with each of these two viruses.

TURNIP YELLOW MOSAIC VIRUS

TYMV is a spherical RNA virus having a molecular weight of 5.6×10^6 daltons. Till now it was believed that the entire genetic information of this virus was contained on a single-stranded RNA molecule of 2×10^6 daltons. For some time now it has been known that the 3'OH end of the virus RNA can be aminoacylated by valine in conditions very similar to those used for aminoacylation of a true tRNA (1). This observation suggests that the 3'terminal part of the RNA sequence has a tRNA-like structure. It was of course of interest to establish the nucleotide sequence of this region of the RNA and we have recently done this in our laboratory.

Primary Structure of a 3'End Fragment of 159 Nucleotides

Our principal goal has been to isolate and sequence a fragment from the 3'end of TYMV RNA of a length similar to that of a true tRNA. With this end in mind we performed a partial T_1 RNase digest of highly labeled ^{32}P TYMV RNA and separated the breakdown products by gel electrophoresis. In order to identify those fragments containing the 3'end of the viral RNA we completely digested the RNA in each band with U_2 RNase. At acid pH the 3'terminal oligonucleotide of the whole RNA is positively charged and migrates to the cathode in contrast to all the other oligonucleotides. In this way three fragments containing the 3'terminus were identified, the two shorter being derived from the longest. The longest fragment was 159 nucleotides in length and was sequenced by conventional methods. Fig. 1 shows the primary sequence of the fragment. Concerning the sequence, two points are worth noting.

Localisation of the coat protein gene. Five nonsense codons (UAA) and (UAG) are found at irregular intervals within the sequence but in phase. This suggests that the nucleotides preceding the first UAA could code for the carboxy-terminal part of a viral protein (fig. 1). Examination of the coding capacity of the first

```
           -150              -140              -130              -120              -110
  G U A U C A G G A A C U C U C U C G A U G C A C U C U C C G C U C A U C A C G G A C A C U U C C A C C U A A --
```
```
          -100              -90               -80               -70               -60
  --G U(U,C)U C G A U C U U U A A A A U C G U U A G C U C G C C A G U U A G C G A G G U C U G U C C C C A C A C--
```
```
      -50               -40               -30               -20               -10
  --G A C A G A U A A U C G G G U G C A A C(U,C,C)C G C(C,C,C,U,C,U,U,C,C)G(AG,G,G)U C A U C G G A A C C OH
```

Figure 1. Nucleotide sequence of the 159 nucleotides at the 3'OH end of TYMV RNA. The five in-phase termination signals are boxed.

51 nucleotides at the 5'end of the fragment give a sequence identical to that expected for the part of the coat protein cistron coding for the last 17 amino acids of the coat protein (fig. 2). We can conclude that the coat protein cistron is located very close to the 3'end of the viral RNA. The untranslated part of the 3'terminus of TYMV RNA is restricted to the last 108 nucleotides.

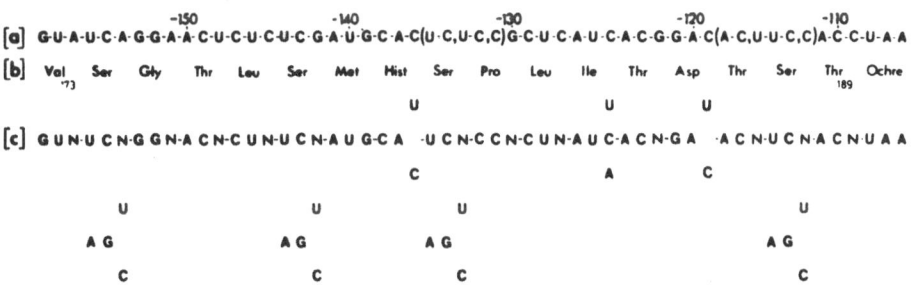

Figure 2. Experimentally determined nucleotide sequence of the 5'-terminal 54 nucleotides of the 3'end fragment (a) and the amino acid sequence of the carboxyl-terminal portion (val[173] to thr[189]) of the coat protein (b). The nucleotide sequence of the corresponding portion of the coat protein cistron as determined by the genetic code is shown in (c). The complete nucleotide sequence of the terminal portion of the cistron, deduced from consideration of (a) and (c), is given in (d).

Folding of the last 108 nucleotides.   As is shown in Fig. 1 four oligonucleotides could not be resolved by our techniques, [U(C,U)UCG , C(C,C,C,U,C,U,U,CC)G , CAAC(U,C,C)CG and G(AG,G,G)U] . However, by employing other methods for isolating the 3'OH fragment and for sequencing it, Silberklang et al (2) were able to establish the complete pancreatic and T_1 oligonucleotides catalogue of the 3'-terminal region. The results obtained are in close accord with our data and give the sequence of our four unsolved oligonucleotides. This allows us to present a complete primary structure for the 3'terminus using the overlaps we have obtained from partial pancreatic and T_1 digest of the fragment. Fig. 3 and 4 give two possible secondary structures for this portion of the chain. With reference to the amino acceptor properties of the terminus the following remarks can be made :

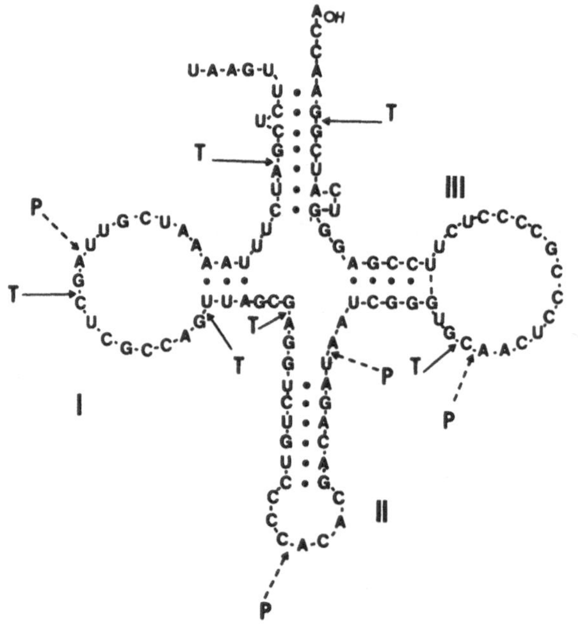

Figure 3. A possible secondary structure (model I) for the 108 untranslated nucleotides at the 3'-end of TYMV RNA. Arrows indicate the prefered sites of action for T_1 RNase and pancreatic RNase.

The fragment can be folded into a cloverleaf type of structure but the cloverleaf still has many aspects different from a true tRNA, particularly its size which is much too large (108 nucleotides instead of 75 to 95 for tRNA's).

No modified bases are present on the fragment. This seems to be a general rule for viral RNA showing amino acid accepting properties ; indeed none of the 3'terminal fragments from any of the viruses that we have studied contain modified bases.

Virtually all tRNAs of eucaryotic or procaryotic origin share certain common sequences particularly the sequence GTψCR (R is very often G) characteristic of the TψC loop. No such sequence is found in the TYMV RNA fragment. This may be regarded as surprising in view of the reported results of Haenni et al (3) showing that the tRNA-like structure of TYMV RNA is able to transfer valine during in vitro protein synthesis if we assume that such a sequence plays the same role in the reaction of euca-

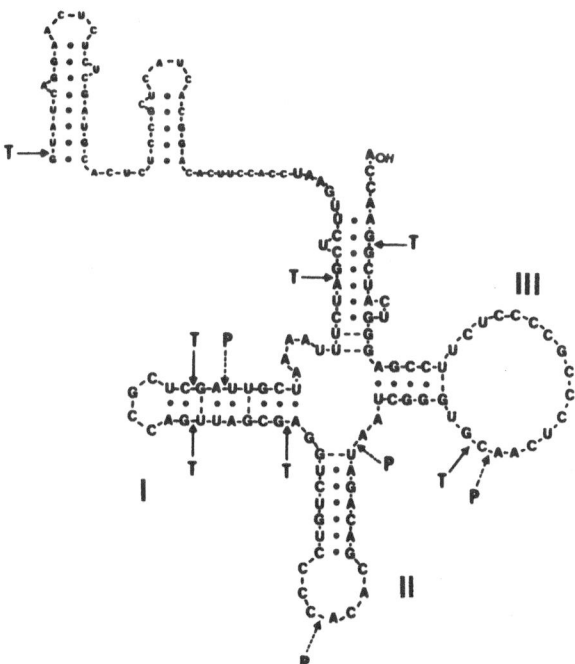

Figure 4. Another possible secondary structure (model II) for the 3'terminus of TYMV RNA.

ryotic tRNA's with ribosomes as in procaryotic systems.

In contrast with the negative characteristics described above, there are certain features in common between the proposed 3'terminal structure and eucaryotic tRNA^{val} (from Torula utilis, for example).

Concerning the sequences common to all tRNA^{val}, the 3'terminal oligonucleotide is ...ACCA$_{OH}$ in all cases. The sequence AGDDGGDCAU, which is found in the D loop of certain tRNA^{val}, is to be found at positions −8 to −15, except that the two neighboring D residues are missing and, of course, the modified base has been replaced by its normal equivalent. However, whatever the model chosen, this sequence is not an integral part of a loop structure (fig. 3). Turning to other aspects of the cloverleaf structures both models have three loops, three arms and one stem which can be likened to the D−arm (loop I), anticodon arm (loop II), TψC arm (loop III) and acceptor stem of true tRNA's. The extra-loop is missing or vestigial but stems I and III and the acceptor arm have the same number of bases pairs as in the type II tRNA (4) while the anticodon stem has only one base pair more that a true anticodon stem. In the anticodon loop the CAC valine anticodon has the predicted position in a seven base loop. It is worth noting that the anticodon arm is very similar to the anticodon arm of known eucaryotic tRNA^{val}'s, suggesting that it, probably in conjunction with the terminal ACCA$_{OH}$, plays an important role in the recognition of the structure by valyl−tRNA synthetase.

Fig. 4 shows another possibility of folding. It differs in the base pairing of arm I and of the acceptor stem and it is less closely related to a tRNA structure than model I. The main interest of model II is the repetition of the sequence CGAGG and AAUCG in two base−paired blocks. The significance of this feature, if it has one, is completely unknown. It is worth noting that the part of the isolated fragment folded in a cloverleaf structure is able to be aminoacylated when it is isolated from whole TYMV RNA by means of ribonuclease P, one of the enzymes responsible for processing tRNA precursors (5). This proves that the amino acid acceptor capacity involves only the 109 last nucleotides from the 3'end.

Recently, in collaboration with R. Giégé it has been demonstrated that the affinity of purified yeast valyl−tRNA synthetase for

the tRNA-like structure of TYMV RNA is comparable to that of
a true tRNA^{val}. The same observation applies for the rate of
aminoacylation. In other words the interaction of TYMV RNA with
valyl-tRNA synthetase is not at all like the mischarging of a heterologous tRNA (such as tRNA^{phe}) by this enzyme but is like that
of an authentic tRNA^{val} .(Giégé et al, to be published). This important observation suggests strongly that the 3'terminal TYMV
RNA fragment has a spatial conformation which closely resembles tRNA^{val} in certain critical aspects. Which parts of the tertiary structure play a crucial role in the recognition process is
not known ; it is however worth noting that though no great sequences analogy between valyl-tRNA and the TYMV RNA 3'end
are observed, the folding of the molecule can mimic, at least in
the critical region, that of a true tRNA. On the basis of the sequence homology which does exist, we believe that the terminal
ACCA$_{OH}$ plays a major role and probably also the anticodon loop.
Attempts to fold the "cloverleaf" part of the fragment in a L sha-
ped tertiary structure have shown that this was possible and that
the extra nucleotides localized in the I and III loops do not hinder
this type of folding. A more precise identification of the parts of
the molecule which interact with valyl-tRNA synthetase will be of
great interest.

TYMV RNA is a "Processed" Genome

A recent careful study of the acceptor properties of TYMV
RNA in the presence of highly purified yeast valyl-tRNA synthetase has shown that, if the number of TYMV RNA molecules in
the incubation medium was calculated on the bases of a molecular weight of 2×10^6 daltons, $1.4 - 1.6$ molecules of valine were
bound for each viral RNA molecule. This unexpected finding induced us to determine whether or not the TYMV RNA preparation
was homogenous. Indeed, it is known that in the case of Cowpea
strain of TMV 2 components are found. The heavy component
contains the entire genetic information and has a molecular
weight of 2×10^6 daltons ; the lighter component has a molecular
weight of about 300 000 daltons. We wondered if the situation with
TYMV RNA might be similar to that of the Cowpea strain of TMV.
Analysis of whole TYMV RNA by means of polyacrylamide gel
electrophoresis (PAGE) shows a major product having a molecular weight of 2×10^6 daltons and small quantities of various intermediate products with molecular weights ranging from 200 000

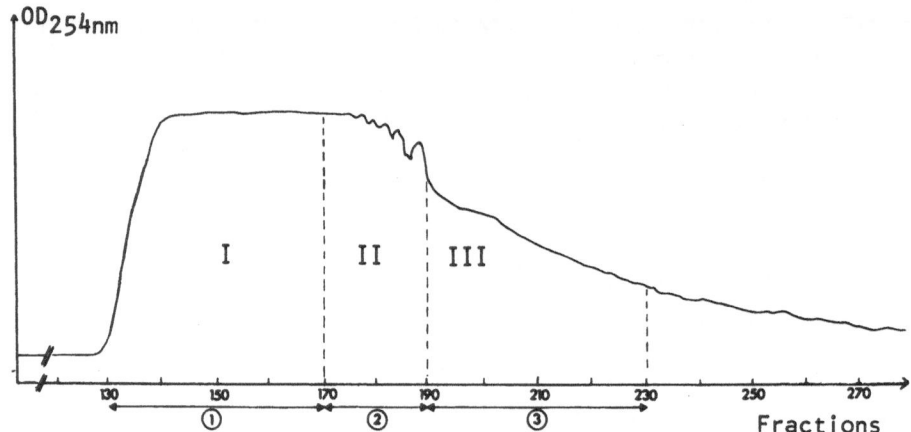

Figure 5. Purification of light (fraction III) and heavy (fraction I) RNAs of TYMV by gel filtration in a column of Ultrogel AcA22. Outflow was 9 ml/h.

to 1.5×10^6 daltons. No other major band was observed. Next we tried to fractionate the whole TYMV RNA after heating in the presence of EDTA by passage on an acrylamide-agarose column. Fig. 5 shows the elution profile obtained. The three indicated fractions were analyzed by mean of PAGE and it is clear (fig. 6) that fraction I consists of 2×10^6 dalton material and fraction III of a 300 000 dalton product which is slightly contaminated by products of higher molecular weight.

In vitro translation of whole TYMV RNA in the wheat germ embryo system gives rise (fig. 7 a) to coat protein and to a 35 000 dalton product. Small amounts of a protein of about 100 000 daltons are also made (fig. 7 a). The same whole RNA, heaten in the presence of EDTA gives an increased amount of coat protein and heavy proteins with a molecular weight of more than 165 000 (fig. 7 b). In contrast, the purified heavy RNA (after passage through the gel filtration column) codes only for a very small amount of 20 000 daltons (the nature of this protein, which comigrates with the coat protein, was not determined as it was present in too small amount). The 300 000 dalton RNA is very efficently translated to give a protein comigrating with the viral coat protein (fig. 7 c). Serological and tryptic fingerprints ana-

TYMV RNA

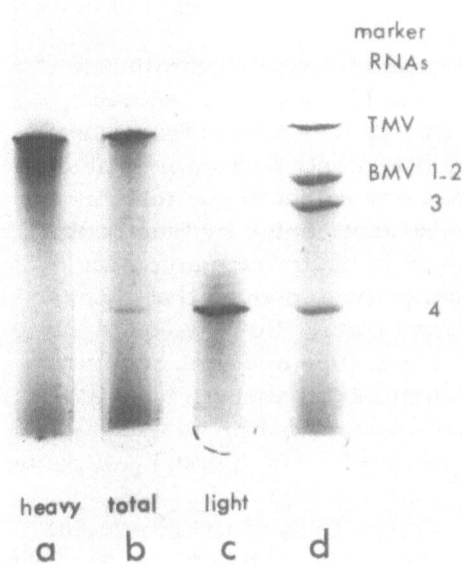

marker
RNAs

TMV
BMV 1.2
      3

      4

heavy total light
  a   b   c   d

Figure 6. Polyacrylamide
agarose (2.4 % - 0.5 %) gel
analysis of TYMV RNAs
purified by gel filtration
a. Heavy TYMV RNA
   (fraction I)
b. Total nonfractionated
   TYMV RNA
c. Light TYMV RNA
   (fraction III)
d. Marker RNAs

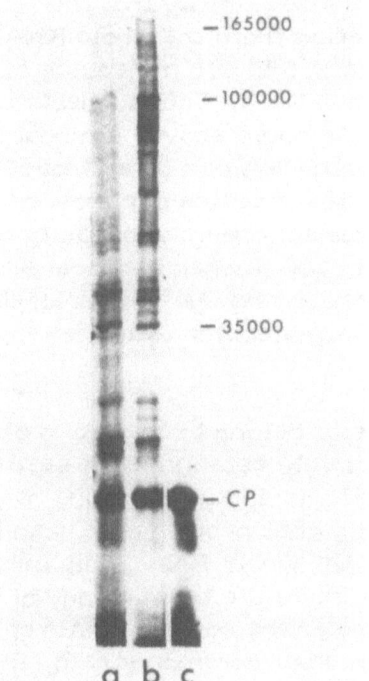

— 165000

— 100000

— 35000

— CP

  a b c

Figure 7. Polyacrylamide gel
analysis of in vitro translation products of unfractionated TYMV RNA (a),
unfractionated TYMV RNA
after heating to 60° (b) and
of purified light RNA (c)
CP = coat protein

lysis have proved that this translation product is indeed the coat protein. Thus the light RNA is a very efficient messenger for coat protein.

We conclude that the TYMV genome is probably similar to the TMV Cowpea strain genome. One possible source of artefact, however, must be eliminated. In effect, it is known from earlier work (6) that TYMV RNA is easily fragmented into small pieces by heat or EDTA treatment. Hence, one might argue that 300 000 daltons fragment could conceivabl arise by fragmentation of the RNA during purification or subsequent manipulation of the virus. Several lines of evidence have proved that this is not the case. First, both the light and heavy RNAs have the same $ACCA_{OH}$ 3'terminal oligonucleotide and they are aminoacylated by valine to similar extents. This indicates that both have a 3'-terminal tRNA-like structure which is identical or at least very similar. Furthermore, both are "capped". The 5'end T_1 oligonucleotide of the large RNA is $m^7G^{5'}ppp^{5'}Gp$ and that of the light RNA is heterogenous : $m^7G^{5'}ppp^{5'}Gp$ for 67 % of the molecules and $m^7G^{5'}ppp^{5'}Ap$ for 33 % of the molecules. The reason for this heterogeneity of the cap is not known. The presence of "cap" at the 5'end of the small RNA rules out the possibility that this fragment arises by accidental degradation from the whole RNA. In addition it is worth noting that about 0.5 to 0.6 molecules of light RNA are present per mole of heavy RNA. This suggests a ratio of 1:2 for light and heavy RNA. As noted above, aminoacylation experiments have proved that both RNA can be aminoacylated to the same extent (0.6 to 0.7 mole of valine per mole of each RNA). This explains why the apparent charging capacity of TYMV RNA molecule was greater than 100 % when the viral RNA suspension was considered as homogenous and the number of its molecules determined on the basis of a molecular weight of 2×10^6 daltons.

In conclusion, TYMV RNA appears to belong to the group of processed viruses. Although not yet firmly established it seems likely that the whole genetic information is contained in the heavy RNA, but that certain genes (especially coat protein gene) can be translated only from a small piece of the whole RNA. This piece, which may be termed the LMC according to the terminology of Zaitlin (7), can be coated as in the case of the cowpea strain of TMV, in contrast to the situation with TMV common strain. In this regard, TYMV seems to resemble cowpea strain.

TOBACCO MOSAIC VIRUS

Sequences work with TMV RNA was devoted to resolution of two major problems : (1) the structure of the amino acid accepting extremity of the TMV RNA and (2) the characterization of the initiation site for the in vitro reconstitution of TMV.

Primary Structure of the 3'End of the TMV RNA

Using methods described for TYMV it was possible to isolate a fragment of 71 nucleotides containing the 3'terminal oligonucleotide..$CCCA_{OH}$. The sequence of this fragment is shown in fig. 8.

Several points should be noted in view of the results reported above for TYMV.

As for TYMV no modified bases are present.

From the point of view of the sequence, the 3'terminal oligonucleotide $CCCA_{OH}$ is identical to the histidyl–tRNA of Salmonella typhimurium. Furthermore we find the sequence GGUAG

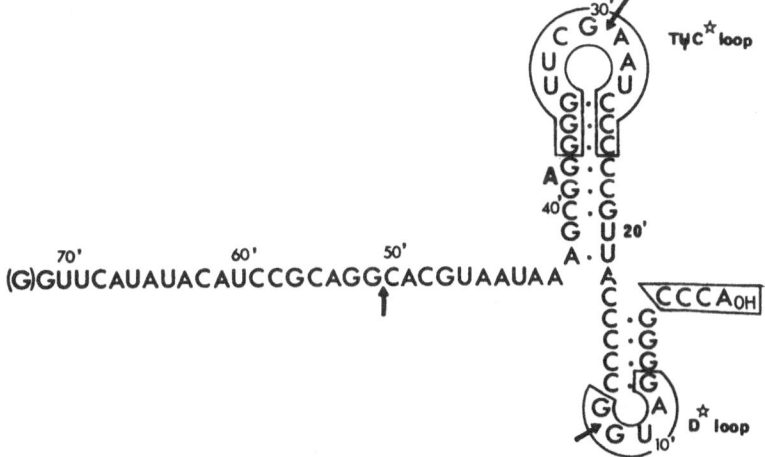

Figure 8. Possible secondary structure for the 3'end of TMV RNA.

which is a part of the D-loop of many tRNA. The sequence
...GGGUUCGAAUCCC... is identical to that found in the TψC
loop of the true histidyl-tRNA (fig. 9). It is worth noting that in
contrast with the TYMV tRNA-like structure, the sequence
...GUUCG.. (analogue to TψCG) is present.

No anticodon was found in the sequenced fragment.

The established structure can not be folded in a cloverleaf
structure although the length of the sequence is close to that of
many true tRNA's.

If we compare it to the TYMV tRNA-like structure it seems
likely that the sequenced fragment may be too short to contain the
whole nucleotide chain which can be folded in a cloverleaf structure and which is able to be charged by histidyl-tRNA synthetase.
Experiments are now in progress to determine whether or not the
71 nucleotide fragment can be charged. Fig. 10 shows the sequence of 74 nucleotides from the 3'end of GTAMV RNA, a virus
strain belonging to the TMV group but which is only distantly
related to TMV common strain. It is worth noting that the only
parts of this sequence which are common to the corresponding
parts of the common strain are those which also show sequence

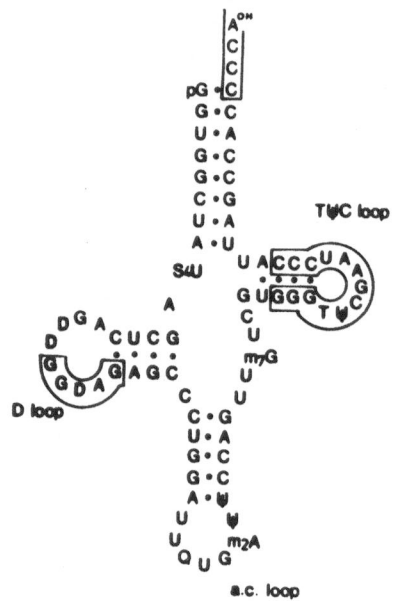

Figure 9. tRNA^hist of
S. typhimurium (8).

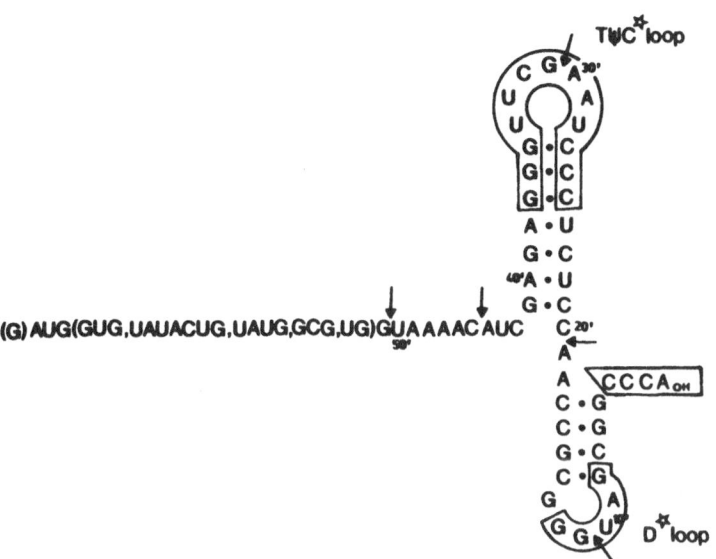

Figure 10. Possible model for the secondary structure of the
3'end fragment (n = 74 nucleotides) of GTAMV RNA.

analogy with the histidyl tRNA of Salmonella typhimurium. Both
strains of TMV can be aminoacylated by histidine suggesting
strongly that these conserved sequences are important in the recognition by histidyl-tRNA synthetase.

TMV RNA Sequences Recognized by TMV Protein in the
in Vitro Reconstitution Process

In 1969 Stussi et al (9) have discovered that, during the
in vitro reconstitution of TMV, incomplete coating of viral RNA
molecules occurs. The results obtained suggested strongly that
the TMV reconstitution was a polar process. In 1971 Thouvenel
et al (10), Butler and Klug (11) and Ohno et al (12) showed independently that the 5'end of TMV RNA was recognized by the TMV
protein disk. This observation suggested the presence of a peculiar nucleotide sequence having a great affinity for the coat protein at the 5'end. The determination of this recognition sequence
is evidently of a major interest in understanding the specific

mechanism of recognition of an RNA sequence by a protein. Two
methods may be used to isolate the nucleation sequence. The
first is to carry out limited reconstitution in order to isolate very
small RNA fragments containing the initiation site. The second is
to perform partial endonuclease digestion of TMV RNA and to try
to select RNA fragments having a great affinity for the TMV
protein disk in the conditions of reconstitution. Both techniques
have advantages and disadvantages. The attempts we have made
to perform partial reconstitution giving rise to RNA fragments of
limited size were disappointing. Generally we obtain fragments
having lengths of 300 to 600 nucleotides, too long to be easily sequenced. Thus we used the technique of selecting reactive RNA
fragments from a partial digest. The disadvantage of this approach is, of course, that other lines of evidence must be provided in order to show that the selected fragment comes from the
initiation site.

Isolation and sequencing of a fragment present in a T_1 partial
digest of TMV RNA and selected by TMV protein.   We have
already reported in detail the results of experiments in which
disks are reacted with a T_1 RNase partial digest of TMV RNA.
Fig. 11 shows that when TMV protein is added in the conditions
of reconstitution to a T_1 partial digest of TMV RNA several RNA
fragments are selected, but all these fragments (except band 6)
are degradation products of the band 1, which is 105 nucleotides
in length. No other fragments were selected. In particular it is
of interest to note that fragments containing the cap are not trap-
ped, thus proving that the selected fragment does not arise from
the 5'end of the viral RNA and is not the initiation site if we consider that reconstitution starts from the 5'end. The fragment does
not contain any of the sequence present at the 3'end of the TMV
RNA and hence it must arise from the interior of the RNA molecule. The sequence of this fragment was established and its possible secondary structure is given in fig. 12. In view of the origin of this fragment it was examined for its coding capacity and
was found to be a part of the coat protein cistron. Band 2 and 4
were not pure but were contaminated with fragments which came
from the part of the coat protein cistron just to the left of the
105 nucleotide fragment. Their sequence was determined with
the aid of the coat protein sequence. Fig. 13 gives the secondary
structure of a fragment of the coat protein cistron coding from
the amino acids 53 to 130 of the TMV protein subunit. It is worth
noting that the fragment corresponding to band 1 (105 nucleotides)
is rich in A and that its secondary structure is a very stable one.

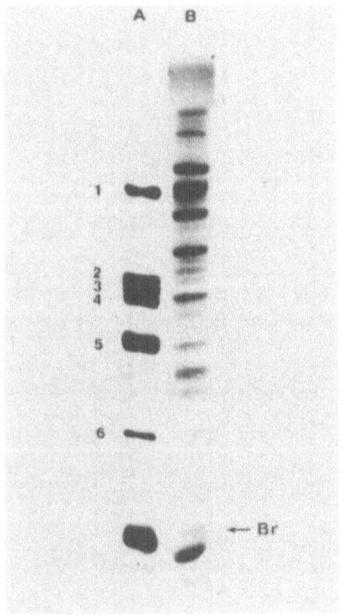

Figure 11. Polyacrylamide gel electrophoresis of a partial T$_1$ ribonuclease digest of TMV RNA before (B) and after (A) reconstitution with 25 S TMV protein.

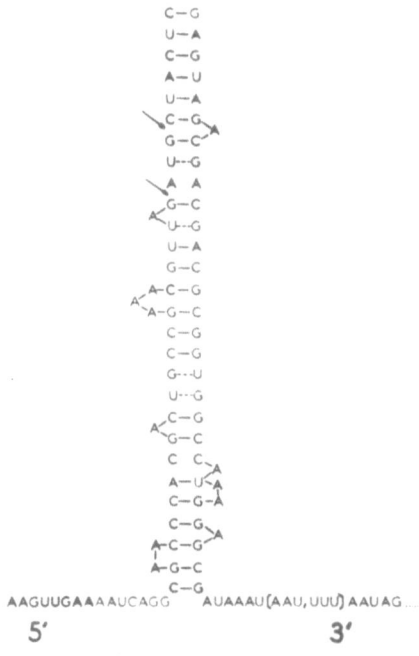

Figure 12. Possible secondary structure of the fragment 1 (n = 105 nucleotides).

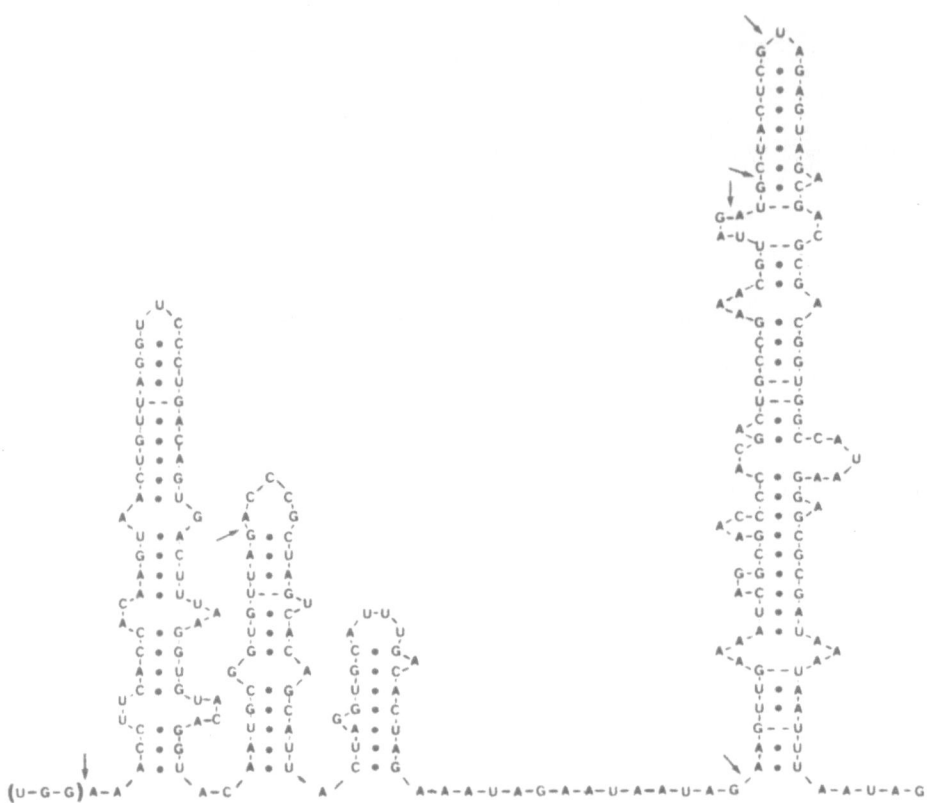

Figure 13. Possible secondary structure of the fragment of
the coat protein cistron which codes for amino
acids 53 to 130 of the TMV coat protein.

The reason why this structure is selected by the TMV protein will
be discussed below. The realisation that this sequence came from
the coat protein cistron allowed us to locate the coat protein cistron on the whole RNA. Stripping of TMV RNA by DMSO
(Lebeurier et al, to be published) starts from the 3'end and appropriate concentrations of DMSO can liberate 1500 to 2000 nucleotides from this end without stripping of the 5'end. It is possible to test the free RNA part of partially uncoated particles for
the presence of the nucleotides sequence corresponding to the
coat protein cistron and also the RNA extracted from the remaining uncoated part of the virus. Using ^{32}P labeled virus it was
possible to identify the coat protein in the free RNA and never

in the coated RNA from the partly uncoated TMV particles. This allowed us to locate the coat protein cistron within 2000 nucleotides of the 3'end. These results were confirmed later by Hunter et al using other methods (13).

We have also studied the fragments of a partial pancreatic RNase digest of TMV RNA which are captured by the disk. In this case we were able to isolate a unique RNA fragment which was selected by the viral coat protein in the conditions of reconstitution. This fragment was sequenced and its structure is given in fig. 14. Several attempts were made to determine where the fragment is located on the RNA molecule. The coding capacity of the fragment was examined and only one frame (among the 3 possible) has the capacity to code for a definite polypeptide. The comparison with the known coat protein sequence shows that it is not a part of the coat protein cistron. Using partially reconstituted and partly stripped particles and using too the knowledge of the fingerprint of the first 250 nucleotides from the 5'end of the RNA molecule we were able to conclude that the fragment does not contain the "cap" and has no sequence in common with the first 250 nucleotides of TMV RNA. Hence, it is surely within the interior of the RNA molecule. In the partly reconstituted particles it is always in the coated part of the RNA, this in contrast with the coat protein cistron which is always in the part of the RNA which is unencapsidated in incompletely coated particles. This observation suggests strongly (a) that this fragment is further from the 3'end than the coat protein cistron (b) that, under the conditions of reconstitution used, the coat protein cistron does not play the role of initiation site but that the pancreatic RNase fragment could play this role (c) that if the pancreatic RNase fragment is the initiation site it is located relatively far from both ends of the TMV RNA molecule. This could suggest that the in vitro reconstitution of TMV does not begin from one end but from an inside part of the molecule. Thus the problem remains : does the pancreatic RNase fragment contain the initiation site ? Using the partial reconstitution method, Zimmern (14) has isolated 3 fragments which contain oligonucleotides identical to those contained in our P_1 fragment. Using our overlaps it is possible to show that the Zimmern's fragments can be organized to give our sequence. Under these circumstances we can conclude that the pancreatic RNase fragment is (or contains) the initiation site and that TMV reconstitution is initiated by an oligonucleotide sequence or by a particular tertiary structure (or more likely by both) contained within the interior of the RNA molecule.

Figure 14. Possible secondary structure for the viral assembly initiation region of TMV RNA.

Whatever the exact location of the initiation site is, it is sure that elongation proceeds in both directions 3'—5' and 5'—3'. If we consider that in the partially reconstituted particles the free parts of the RNA always contain the 3'end, the coat protein cistron and the "cap" we would conclude that in condition where the stoechiometry between RNA and protein exists, a rapid elongation occurs from the initiation site to a point far from either end. In complete morphogenesis, coating of the remaining free part of the RNA presumably proceeds more slowly. Perhaps in the first step disk protein plays the major role and in the second step only the free 4 S protein. In any event, the problem of the mechanism of elongation has to be reconsidered.

Which specific characteristics exist in both the T_1 RNase and pancreatic RNase fragments which confer reactivity toward the disk ? This is an important problem especially if we consider that both fragments have the same gross affinity for the TMV protein disk but that in vitro only the P1 fragment plays the role of an initiation site (Jonard et al, to be published).

It is clear that both fragments are relatively rich in purine and especially in A. This is especially the case for the pancreatic RNase fragment where blocks of 11 purines (fig. 14) are to be found not far from either ends of the fragment. However, no blocks of homologous sequence are shared by the two sequences even if we consider that 3 or 4 nucleotides constitute a block However, two characteristics are shared by both fragments. (1) A very stable hairpin structure and (2) a periodicity of order three in purines. The order of the periodicity is important since three nucleotides interact with each protein subunit. We can see that whatever the frame used to consider the periodicity of purine we can found a purine at the same position for each subunit. This could explain why other fragments which does not show this periodicity are not recognized.

It is now evident that since the initiation site is in the interior of the RNA no free extremities can play a role (although this may not be the case for recognition of short isolated fragments). In any event, the hairpin structure and especially the loop part of this structure may be important. It is worth noting, however, that the loop sequence of the pancreatic and T_1 RNase selected fragments are different.

Studies of the reactivity of subfragments of the pancreatic RNase selected sequence suggest that the part of the sequence playing a critical role is contained in the middle part of the fragment. This is an indirect proof that the extremity of the sequence does not play a role in the recognition process.

The above experiments raise almost as many questions as they have answered. One of the most important unanswered problems concerns the conformational change of the protein disk to trap the initiation site and to give a helical structure. Many of these questions are discussed in detail in a paper in preparation.

References

(1) Pinck, M., Yot, P., Chapeville, F. and Duranton, H.M. (1970) Nature 226, 954–956.

(2) Silberklang, M., Prochiantz, A., Haenni, A.L. and RajBhandary, U.L., submitted to the Editor of Eur. J. Biochem.

(3) Haenni, A.L., Prochiantz, A., Bernard, O. and Chapeville, F. (1973) Nature New Biology 241, 166–168.

(4) Levitt, M. (1969) Nature 224, 759–763.

(5) Prochiantz, A. and Haenni, A.L. (1973) Nature New Biology 241, 168–170.

(6) Hirth, L., Horn, P. and Strazielle, C. (1965) J. Mol. Biol. 13, 720–734.

(7) Bruening, G., Beachy, R.N., Scalla, R. and Zaitlin, M. (1976) Virology 71, 498–517.

(8) Singer, C. and Smith, G.R. (1972) J. Biol. Chem. 247, 2989

(9) Stussi, C., Lebeurier, G. and Hirth, L. (1969) Virology 38, 16–25.

(10) Thouvenel, J.C., Guilley, H., Stussi, C. and Hirth, L. (1971) FEBS Letters 16, 204–206.

(11) Butler, P.J.G. and Klug, A. (1971) Nature New Biology 229, 47–50.

(12) Ohno, T., Nozu, Y. and Okada, Y. (1971) Virology 44, 510–516.

(13) Hunter, T.R., Hunt, T., Knowland, J. and Zimmern, D. (1976) Nature 260, 759–764.

(14) Zimmern, D. (1976) Phil. Trans. Roy. Soc. B. (In press).

VIROIDS IN AGRICULTURE

T. O. Diener

Plant Virology Laboratory, Plant Protection Institute
Agricultural Research Service, U.S. Department of
Agriculture, Beltsville, MD 20705 U.S.A.

INTRODUCTION

The term "viroid" has been introduced to denote a recently
recognized class of subviral plant pathogens (7). Presently known
viroids consist solely of a short strand of RNA with a molecular
weight of about 75,000 to 125,000 daltons. Introduction of this
low-molecular-weight RNA into susceptible hosts leads to replication
of the RNA and, in some hosts, to disease.

The first viroid came to light during efforts to purifiy and
characterize the agent of the potato spindle tuber disease (PSTV),
a disease which, for many years, had been assumed to be of viral
etiology (15). Diener and Raymer (13) reported that the infectious
agent of this disease is a free RNA and that virus particles,
apparently, are not present in infected tissue. Later, sedimentation
and gel electrophoretic analyses conclusively demonstrated that the
infectious RNA has a very low molecular weight (7) and that the
agent, therefore, differs basically from conventional viruses.

Four additional plant diseases, citrus exocortis (33), chrysanthemum stunt (12), chrysanthemum chlorotic mottle (28) and cucumber
pale fruit (39, and Peters, unpublished results), are now known also
to be caused by low-molecular-weight RNAs; i.e., by viroids.

The recognition of viroids as a newly identified class of
pathogens raises several questions that have potentially important
implications for microbiology, molecular biology, and plant pathology,
as well as for veterinary and human medicine.

PROPERTIES OF VIROIDS

Sedimentation Properties and Nuclease Sensitivity

Diener and Raymer (13) showed that most of the infectious·
material in crude extracts prepared from potato or tomato leaves
affected with the potato spindle tuber disease sediments in sucrose
gradients at a very low rate (about 10S). Treatment of crude
extracts with phenol affected neither infectivity nor the sedimentation properties of the agent. Incubation of extracts with nuclease
revealed that the agent is sensitive to ribonuclease, but not to
deoxyribonuclease. In view of these findings, the authors proposed
the agent to be a free nucleic acid.

Somewhat similar results were later reported by Singh and
Bagnall (35) who also worked with the potato spindle tuber disease,
by Semancik and Weathers (32) with citrus exocortis disease, and by
Lawson (23) with chrysanthemum stunt disease.

Absence of virions. Although there was little doubt that the
slowly sedimenting infectious material was, in each case, free RNA,
the question arose as to whether this RNA exists as such in situ or
whether it is released from conventional virus particles during
extraction. A systematic study of this question led to results
that are incompatible with the concept that conventional viral
nucleoprotein particles exist in PSTV-infected tissue (6). Furthermore, comparisons of proteins isolated from PSTV-infected tissue
with those isolated from healthy tissue gave no evidence for the
synthesis in infected leaves of proteins that could be construed
as viral coat proteins under conditions such that coat protein of
defective strains of tobacco mosaic virus could readily be
demonstrated (40).

Subcellular location. Isolation of subcellular particles from
PSTV-infected tissue revealed that appreciable infectivity is presen·
only in the original tissue debris and in the fraction containing
nuclei. Chloroplasts, mitochondria, ribosomes, and the "soluble"
fraction contain no more than traces of infectivity. Furthermore,
when chromatin was isolated from infected tissue, most infectivity
was associated with it and could be extracted with phosphate buffer
as free RNA (6). These and other experiments suggest that, in situ,
PSTV is associated with the nuclei and particularly with the
chromatin of infected cells.

Recognition of the low molecular weight of viroids. The low
sedimentation rate of PSTV is consistent with a viral genome of
conventional size ($> 10^6$ daltons) only if the RNA is double- or
multi-stranded. Early experiments indeed indicated that the RNA

might be double-stranded (14), but later results showed that its chromatographic properties are not compatible with this hypothesis (5).

Evidently, determination of the molecular weights of viroids is of great importance in elucidating their structure. This determination is difficult, because the agents occur in infected tissue in very small amounts and are, therefore, difficult to separate from host RNA and to purify in amounts sufficient for conventional biophysical analyses.

A concept elaborated by Loening (24) made it feasible to determine the molecular weight of PSTV, using infectivity as the sole parameter. Combined sedimentation and gel electrophoretic analyses conclusively showed that the infectious RNA has a very low molecular weight (7). A value of 5×10^4 daltons was compatible with the experimental results (7). This conclusion was confirmed by the ability of PSTV to penetrate into polyacrylamide gels of high concentration (small pore size), from which high molecular weight RNA molecules are excluded (16).

On the basis of its electrophoretic mobility in 5% polyacrylamide gels, Semancik and Weathers (33) came to the conclusion that the agent causing citrus exocortis disease is also a low-molecular-weight RNA. They estimated that the RNA has a molecular weight of 1.25×10^5 daltons. Sänger (29), on the other hand, estimated that the exocortis disease agent has a molecular weight of 5 to 6×10^4 daltons.

Diener and Lawson (12) showed by a combination of isokinetic density-gradient centrifugation and electrophoresis in 20% polyacrylamide gels that the agent of chrysanthemum stunt disease is a low-molecular-weight RNA similar to, but distinct from, PSTV.

Purification of PSTV. In all experiments so far described, PSTV was identified solely by its biological activity; and no clearly and consistently recognizable ultraviolet light-absorbing component was correlated with infectivity distribution in sucrose gradients or in polyacrylamide gels. Detailed characterization of PSTV requires its isolation in amounts sufficient for conventional biophysical and biochemical analyses. Large-scale isolation of PSTV, together with improvements in separation techniques, have made this goal attainable (8).

As shown in Fig. 1(A), electrophoresis in 20% polyacrylamide gels of highly concentrated, low-molecular-weight RNA preparations from healthy plants discloses aside from 5S RNA, at least three RNA species of low molecular weight (I, III, and IV). Electrophoresis of identically prepared samples from PSTV-infected plants reveals the

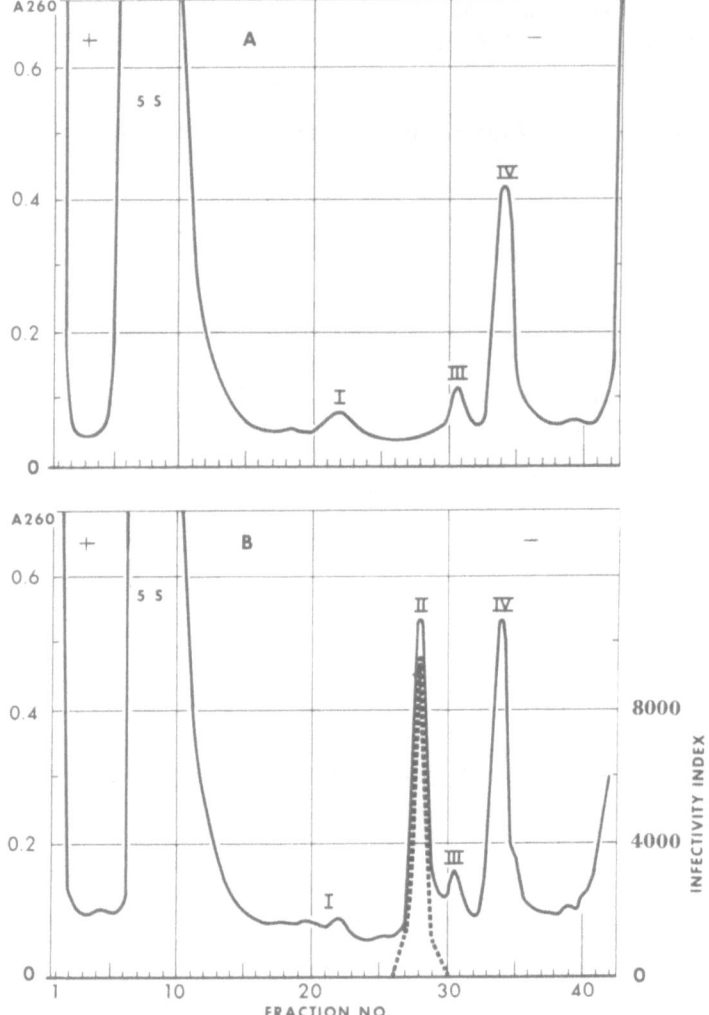

Fig. 1. (A) Ultraviolet light-absorption profile of RNA prepara-
              tion from healthy tomato leaves after electrophoresis
              in a 20% polyacrylamide gel for 7.5 hrs at 4°C (5 mA
              per tube, constant current). (From ref. 8.)

         (B) Ultraviolet light-absorption (———) and infectivity
              distribution (-----) of RNA preparation from PSTV-
              infected tomato leaves after electrophoresis in a 20%
              polyacrylamide gel (same conditions as in Fig. 1).
              5S = 5S ribosomal RNA; I, III, IV = unidentified minor
              components of cellular RNA; II = PSTV; A_{260} = Absorb-
              ance at 260 nm. Electrophoretic movememt from right
              to left. (From Ref. 8.)

same components; namely 5S RNA and components I, III, and IV; but, in addition, another prominent ultraviolet-light-absorbing component, II, is evident [Fig. 1(B)]. Bioassays of individual gel slices demonstrated that infectivity coincides with component II [Fig. 1(B)]. This coincidence, the high level of infectivity, and the fact that component II is not recognizable in preparations from healthy leaves constitutes strong evidence that component II is PSTV.

Pure PSTV has been prepared using electrophoresis in 20% poly-acrylamide gels as the last step in the purifcation scheme, followed by elution from gel slices and reconcentration of the RNA by chromatography on hydroxyapatite (11). Quantities so far produced have been sufficient to determine several properties of the RNA by conventional means.

Thermal denaturation properties. The total hyperchromicity shift of PSTV in 0.01 x SSC (0.15 M sodium chloride-0.015 M in sodium citrate, pH 7.0) is about 24% and the T_m about 50° C (8). The thermal denaturation curve indicates that PSTV is not a regularly base-paired structure, such as double-stranded RNA, since in this case, denaturation would be expected to occur over a narrower temperature range and at higher temperatures (25). PSTV could, however, be an irregularly base-paired, single-stranded RNA molecule, similar to transfer RNA, in which single-stranded regions alternate with base-paired regions.

Molecular weight of PSTV. With the availability of purified PSTV, a redetermination of its molecular weight based, not on its biological activity, but on its physical properties, became possible. For this purpose, a method described by Boedtker (2) appeared particularly promising, as it permits the determination of the molecular weight of an RNA independent of its conformation. Application of this method to PSTV led to a molecular weight estimate of 7.5 to 8.5 x 10^4 daltons (17).

Visualization of PSTV. In view of the purity of available PSTV preparations, it appeared feasible to visualize PSTV by electron microscopy and to determine its molecular weight by direct length measurements of the RNA in electron micrographs. This, indeed, proved feasible (37).

Fig. 2 shows an electron micrograph of a mixture of a double-stranded DNA; namely, coliphage T7-DNA, and PSTV. Length measurements indicated that T7-DNA is about 280 times longer than PSTV. It is also apparent that the width of PSTV is similar to that of T7-DNA. If one assumes a molecular weight of 25 x 10^6 daltons for T7-DNA and assumes that PSTV in urea is formed by two more or less base-paired strands (either as a hairpin or a double helix), then one obtains a molecular weight estimate for PSTV of 8.9 x 10^4 daltons.

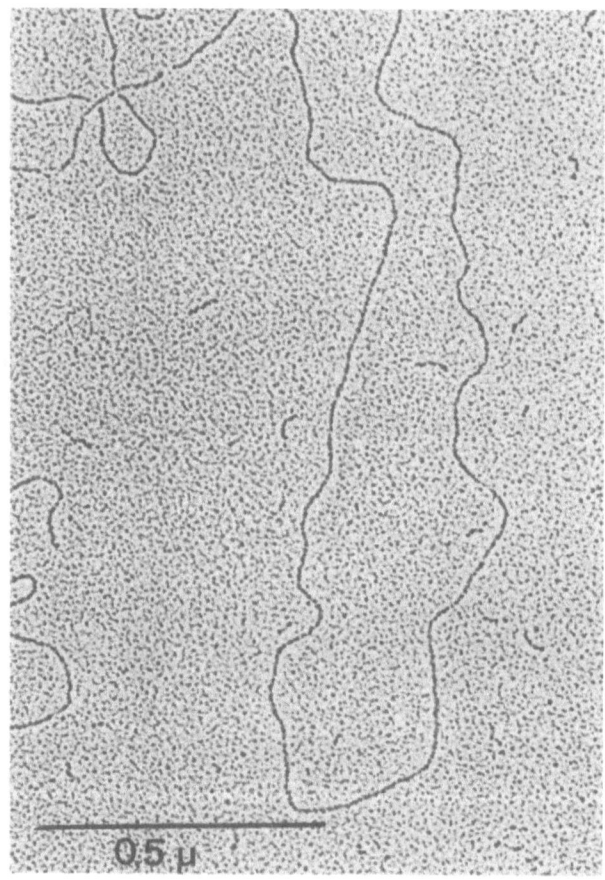

Fig. 2. Electron micrograph of PSTV mixed with a double-stranded
DNA, coliphage T7-DNA. Native T7-DNA (0.8 μg/ml) was mixed
with purified PSTV (0.4 μg/ml) previously heated for 10 min
at 63°C in the presence of 8 M urear, followed by quenching
in ice water. Note that double-stranded T7-DNA and PSTV
have similar widths. (Courtesy of T. Koller and J. M. Sogo,
Swiss Federal Institute of Technology, Zurich.)

In other experiments, mixtures of PSTV and a single-stranded viral RNA were examined and measured. PSTV appeared thicker than this single-stranded viral RNA, and from length comparisons, a molecular weight of 7.9×10^4 daltons was obtained (37).

The molecular weight estimates obtained by electron microscopy are, therefore, in excellent agreement with the values obtained from analysis of heat-denatured, formylated PSTV in polyacrylamide gels.

Inactivation by ultraviolet light. In view of the low molecular weight of PSTV, it was of interest to determine its sensitivity to irradiation with ultraviolet light. Although one might expect that a small molecule, such as PSTV, would be considerably more resistant to ultraviolet irradiation than a conventionally sized viral RNA or DNA, the effect of size on ultraviolet sensitivity of nucleic acids is not well understood (1).

Exposure of purified PSTV, of tobacco ringspot virus (TRSV), and of its satellite (SAT) (30) to ultraviolet radiation of 254 nm showed that the inactivation dose for PSTV and SAT is 70 to 90 times as large as that for TRSV (19). Although other explanations are possible, this marked difference in sensitivity to ultraviolet radiation is probably a consequence of the smaller size (smaller target volume) of PSTV and SAT-RNA, as compared with TRSV-RNA.

Composition. In view of these results, little doubt exists that the infectious RNA has a very low molecular weight. It is conceivable, however, that PSTV is not a single molecular species, but rather a population of several RNA molecules of similar length, with different nucleotide sequences, which together may constitute a viral genome of more or less conventional size. As discussed previously (9, 10), existing knowledge does not support this model, and the high resistance of PSTV to UV radiation (19) may be interpreted as further evidence for the smallness of the PSTV genome.

More definitive results have recently been obtained by two-dimensional fingerprinting of highly purified PSTV and citrus exocortis viroid (CEV) labeled in vitro with ^{125}I (4). These analyses demonstrated that each of these RNAs has a complexity compatible with the size estimate of 250 to 350 nucleotides, thus supporting the concept that each viroid consists of a single RNA species of defined sequence. These studies also demonstrated that PSTV and CEV do not have the same primary sequence, a result which contradicts conclusions drawn by some workers from biological experiments which suggested that the two viroids were independent isolates of the same pathogen (31, 34, 36).

<u>Messenger RNA and aminoacylation properties</u>. In <u>vitro</u> experiments with several cell-free protein synthesizing systems indicated that neither purified PSTV (3) nor CEV (22) act as messenger RNAs in these systems. With CEV, it has been shown that the RNA does not serve as an amino acid acceptor (22).

REPLICATION OF VIROIDS

By what mechanisms RNAs of such a small size are replicated in susceptible host cells is, at present, unknown. In view of the small amount of genetic information that PSTV introduces into its host cells, one might plausibly assume PSTV to be analogous to a satellite RNA that requires a helper virus for its own replication. However, efforts to demonstrate the presence of such a helper virus in uninoculated tomato plants gave negative results (7). In view of these and other results (7, 20), it appears most unlikely that a conventional helper virus is necessary for the replication of PSTV. Hence, PSTV, in spite of its small size, appears to replicate autonoumously in susceptible cells. Because the RNA can code for only 80 to 120 amino acids and because it is not translated in cell-free protein-synthesizing systems, it appears that PSTV is replicated by preexisting host enzymes.

Two schemes of PSTV replication are most readily compatible with present views on cellular and viral RNA synthesis. The first scheme postulates that PSTV is replicated on a DNA template, which either is already present in uninfected hosts in repressed form, or is synthesized as a consequence of infection with PSTV. The second scheme postulates that PSTV is replicated independently of DNA; i.e., that its replication is analogous to that of many viral RNAs, proceeding via a complementary RNA transcribed from PSTV.

A distinction between these two schemes should be made possible by the use of drugs that specifically inhibit RNA transcription from DNA templates. Such drugs should inhibit PSTV replication if the first scheme is correct. We used actinomycin D because this antibiotic inhibits cellular RNA synthesis in plant cells but does not seriously interfere with the replication of several plant viral RNAs.

Leaf strips from healthy or PSTV-infected tomato plants were incubated in solutions containing [^{3}H]uracil or [^{3}H]UTP. Extraction of nucleic acids and analysis by polyacrylamide gel electrophoresis revealed ^{3}H incorporation into a component with electrophoretic mobility identical with PSTV in extracts from infected, but not in extracts from healthy leaves. No ^{3}H incorporation into PSTV could be detected when leaf strips were pretreated with actinomycin D under conditions which reduced uptake of [^{3}H] uracil only 10 to 12%

but inhibited cellular RNA transcription 87 to 99%. These results
suggest that, in infected tomato leaves, PSTV replication may require
the continued synthesis of one or more cellular RNA species or that
PSTV replication may proceed via a DNA intermediate (18).

Similar results were obtained with an in vitro RNA synthesizing
system, using purified cell nuclei from healthy or PSTV-infected
tomato leaves as an enzyme source. Isolation of low-molecular-weight
RNAs from the in vitro reaction mixtures and analysis by gel electrophoresis revealed that PSTV replication is sensitive to incubation
with actinomycin D (38). These results also demonstrate that PSTV
is replicated in the nucleus of infected cells. The observed sensitivity of PSTV replication to actinomycin D, in both in vitro and
in vivo systems suggests that PSTV may be transcribed from a DNA
template.

Recent experiments (21), in which purified, ^{125}I-labelled PSTV
was used as a hybridization probe to detect sequences complementary
to PSTV in DNA isolated from healthy and PSTV-infected host plants
give support to this hypothesis. These experiments showed that
infrequent DNA sequences complementary to PSTV exist in both uninfected and infected tomato cells. DNA titration experiments
revealed that at least 60% of the PSTV molecule is represented by
complementary sequences in DNA of several normal solanaceous host
species and that phylogenetically diverse plants contain sequences
related to less of the PSTV (21). PSTV-infected tomato plants do
not possess new PSTV sequences at detectable levels (21). These
results support the hypothesis that synthesis of PSTV, and presumably
of other viroids, is a normal capacity of the host genome which,
however, is completely repressed in normal plants. If so, viroids
act as derepressors of this latent capacity and could, therefore,
be considered as abnormal regulatory molecules.

CONCLUSIONS

With the knowledge now at hand, it is evident that viroids
constitute a novel class of pathogens which are clearly distinct
in several respects from conventional viruses. Aside from the lack
of a protein coat and the smallness of their genomes, viroids differ
from viruses in that infection with viroids apparently does not lead
to the synthesis of novel, pathogen-specific proteins. Also, PSTV
is the first known RNA pathogen of plants which possesses complementarity to its host DNA. Having these properties, viroids lend
themselves admirably to studies, on the molecular level, of the
mechanisms of pathogenesis and control of gene regulation in
eukaryotic cells.

From a practical standpoint it can be predicted that many infectious diseases of plants and possibly of animals, the etiology of which is now obscure, will be found to be caused by agents similar to presently known viroids. Some of these plant diseases may be of considerable economic importance. A case in point is cadang-cadang disease of coconut palms, which poses a serious threat to the coconut industry, particularly in the Philippines. Although information is incomplete, recent results demonstrated the association of a viroid-like, low-molecular-weight RNA with extracts from diseased tissue and its absence in extracts from healthy tissue, suggesting that cadang-cadang disease may be of viroid etiology (27).

Methods developed for the study and purification of viroids have already been adapted as a valuable diagnostic aid for indexing elite or basic potato seed stocks in certification programs (26).

LITERATURE CITED

1. ADAMS, D. H. 1970. The nature of the scrapie agent. A review of recent progress. *Pathol. et Biol.* 18:559-577.

2. BOEDTKER, H. 1971. Conformation independent molecular weight determination of RNA by gel electrophoresis. *Biochim. Biophys. Acta* 240:448-453.

3. DAVIES, J. W., P. KAESBERG, and T. O. DIENER. 1974. Potato spindle tuber viroid. XII. An investigation of viroid RNA as a messenger for protein synthesis. *Virology* 61:281-286.

4. DICKSON, E., W. PRENSKY, and H. D. ROBERTSON. 1975. Comparative studies of two viroids: Analysis of potato spindle tuber and citrus exocortis viroids by RNA fingerprinting and polyacrylamide gel electrophoresis. *Virology* 68:309-316.

5. DIENER, T. O. 1971. A plant virus with properties of a free ribonucleic acid: Potato spindle tuber virus. Pages 433-478, *In:* Comparative Virology (K. Maramorosch and E. Kurstak, eds.), Academic Press, New York.

6. DIENER, T. O. 1971. Potato spindle tuber virus. III. Subcellular location of PSTV-RNA and the question of whether virions exist in extracts or in situ. *Virology* 43:75-89.

7. DIENER, T. O. 1971. Potato spindle tuber "virus." IV. A replicating, low molecular weight RNA. *Virology* 45:411-428.

8. DIENER, T. O. 1972. Potato spindle tuber viroid. VIII. Corre-
    lation of infectivity with a ultraviolet light-absorbing compon-
    ent and thermal denaturation properties of the RNA. *Virology*
    50:606-609.

9. DIENER, T. O. 1972. Viroids. *Advan. Virus Res.* 17:295-313.

10. DIENER, T. O. 1973. Potato spindle tuber viroid: A novel
    type of pathogen. *Perspect. Virol.* 8:7-30.

11. DIENER, T. O. 1973. A method for the purification and recom-
    mendation of nucleic acids eluted or extracted from polyacryla-
    mide gels. *Anal. Biochem.* 55:317-320.

12. DIENER, T. O. and R. H. LAWSON. 1973. Chrysanthemum stunt:
    A viroid disease. *Virology* 51:94-101.

13. DIENER, T. O. and W. B. RAYMER. 1967. Potato spindle tuber
    virus: A plant virus with properties of a free nucleic acid.
    Science 158:378-381.

14. DIENER, T. O. and W. B. RAYMER. 1969. Potato spindle tuber
    virus: A plant virus with properties of a free nucleic acid.
    II. Characterization and partial purification. *Virology* 37:
    351-366.

15. DIENER, T. O. and W. B. RAYMER. 1971. Potato spindle tuber
    "virus." *Descriptions of Plant Viruses. No. 66.* Commonwealth
    Mycol. Inst. Assoc. Applied Biol., Kew, Surrey, England. 4 pp.

16. DIENER, T. O. and D. R. SMITH. 1971. Potato spindle tuber
    viroid. VI. Monodisperse distribution after electrophoresis
    in 20% polyacrylamide gels. *Virology* 46:498-499.

17. DIENER, T. O. and D. R. SMITH. 1973. Potato spindle tuber
    viroid. IX. Molecular weight determination by gel electrophoresis
    of formylated RNA. *Virology* 53:359-365.

18. DIENER, T. O. and D. R. SMITH. 1975. Potato spindle tuber
    viroid. XIII. Inhibition of replication by actinomycin D.
    Virology 63:421-427.

19. DIENER, T. O., I. R. SCHNEIDER, and D. R. SMITH. 1974. Potato
    spindle tuber viroid. XI. A comparison of the ultraviolet light
    sensitivities of PSTV, tobacco ringspot virus, and its satellite.
    Virology 57:577-581.

20. DIENER, T. O., D. R. SMITH, and M. J. O'BRIEN. 1972. Potato spindle tuber viroid. VII. Susceptibility of several solanaceous plant species to infection with low molecular weight RNA. *Virology* 48:844-846.

21. HADIDI, A., D. M. JONES, D. H. GILLESPIE, F. WONG-STAAL, and T. O. DIENER. 1976. Hybridization of potato spindle tuber viroid to cellular DNA of normal plants. *Proc. Nat. Acad. Sci. U.S.A.* 73:2453-2457.

22. HALL, T. C., R. K. WEPPRICH, J. W. DAVIES, L. G. WEATHERS, and J. S. SEMANCIK. 1974. Functional distinctions between the ribonucleic acids from citrus exocortis viroid and plant viruses: Cell-free translation and aminoacylation reactions. *Virology* 61:486-492.

23. LAWSON, R. H. 1968. Some properties of chrysanthemum stunt virus. *Phytopathology* 58:885.

24. LOENING, U. E. 1967. The fractionation of high-molecular-weight ribonucleic acid by polyacrylamide-gel electrophoresis. *Biochem. J.* 102:251-257.

25. MIURA, K. I., I. KIMURA, and N. Suzuki. 1966. Double-stranded ribonucleic acid from rice dwarf virus. *Virology* 28:571-579.

26. MORRIS, T. J. and N. S. Wright. 1975. Detection on polyacrylamide gel of a diagnostic nucleic acid from tissue infected with potato spindle tuber viroid. *Amer. Potato J.* 52:57-63.

27. RANDLES, J. W. 1975. Association of two ribonucleic acid species with cadang-cadang disease of coconut palm. *Phytopathology* 65:163-167.

28. ROMAINE, C. P., and R. K. HORST. 1975. Suggested viroid etiology for chrysanthemum chlorotic mottle disease. *Virology* 64:86-95.

29. Sänger, H. L. 1972. An infectious and replicating RNA of low molecular weight: The agent of exocortis disease of citrus. *Advan. in Biosci.* 8:103-116.

30. SCHNEIDER, I. R. 1969. Satellite-like particle of tobacco ringspot virus that resembles tobacco ringspot virus. *Science* 166:1627-1629.

31. SEMANCIK, J. S., D. S. MAGNUSON, and L. G. WEATHERS. 1973. Potato spindle tuber disease produced by pathogenic RNA from citrus exocortis disease: Evidence for the identity of the causal agents. *Virology* 52:292-294.

32. SEMANCIK, J. S. and L. G. WEATHERS. 1968. Exocortis virus of citrus: Association of infectivity with nucleic acid preparations. *Virology* 36:326-328.

33. SEMANCIK, J. S. and L. G. WEATHERS. 1972. Exocortis disease: Evidence for a new species of "infectious" low molecular weight RNA in plants. *Nature New Biology* 237:242-244.

34. SEMANCIK, J. S. and L. G. WEATHERS. 1972. Pathogenic 10 S RNA from exocortis disease recovered from tomato bunchy-top plants similar to potato spindle tuber virus infection. *Virology* 49: 622-625.

35. SINGH, R. P. and R. H. BAGNALL. 1968. Infectious nucleic acid from host tissues infected with potato spindle tuber virus. *Phytopathology* 58:696-699.

36. SINGH, R. P. and M. C. CLARK. 1973. Similarity of host response to both potato spindle tuber and citrus exocortis viruses. *FAO Plant Protect. Bull.* 21:121-125.

37. SOGO, J. M., T. KOLLER, and T. O. DIENER. 1973. Potato spindle tuber viroid. X. Visualization and size determination by electron microscopy. *Virology* 55:70-80.

38. TAKAHASHI, T. and T. O. DIENER. 1975. Potato spindle tuber viroid. XIV. Replication in nuclei isolated from infected leaves. *Virology* 64:106-114.

39. VAN DORST, H. J. M. and D. PETERS. 1974. Some biological observations on pale fruit, a viroid-incited disease of cucumber. *Neth. J. Plant Path.* 80:85-96.

40. ZAITLIN, M. and V. HARIHARASUBRAMANIAN. 1972. Gel electrophoretic analysis of proteins from plants infected with tobacco mosaic viruses. *Virology* 47:296-305.

AMINOACYLATION OF VIRAL RNAs

M. PINCK, M. GENEVAUX, P. LESTIENNE and H. DURANTON

E.R.A. 104, C.N.R.S.

Institut de Botanique 28, Rue Goethe 67083-STRASBOURG

I - INTRODUCTION

In protein biosynthesis a primary step is the aminoacylation of tRNAs .

a) Activation of aminoacid

$$\text{Aminoacid+ATP+Enz} \xrightleftharpoons{Mg^{++}} \text{(aminoacyl.AMP.Enz)+PPi}$$

b) Formation of aminoacyl-tRNA

$$\text{(aminoacyl.AMP.Enz.)+tRNA} \xrightleftharpoons{Mg^{++}} \text{aminoacyl.tRNA+Enz+AMP}$$

The tRNAs are ribonucleic acids of low molecular weight: 25,000 - 30,000 daltons.

The question is : can we aminoacylate viral RNAs ? The answer is yes as you will see from this paper.

II - AMINOACYLATION OF TURNIP YELLOW MOSAIC VIRUS RNA (TYMV RNA)

TYMV RNA has a molecular weight of 2.10^6 daltons and a sedimentation constant of about 23S. M. PINCK et al. (1.2) showed that this RNA binds valine when incubated with ATP and *E. coli* cell-free extracts devoid of nucleic acids.

The requirements of this reaction are shown in table 1.

TABLE 1

REQUIREMENTS FOR THE BINDING OF VALINE TO TYMV-RNA

Conditions	^{14}C-Valine bound (pmoles)
Complete system	37
− Proteins	< 1
− ATP	< 1
− TYMV-RNA	< 1

The complete system is required to obtain binding of valine to cold TCA precipitable material. Neither GTP, nor CTP, nor
UTP can replace ATP in this reaction.

As can be seen in table 1, 37 pmol of valine were bound
per 50 pmol of TYMV RNA, which is a ratio of 0.7 : 1. This ratio
fluctuated between 0.3 and 0.8, depending on the RNA preparations,
but it never exceeded one. The incubation of RNA with a mixture of
labeled amino acids, followed, by alkaline hydrolysis and chromatography of the radioactive products bound to the RNA showed that
valine was the only aminoacid retained by TYMV RNA. The reaction
of valine binding to TYMV RNA is complete in 15 minutes at 30°. The
half-life of valyl-RNA of TYMV is similar to that of valyl-tRNA :
60 minutes at pH 8.6. Another similarity between valyl-RNA of TYMV
and valyl-tRNA is the fact that the NH_2 group of valine bound to

RNA of TYMV is free. This was shown by chemical acetylation to
yield acetyl valyl-RNA. This compound is readily hydrolyzed by
N-acyl aminoacyl-tRNA hydrolase, liberating acetyl valine, as is
also the case for acetyl valyl-tRNA (3).

Furthermore, chromatography of pancreatic RNase or snake
venom phosphodiesterase digests of ^{14}C-valyl-RNA of TYMV or of

[14]C-valyl-tRNA, yield radioactive compounds which correspond to valyl-adenosine or valyl-AMP respectively. This indicate that in the case of valyl-RNA of TYMV valine is bound to the 3'-terminal adenosine of the RNA molecule (1).

In order to elucidate whether, valyl-tRNA synthetase, the enzyme responsible for the binding of valine to tRNA^{val}, was also responsible for the binding of valine to TYMV RNA, purified valyl-tRNA synthetase was used. Surprisingly, in these conditions, valine binding was not observed. However when ATP (CTP) tRNA nucleotidyl transferase was included in the incubation mixture with valyl-tRNA synthetase a terminal adenosine was positioned on the 3'-end of the RNA, and binding of valine ensued (4). Thus, these two enzymes, valyl-tRNA synthetase and ATP (CTP) tRNA nucleotidyl transferase, as well as the hydrolase mentioned above are all enzymes which specifically react with tRNAs, and also recognize RNA of TYMV.

Because of the difficulty of preparing purified valyl-tRNA synthetase and ATP (CTP) tRNA nucleotidyl transferase, all subsequent experiments were performed with crude *E. coli* extracts freed of nucleic acids by chromatography on DEAE cellulose.

The great similarity that exists between valyl-RNA of TYMV and valyl-tRNA made it necessary to examine whether valine could be bound to host tRNA entrapped within the virus particle during the encapsidation process. It has, indeed, been shown that various tRNA molecules are recovered within avian myeloblastosis virus.

Therefore, the RNA of TYMV bearing valine was analysed by sucrose gradient centrifugation. About 80% of the valine was recovered with 23 S RNA, and most of the remaining 20% was bound to slowly sedimenting material (1). When the 23 S material was further examined by filtration through Sephadex G-75, it was not retarded on the column as compared to valyl-tRNA included as marker (3).

However, these experiments did not exclude the possibility that putative host valyl-tRNA entrapped within the virus particle could somehow be bound to the viral genome by protein-mediated linkages, ionic bonds or hydrogen bonds. We have made a thorough investigation of these possibilities, but none of our attempts have led to the liberation of a valyl-tRNA molecule of host origin.

Chromatography on Sephadex G-75 was employed to analyse TYMV RNA previously incubated with valine, after the following treatments :

- Radioactive RNA was treated with DMSO followed by formaldehyde. DMSO and formaldehyde break any hydrogen bond between a viral RNA and a tRNA (3).

Chromatography was also carried out in the presence of SDS and EDTA. SDS splits the bonds involving a protein, while EDTA breaks any bonds formed by a complex ion.

Our results demonstrate that TYMV RNA is involved in the esterification. The possibility of a host tRNA linked to the viral RNA by a hydrogen bond, a protein bridge, or a complex ionic bond is eliminated.

We have also analysed the valyl-oligonucleotides obtained by RNAse T_1 digestion of cabbage valyl-tRNA and compared them to the material obtained with TYMV valyl-RNA. Analysis of the products by DEAE cellulose chromatography in the presence of urea and an ammonium acetate gradient shows that two valyl-oligonucleotide fragments were recovered from cabbage valyl-tRNA. With valyl-RNA of TYMV, only one fragment was observed and it was distinct from the two valyl-oligonucleotides originating from valyl-tRNA (3).

II - UBIQUITY OF THE PHENOMENON

RNAs of other viruses from the tymovirus group have been tested. These viruses are : okra mosaic virus, eggplant mosaic virus (EMV), wild cucumber mosaic virus, cacao yellow mosaic virus ; belladona mottle virus and dulcamara mottle virus (5). All these viral RNAs have tRNA-like accepting properties. This particular function of a part of the viral genome seems to be a frequent phenomenon for plant viruses. Indeed other authors have shown that tobacco mosaic virus can be charged with histidine by a yeast aminoacyl tRNA synthetase preparation (6), and brome mosaic virus RNA charged with tyrosine by a bean aminoacyl tRNA synthetase fraction (7). This property was also demonstrated for an animal virus in the case of mengo virus RNA, which can be charged with histidine (8).

III - COMPLEXITY OF THE PROBLEM

Recently we have developed our research on EMV RNA. It has the property to accept valine but also has a 4 S "companion" RNA (9).

EMV RNA was incubated in the presence of radioactive amino acids and analysed by Sephadex filtration as described previously. The amino acids bound to the RNAs were released by alkaline hydrolysis and analysed by thin layer chromatography. The radioactive material bound to EMV RNA was identified as valine and lysine.

Protein-mediated linkages and hydrogen or ionic bonds were studied by SDS, urea, DMSO and EDTA treatments .We have shown that urea DMSO and EDTA treatments release (^{3}H) lysyl-RNA from EMV RNA. It is striking that binding of ^{14}C valine to EMV RNA is not modified by any of these treatments.

Thus, valine is linked to the genome of the virus but lysine is linked to a 4 S "companion" RNA.

We have analysed the lysyl-oligonucleotides obtained by RNAse T$_1$ digestion of *Datura stramonium* lysyl-tRNA, and we have compared them with the oligonucleotides obtained from purified EMV$_4$ "companion" RNA. The results obtained show that two major (^{14}C) lysyl-oligonucleotide peaks are obtained from infected or healthy *Datura stramonium* lysyl-tRNA (9). This proves that there are at least two different isoacceptor tRNAs for lysine in *Datura stramonium*. Only one lysyl-oligonucleotide is found with lysyl "companion" RNA of EMV. It corresponds to the first peak eluted from *Datura* lysyl-tRNA. However, this does not mean that lysyl "companion" RNA is similar with one of the lysyl tRNAs of the *Datura* (9).

It is known that EMV typically has two types of particles. The heavy particles (bottom component) only contain the genome and the linked "companion" RNA but not free tRNAs. The light particles (top component), which are considered as "empty" protein shells in the tymovirus group contain 2.5 per cent of 4 S RNAs (10). The heavy RNA is amino acylated by valine and lysine. The RNA extracted from the light particles accepted alanine arginine, histidine, leucine, lysine, phenylalanine, tyrosine and valine (10).

The presence of the "companion" RNA and of the 4 S RNAs was not found to be dependent on the host plant choosen to culture the virus. Identical results were obtained with EMV RNA extracted from *Datura stramonium*, tomato (*Lycopersicum esculentum*), tobacco (*Nicotiana clevelandii*) and two Chenopodium species (*C. amaranticolor* and *C. quinoa*) (11-12).

IV - CELLULAR LOCALISATION OF THE VALYL-tRNA SYNTHETASE WHICH IS ABLE TO AMINOACYLATE VIRAL RNA

We know from the work of BOVE's group that the double standed RNA of TYMV is associate with polyplast (13). The viral replicase has been partially purified from chloroplasts of infected leaves. It is bound to the membrane. On the other hand, autoradiographic studies have indicated the possible existence, in the nucleus of a TYMV-induced RNA synthesis that is resistant to actinomycin D (14).

Thus, if a valyl-tRNA synthetase specific for tymovirus RNA was found to be present in a specific compartment (chloroplast, cytoplasm or mitochondrion) this could be an argument for the localisation of viral replication, especially if aminoacylation of the 3' end of viral RNA is really involved in its replication.

Studies were performed on tomato (host plant) and bean (non-host plant) in order to analyse the valyl-tRNA synthetase of the different cell compartments (15 - 16). The same profile of activity was obtained with these two plants ; however, bean extracts were studied in more detail since it is easier to obtain mitochondria and hypocotyls. Fig. 1 summarises our results.

1) Valyl-tRNA synthetase eluted in peak I corresponds to the chloroplast enzyme. It was not possible to aminoacylate EMV RNA significantly with this enzyme, although tRNA of *E. coli* could be aminoacylated.

2) Valyl-tRNA synthetase eluted in peak II is present in mitochondria purified according to GUILLEMAUT et al. (17), and also in small quantity in young dark grown hypocotyls. It can aminoacylate EMV RNA but not *E. coli* tRNA.

3) Valyl-tRNA synthetase eluted in peak III is present in leaves, hypocotyls and mitochondria. Two enzymes are probably present in this peak : one of mitochondrial origin and the other cytoplasmic, as shown by sensitivity to potassium phosphate.

They do not aminoacylate *E. coli* tRNA, and poorly charge EMV RNA.

The total enzyme preparation from bean mitochondria aminoacylated RNAs from eggplant mosaic virus, turnip yellow mosaic virus, okra mosaic virus, wild cucumber mosaic virus with valine and brome grass mosaic virus with tyrosine.

From these results, we can conclude that :
a) The presence of aminoacyl-tRNA synthetase which can recognize viral RNA does not confer host specificity since acylating enzymes can be obtained from plants which do not function as hosts.

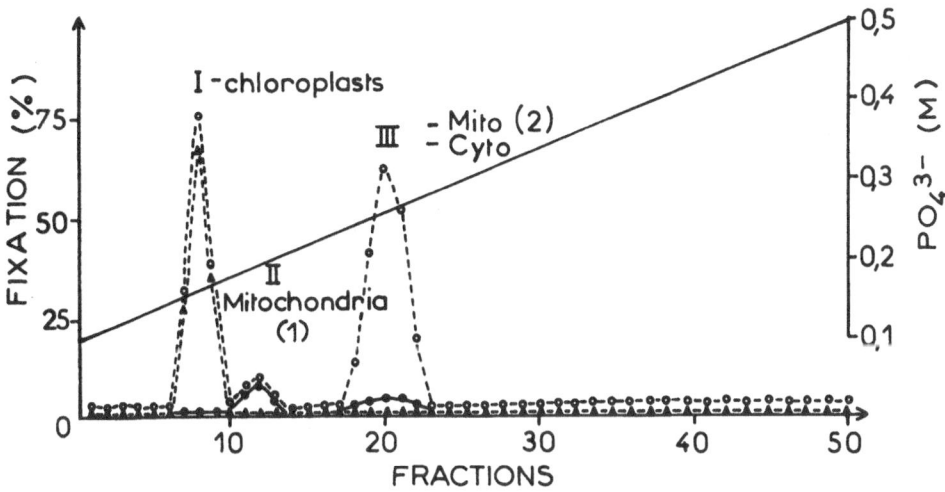

FIGURE 1

        Chromatography of valyl-tRNA synthetases from *phaseolus vulgaris* on hydroxyl apatite column (1 cm x 10 cm) at pH 7.5. The column was equilibrated with 0.1 M potassium phosphate buffer pH 7.5, 1 mM $MgCl_2$, 1 mM 2 β Mercapto ethanol, 10% glycerol. Proteins bound to the column were eluted in a 0.1-0.5 M potassium phosphate gradient (total volume = 100 ml) in the same buffer. 2 ml fractions were collected at a flow rate of 6 ml/h. Valyl-tRNA synthetase activities were determined as follows.

        The incubation mixture contained in a final volume of 100 /ul : 10 /umoles of Hepes (pH 7.5) ; 1 /umol (for tRNA) or 1.5 /umol (for EMV RNA) $MgCl_2$; 1 /umol (for tRNA) or 0.75 /umol (for EMV RNA) ATP ; 0.25 /umol of 2 β mercaptoethanol, 3 nmol of ^{14}C valine (280 mci/mmol) ; 300 pmol of tRNA or 16 pmol of EMV RNA, and samples of 5 /ul of each fraction. After incubation for 20 min at 37°C, the reactions were stopped by addition of 5% (w/v) cold TCA and the resulting precipitates extensively washed on "millipore" filters (0.45 /um), and counted in toluene-PPO (5 g/l.). Enzymatic activity is shown for EMV RNA (●——●), hypocotyl tRNA (o----o) and *E. coli* tRNA (▲----▲).

b) The mitochondrion seems to be the compartment containing valyl-tRNA synthetase which is able to aminoacylate RNA of viruses belonging, at least, to the tymovirus group.

Those results have to be considered as preliminary. We intend to work further in this area.

V - CONCLUSION

The biological function of the amino-accepting activity of RNA from the tymoviruses and from other viruses is still unknown, but one can conjecture that the aminoacylation of viral RNA might be involved in the regulation of replication (18) (19) or translation (20) (21). The valine-accepting structure seems to be replicated, and for EMV the lysine-accepting structure ("companion RNA") is bound to the replicative form (22).

Further studies will be required to determine the role of these amino acid accepting RNAs.

REFERENCES

1) PINCK M., YOT P., CHAPEVILLE F. and DURANTON H.M. - Nature 226, 954, 1970.
2) PINCK M. - Thèse doctorat d'Etat, STRASBOURG (1970).
3) YOT P., PINCK M., HAENNI A.L., DURANTON H.M. and CHAPEVILLE F. - Proc. US Nat. Acad. Sci. 67, 1345 (1970).
4) LITVAK S., CARRE D.S. and CHAPEVILLE F. - FEBS Letters 11, 316, (1970).
5) PINCK M., CHAN S.K., GENEVAUX M., HIRTH L. and DURANTON H. - Biochimie 54, 1093 (1972).
6) OBERG B. and PHILIPSON L. - Biochem. Biophys. Res. Communi. 48, 927 (1972).
7) HALL T.C., SHIH D.S. and KAESBERG P. - Biochem. J. 129, 969, (1972).
8) SALOMON K. and LITTAUER U.Z. - Nature 249, 32 (1974).
9) PINCK M., GENEVAUX M. and DURANTON H. - Biochimie 56, 423 (1974).
10)BOULEY J.P., BRIAND J.P., GENEVAUX M., PINCK M. and WITZ J. - Virology 69, 775 (1976).
11)GENEVAUX M., PINCK M., LESTIENNE P. and DURANTON H.M. - Proceedings of the Tenth FEBS Meeting (1975).
12)GENEVAUX M., PINCK M. and DURANTON H.M. - Ann. Microbiol. (Inst. Pasteur) 127, 47, 1976.
13)MOUCHES C., BOVE C. and BOVE J.M. - Virology 58, 409 (1974).
14)BREDBROOK J.R. and MATTEWS - Ann. Microbiol. (Inst. Pasteur) 1976, 127 A, 55.
15)BURKARD G., GUILLEMAUT P. and WEIL J.H. - Biochim. Biophys. Acta 224 (1970) 184-198.
16)JEANNIN G. - Thèse de Doctorat d'Etat, STRASBOURG (1976).

17) GUILLEMAUT P., BURKARD G. and WEIL J.H. (1972) - Phytochemistry 11, 2217

18) LITVAK S., TARRAGÓ A., TARRAGÓ-LITVAK L., ALLENDE J.E. - Nature 241, 88 (1973).

19) BASTIN M., HALL T.C. - J. Virol. (submitted).

20) CHEN J.M. and HALL T.C. - Biochemistry 12, 4570 (1973).

21) HAENNI A.L., PROCHIANTZ A., BERNARD O., CHAPEVILLE F. - Nature 241, 166, 1973.

22) PINCK L., GENEVAUX M., BOULEY J.P. and PINCK M. - Virology 63, 589 (1975).

IN VIVO AND IN VITRO TRANSLATION OF THE RNAS OF ALFALFA MOSAIC VIRUS

L. van Vloten-Doting,* J. Bol,* L. Neeleman,* T. Rutgers,*
D. van Dalen,* A. Castel,* L. Bosch,* G. Marbaix, **
G. Huez,** E. Hubert,** and Y. Cleuter**
  *Department of Biochemistry, State University
  Wassenaarseweg 64, Leiden, The Netherlands
 **Department of Molecular Biology, Free University of Brussels
  Brussels, Belgium

SUMMARY

The translation of the purified RNAs of alfalfa mosaic virus, a
virus with a coat protein dependent tripartite genome, was studied
in different systems. On each RNA only one initiation site was
found except on RNA 3. To this RNA two ribosomes could be bound in
the presence of inhibitors of the protein elongation. Both in vivo
and in vitro coat protein was formed under the direction of RNA 3.
The discrepancy between the results of the infectivity and the translation studies, can either be due to an influence of RNA 1 and/or
RNA 2 on the translation of RNA 3, or to a difference in regulation
between wheat germ extracts and oocytes on one hand and tobacco and
bean cells on the other hand. In the E.coli system the translation
of RNA 4 can start either with acetylphenylalanine or with formylmethionine. From a comparison of the translation products we calculated that the initiation codon is located at or beyond position
40 from the 5' terminus.

INTRODUCTION

In a plant cell a very large number of genes is expressed at
the same time. Since the messenger RNAs and the products are both
present in rather small quantities, it is difficult to obtain insight in the mechanisms of translation and regulation of these messengers. One way to simplify this problem is to study the expression of viral RNAs.

The virus on which our group concentrates is alfalfa mosaic
virus, a virus with a tripartite genome. The genome fragments, called RNA 1, 2 and 3 (molecular weights 1.13, 0.82 and 0.70 million

387

daltons, respectively (Heijtink, 1974)), are separately encapsidated
in bacilliform particles. In virus preparations an additional nucleoprotein containing RNA 4 (0.28 million daltons) is found. RNA 4
functions in vitro as an efficient messenger for the viral coat protein (Mohier et al., 1975; Van Vloten-Doting et al., 1975; Thang et
al., 1975).

A mixture of the three genomic RNAs has to be supplemented with
a small amount of coat protein, or its messenger, to initiate the
infection (Bol et al., 1971). This phenomenon is now recognized a-
mong several plant viruses with a tripartite genome (reviewed by
Van Vloten-Doting, 1976 and Van Vloten-Doting and Jaspars, 1976).

From genetic experiments it is evident that the genomic information for the coat protein is located on RNA 3 (Dingjan-Versteegh
et al., 1972). In vivo this information can apparently only come to
expression in the presence of coat protein.

To gain insight into the role of the coat protein we studied the
translation of the AMV-RNAs in different systems. Since it is known
that most eukaryotic RNAs are associated with protein during translation, and these proteins might influence the translation of the
RNA, polyribosomes were isolated from tobacco and supplemented with
all requirements for protein synthesis. The products specified by
the polyribosomes isolated from healthy and infected tobacco were
compared.

The results will be discussed in relation to those obtained
with the translation of the RNAs from bromegrass mosaic virus, since
this virus has a very similar genome constitution but is not dependent upon coat protein for infectivity (Lane and Kaesberg, 1971).

MATERIALS AND METHODS

Preparation of AMV-RNAs, of wheat germ extract and of AMV coat
protein, in vitro protein synthesis, polyacrylamide gel electrophoresis in cylinders, and peptide mapping were performed as described
earlier (Van Vloten-Doting et al., 1975).

Preparation of E.coli cell-free extracts. Cell-free extracts
of E.coli MRE 600 or Q13 were prepared according to Nirenberg and
Matthaei (1961), modified by Voorma et al.(1965).

In vitro protein synthesis in E.coli extracts. The reaction
mixture of 100 µl contained 30 µl S30 (or S100 and ribosomes), 9 µg
leucovorin, 1 mM ATP, 5 mM phosphoenolpyruvate, 0.12 mM GTP, 0.05 mM
unlabeled aminoacids, 12 mM KCl, 50 mM Tris-HCl (pH 7.8), 5 mM 2-
mercaptoethanol , 13 mM $MgCl_2$, RNA 4 and labeled amino acids as indicated in the legends. After incubation for 60 min at 37°, RNase A
(20 µg/ml) and RNase Tl (6 U/ml) were added and the incubation was
prolonged for 15 min.

Injection and labeling of oocytes from Xenopus leavis, preparation of labeled oocytes for SDS-gel electrophoresis, and SDS slab
gel-electrophoresis were performed as described by Knowland (1974).

Isolation of polyribosomes, labeling of leaf material, RNA-RNA hybridization and immune precipitation of the in vitro labeled protein is published elsewhere (Bol et al., 1976).

Binding of AMV (^{3}H)RNA to ribosomes. The binding reactions were carried out in a standard incubation sample of 100 µl. From each of the RNAs 2.5 µg (specific activity of RNA 1 5700 cpm/µg, RNA 2 21,800 cpm/µg, RNA 3 5200 cpm/µg and RNA 4 17,200 cpm/µg) was used. After 10 min at 30° the incubations were stopped by the addition of 200 µl of cold 40 mM Tris pH 8.5, 20 mM KCl and 10 mM Mg-acetate. The samples were applied to 12.5 - 50% sucrose gradients in the same buffer and centrifuged for 3 hr at 36,000 at 4° in a SW-41 rotor. The A260 of the gradients was monitored in a Philips recording spectrophotometer, and the radioactivity of the fractions was determined after addition of 25 volumes of scintillation fluid consisting of a mixture of 1.5 l toluene, 750 ml triton X-100, 225 ml distilled water, 12.37 g PPO (2,5-diphenyl-oxazol) and 225 mg POPOP p-bis-2(5-phenyl-oxazol-benzene).

Preparation of radioactive labeled aminoacyl-tRNA, and binding of aminoacyl-tRNA to ribosomes was carried out as described earlier (Verhoef et al., 1968).

Preparation of N-acetylphenylalanine-tRNA. Ac-(^{14}C)-Phe-tRNA was prepared with acetic anhydride according to Haenni and Chapeville (1966). The percentage of acetylation was estimated after alkaline hydrolysis for 30 min at 37°. The intactness of the ester bond was determined according to Heller (1966).

RESULTS

Figure 1 shows the densitograms of polyacrylamide gels run with a total AMV-RNA preparation or with the purified RNA preparations used as messenger in the experiments described below.

ANALYSIS OF TRANSLATION PRODUCTS

Products directed by RNA 4.

It has been shown earlier that in different eukaryotic systems AMV-RNA 4 can be translated with fidelity into coat protein (Thang et al., 1975; Van Vloten-Doting et al., 1975; Mohier et al., 1975).
Injection of RNA 4 into oocytes resulted in the formation of a product which comigrates with authentic coat protein on SDS polyacrylamide gels (Fig.2 track 2, 3).

Cell-free systems of E.coli also translate RNA 4. The major product is very similar to the coat protein as follows from electrophoresis on SDS polyacrylamide gels (Fig.3B) and fingerprint analysis of the tryptic digest (Fig.4A). The biosynthetic reaction is more readily completed in a crude S 30 E.coli extract (Fig.3B) than in a

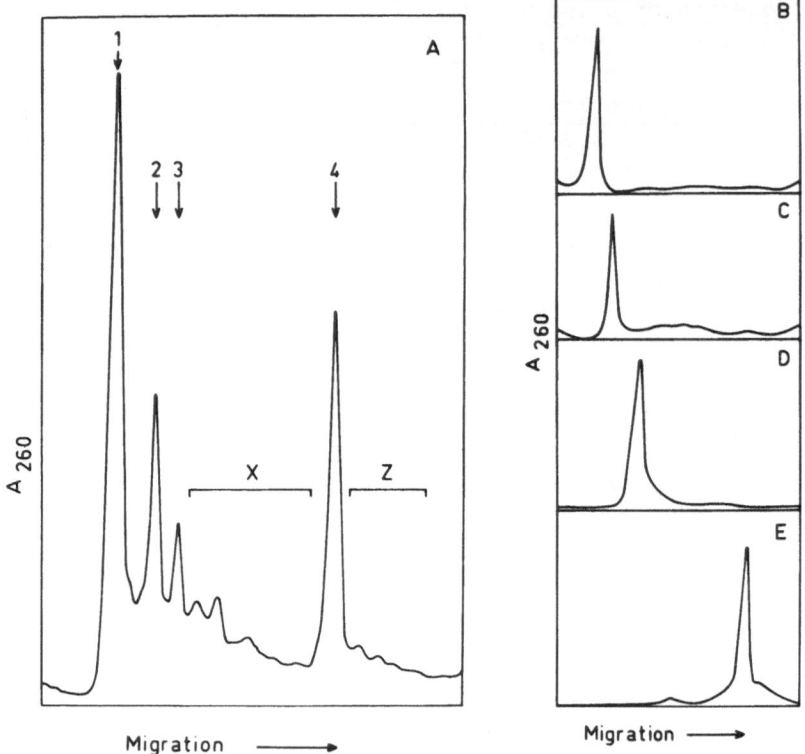

Fig.1. Densitograms of 3% polyacrylamide gels run with a total AMV-RNA preparation (A), purified RNA 1 (B), RNA 2 (C), RNA 3 (D) and RNA 4 (E).

system containing ribosomes and S 100 enzymes (Fig.3A).[*] In the latter case a variety of polypeptides with lower molecular weights dominate. Tryptic analysis shows that they represent N-terminal coat protein fragments of different lengths (results not shown). Under the peptide maps of Figure 4 the relative lengths and the alignment of the tryptic peptides of AMV coat protein is shown (compare Van Beynum, 1975; Kraal et al., 1976; Van Beynum et al., to be published). The N-terminal peptides 1 and 1a are not present in a tryptic digest of the E.coli product (Fig.4A). All others can unambiguously be detected with the exception of the large tryptic peptides 6, 10, 12 and 17 which are difficult to localize even in

---

[*]G. van Dieijen (unpublished results) has shown that also with MS_2RNA a crude extract (S 30) is needed to synthesize coat protein and that S 100 + purified ribosomes is not sufficient. The latter system is only capable to produce coat protein when the brown-yellowish material which floats on the ribosome-pellet after ultra centrifugation, and which is usually discarded, is added back to the system.

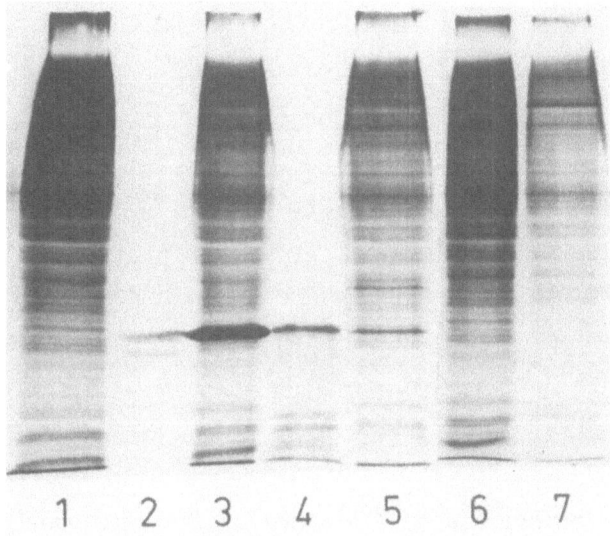

*Fig.2. Gel electrophoretic patterns of products synthesized under
the direction of the different AMV-RNAs.
Into each oocyte 50 nl RNA (1 mg/ml) was injected. 10 oocytes were
incubated for 18 hr in 0,1 ml medium containing 1 mCi/ml (^{3}H)Ala
(50 Ci/mmol).
Track 1: control oocytes, track 2: authentic coat protein, track
3: RNA 4 in oocytes, track 4: RNA 4 in wheat germ, track 5: RNA 3
in oocytes, track 6: control oocytes, track 7: RNA 1 in oocytes.
The samples analyzed on track 6 and 7 were obtained from a different batch of oocytes.*

a digest of the authentic coat protein. The peptides 1 and 1a:
ac-Ser-Ser-Ser-Gln-Lys and ac-Ser-Ser-Ser-Gln-Lys-Lys are present
in the biosynthetic product formed in a cell-free system of wheat
germ (Van Vloten-Doting et al., 1975 and Fig.4B). Since RNA 4 in
the E.coli system can direct the incorporation of N-formylmethionine
it is assumed that the biosynthetic product formed in this system
is initiated by fMet-tRNA. This matter is presently under investigation. It is concluded that RNA 4 is faithfully translated in the
E.coli cell-free system into coat protein with the reservation that
the N-terminus of the biosynthetic product differs from that of the
authentic coat protein.

Products directed by RNA 3.

The main translation product of RNA 3, both in the cell-free
system of wheat germ, reticulocytes and Krebs ascites cells,has an
estimated molecular weight of 35,000 daltons (Van Vloten-Doting et
al.,1975; Mohier et al., 1975). Besides this product two additional

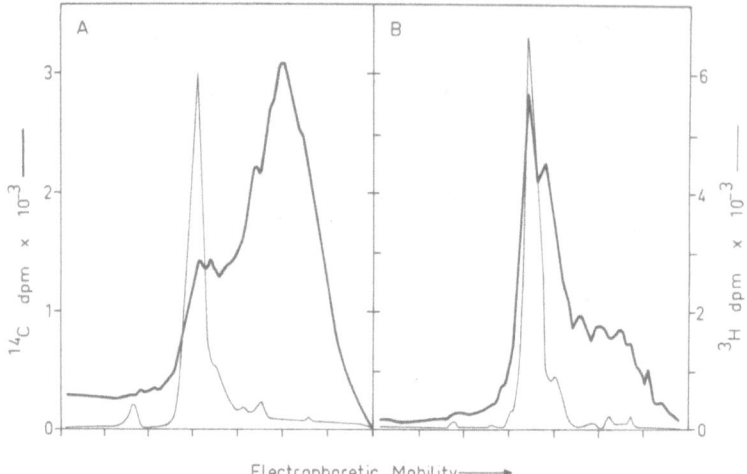

Fig.3. Coelectrophoresis of (^{3}H)-labeled authetic coat protein with (^{14}C)-labeled products synthesized <u>in vitro</u> under the direction of RNA 4; (A) E.<u>coli</u> S 100 extract, supplemented with ribosomes, (B) E.<u>coli</u> S 30 extract. Reaction mixtures (100 µl) contained 20 µg RNA, 0.125 µCi (^{14}C)Lys (348 mCi/mmol) and 0.125 µCi (^{14}C)Arg (312 mCi/ mmol).

products are formed in wheat germ (Van Vloten-Doting, 1976). One of these comigrates with the authentic coat protein, while the other has an estimated molecular weight of 65,000 daltons (compare Fig. 15). The presence of coat protein sequences in these two proteins was demonstrated by the fact that both could be precipitated with AMV antiserum. Further analysis of these proteins is greatly hampered by the fact that RNA 3 is a very inefficient messenger.

Upon injection of RNA 3 into oocytes two proteins were formed, which comigrated with the 35,000 and the coat protein bands formed in the wheat germ cell-free system (Fig.2 track 5). In the region of 70,000 daltons no products were seen.

The three proteins found, can have been formed in one of the following ways :

A) RNA 3 has one initiation site. The RNA is translated into one precursor protein (65,000), which is subsequently cleaved into the 35,000 and the coat protein.

B) RNA 3 has one open and one closed initiation site. Part of RNA 3 is degraded during the incubation, as a result of the degradation the blocked initiation site becomes available for translation.

C) RNA 3 has two initiation sites. Both cistrons are translated simultaneously.

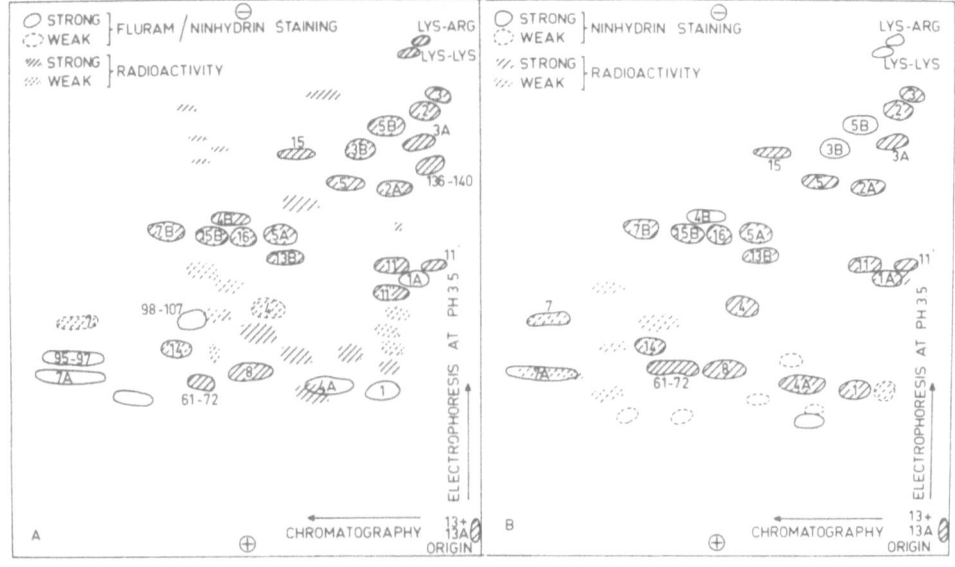

Fig.4. Tryptic "fingerprints" of AMV coat protein mixed before digestion with material synthesized under the direction of RNA 4 in a E.coli extract (S 30) (A, labeled with ^{14}C-Lys and Arg),or in wheat germ extract (B, labeled with ^{14}C-Ser, -Pro, and -Gly). Under the peptide maps the relative length and the alignment of the tryptic peptides is shown (Van Beynum, 1975; Kraal et al.,1976; Van Beynum et al.,to be published). At places with two adjacent basic amino acids two cuts are possible, the resulting splitting products are called 1 and 1^a etc. The trypsin preparation (Serva) used, contained some chymotryptic activity. The resulting splitting products are indicated by 61-72, etc. representing the amino acids starting to count at the N-terminus of the peptide chain.

In both model B and C it has to be assumed that the ribosome occasionally passes along the first termination site, yielding a "read-through" product (65,000).

There are two arguments against model A : the ratio of 35,000 protein to coat protein varies from experiment to experiment, and time studies suggested that the coat protein and the 35,000 protein appeared before the 65,000 protein.

It is very difficult to exclude model B. In oocytes RNAs seem
to be rather stable, since a substantial amount of coat protein was
formed between 48 and 52 hours after injection of RNA 4 (unpublished
results of G.Marbaix, G.Huez and T.Rutgers). However this does not
exclude the possibility that there is a special processing of RNA 3.
In wheat germ a lot of nucleolytic activity must be present since
after an incubation of only 30 min already more than 50% of the RNA
has become TCA soluble (unpublished results).

Products directed by RNA 2 and RNA 1.

In a cell-free system of wheat germ both RNA 2 and RNA 1 promote the synthesis of a large number of products (Fig.5A and B respectively). When the incubation time is extended from 60 min to 120
min the largest product directed by RAN 1 is found to increase in
size from 70,000 daltons to 90,000 daltons (Fig.5B, compare Van
Vloten-Doting et al., 1975). Apparently those RNA molecules which
are engaged in protein synthesis are protected against nucleolytic
activity. The fact that the pattern of the products directed by RNA
1, did not change between 2 hr and 17 hr incubation (results not
shown), suggest that the wheat germ extract contain only a low a-
mount of proteolytic activity, if any.
An extension of the incubation time had much less effect on the
size of the products directed by RNA 2.
The largest product formed under the direction of RNA 2 is
about 70,000 daltons and of RNA 1 90,000 daltons, respectively. Since
products representing all the genetic information would be 80,000 and
110,000 daltons, respectively, it seems likely that the largest products represent complete translates of the RNAs, while the smaller
products represent incomplete chains.
Similar results were found when RNA 2 and RNA 1 were translated
in cell-free systems of reticulocytes or Krebs ascites cells (Mohier
et al., 1975).
Upon injection of RNA 1 into oocytes a major product with an
estimated molecular weight of 110,000 daltons (Fig.2 track 7) was
found. Up to now no products directed by RNA 2 were seen (results
not shown). This can be due either to a failure of the oocytes to
translate RNA 2, or the translation product is removed together with
the yolk protein (Knowland, 1974).
The fact that RNA 2 and RNA 1 can both yield a product which
corresponds to almost the entire genetic information present in these
viral messengers, does not exclude the possibility that one or both
of these RNAs code for more than one protein (compare model B and
C of the previous section).

Products directed by viral polyribosomes.

In Figure 6 the polyribosome profiles obtained from healthy
(Fig.6A) and from infected tobacco leaves (Fig.6B) are shown. The
majority of the polysomes contains 2 to 7 ribosomes, but larger

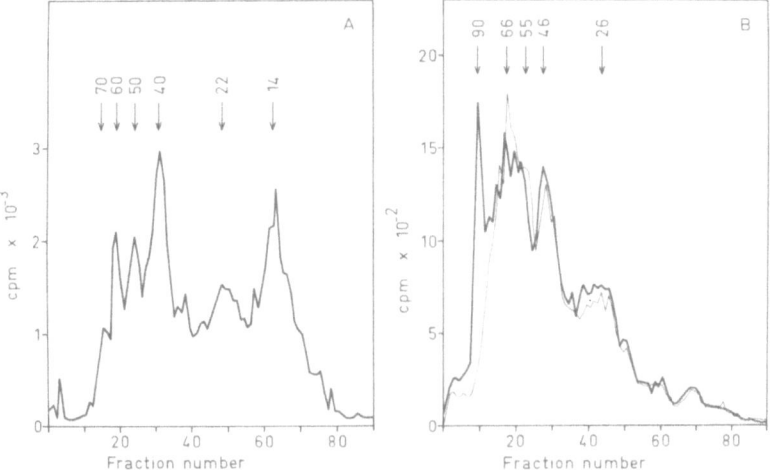

Fig.5. Gel electrophoretic pattern of products synthesized under the direction of RNA 2 (A) and RNA 1 (B) in the wheat germ cell-free system. (A) the reaction mixture (100 μl) contained 2.1 μg RNA 2 and 0.9 μCi (³⁵S) methionine (67 Ci/mmol). Incubation 2 hr at 30°. (B) the reaction mixture (100 μl) contained 4 μg RNA 1, 50 μM spermine and 0.47 μCi (³⁵S) methionine (220 Ci/mmol). After 5 min incubation 4.10⁻⁵ M aurintricarboxylic acid was added and the concentration of KCl was raised from 50 mM to 80 mM, incubation 1 hr (——) and 2 hr (——) at 30°.

structures carrying upto 25 ribosomes are present (see extrapolation in Fig.6B). Upon dissociation of the polyribosomes with EDTA an mRNP is released with a buoyant density of 1.60 g/cm³. A more detailed analysis of the RNA and protein content of the polyribosomes is given by Bol et al.(1976).

To identify the viral mRNAs,(³H) uridine-labeled polyribosomes were isolated from infected and from healthy leaves. Polysomal structures sedimenting faster than monosomes were collected and extracted with phenol and SDS. Figures 7A and B show the patterns of polysomal RNA from infected and healthy leaves, respectively. The polysomal RNA preparations from infected leaves contains RNA species which comigrate with the four major AMV-RNAs and which occur in a similar distribution (compare Fig.1A). In addition two prominent peaks are found in the X-region which are called Xa and Xb and have estimated molecular weights of about 0.6 and 0.5 million daltons, respectively. Polysomal RNA, prepared under comparable conditions from healthy leaves, contains only small amounts of label comigrating with ribosomal RNA (Fig.7B). To demonstrate the presence of viral sequences in polysomal RNA from infected leaves a polyacrylamide gel was run with labeled polysomal RNA, fractionated and the material eluted

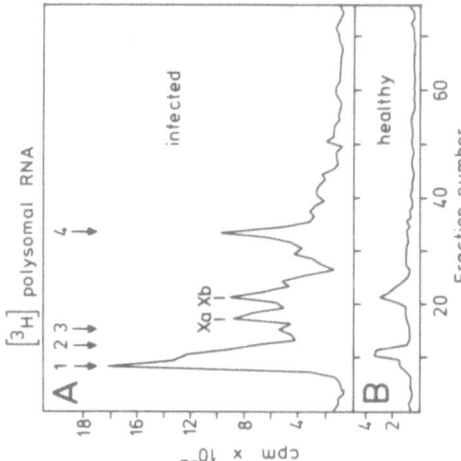

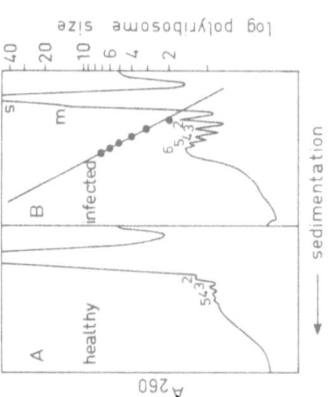

Fig. 6. Sucrose density gradient profiles of polyribosomes extracted from healthy (A) or infected (B) tobacco leaves. The position of AMV particles run in a parallel gradient, is indicated in (B). Nomenclature of the peaks is according to Jackson and Larkins (1976).

Fig. 7. Gel electrophoretic pattern of (³H)uridine-labeled polysomal RNA. Radioactivity pattern of gels run with RNA from polyribosomes isolated from infected (A) and healthy (B) tobacco leaves.

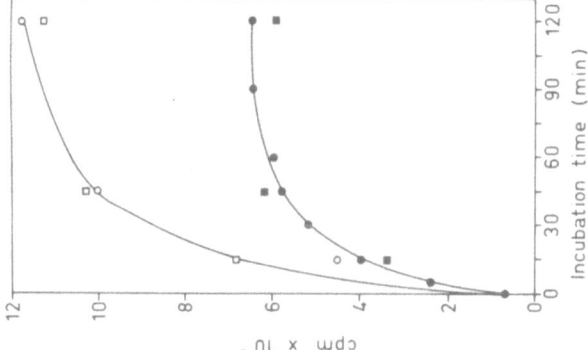

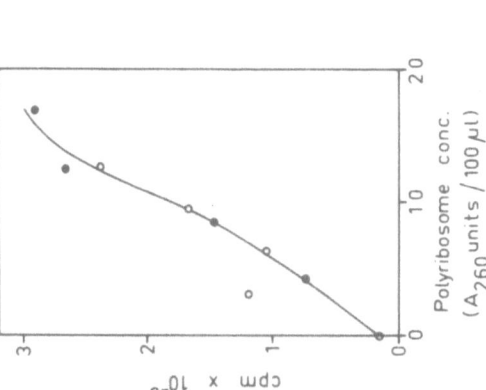

Fig.8. Stimulation of the in vitro amino acid incorporation by increasing amounts of polyribosomes isolated from infected leaves (——) and healthy leaves (——○——). Incubation was for 45 min at 3.2 mM Mg-acetate and 0.1 μCi (14C) Lys (348 mCi/mmol) per incubation mixture (100 μl).

Fig.9. Time course of the in vitro amino acid incorporation directed by polyribosomes in the absence and presence of aurintricarboxylic acid (4 x 10⁻⁵ M ATA). Polyribosomes from infected leaves (——), + ATA (——○——), from healthy leaves (——■——), + ATA (——□——).

from the fractions was annealed to virus-specific ds-RNA. The annealing pattern closely followed the radioactivity pattern, indicating
that labeled polysomal RNA of all size classes contains virus-specific sequences.

Figure 8 shows that addition of increasing amounts of unfractionated polyribosomes isolated from healthy and infected leaves to
a post-ribosomal supernatant from wheat germ and other requirements
for protein synthesis, results in an increasing amino acid incorporation. In figure 9 the time course of the incorporation, determined
in the presence and absence of an inhibitor of the initiation of protein synthesis (aurintricarboxylic acid, Pestka, 1974) is shown. The
incorporation levels off after about 45 min and is not influenced by
the antibiotic. Apparently the incorporation is solely due to elongation of the nascent polypeptide chains.

To identify the polypeptides completed in vitro, polyribosomes
isolated from infected and healthy tobacco leaves were allowed to
complete nascent chains during a 2 hr incubation. The pattern produced by polyribosomes from infected leaves shows a distinct peak
which is absent from the pattern produced by polyribosomes from
healthy material (Fig.10A). The material represented by this peak
comigrates with coat protein and is precipitated by AMV antiserum
(Fig.10B); only residual amounts are present in the supernatant
(Fig.10C).

When polyribosomes were isolated with a low salt buffer, the
mRNP had a higher protein content ($\rho = 1.45$ g/cm^3. This mRNP directed the same products as shown in figure 10).

These results demonstrate that, despite the fact that the viral
mRNP population contains about equimolar amounts of RNA 1 and RNA 4,
the major product directed in vitro by polyribosomes is viral coat
protein. Apparently the in vitro elongation directed by other messengers than RNA 4 is greatly reduced. This can be due to the presence in the mRNP of a viral product which exert a regulatory function in the translation process.

DETERMINATION OF THE NUMBER OF INITIATION SITES

To determine the number of initiation sites on each of the RNAs
we have determined how many ribosomes could be bound to the RNAs in
the presence of inhibitors of the protein elongation. Both fusidic
acid and sparsomycin (Pestka, 1974) were used. Figure 11 shows the
inhibition of the protein synthesis at different concentrations of
the inhibitors.

Figure 12 shows the polysome profiles found in the absence and
the presence of 0.2 mM sparsomycin and 1 mM fusidic acid. In most
experiments only about 10% of the (^{3}H)RNA was bound to the ribosomes,
while the rest of the radioactivity was found at the meniscus. This
is probably due to the high level of nucleolytic activity found in
the wheat germ extracts. The results obtained with RNA 4 (A, B and
C) are consistent with the idea that this RNA is a monocistronic mes-

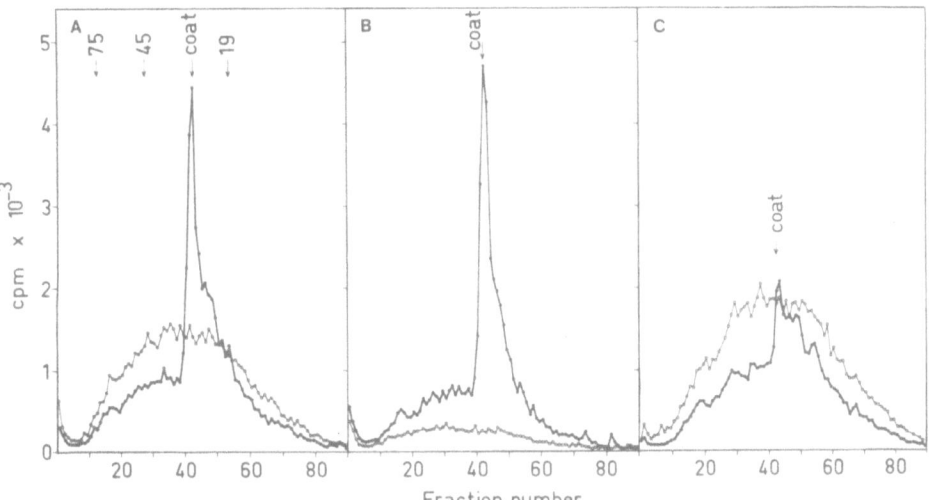

*Fig.10. Gel electrophoretic pattern of products synthezed under
the direction of polyribosomes isolated from infected (——) and
healthy (——) tobacco leaves. (A) unfractionated product, (B)
fraction precipitated with serum raised against AMV, (C) fraction
not precipitated by antiserum. Mixtures of 250 µl, containing 5 mM
Mg-acetate, 2.45 A260 units of polyribosomes and 2 µCi (³⁵S)Met
(60 Ci/mmol), were incubated for 2 hours. The position of coat
protein run in a parallel gel is indicated.In (A) the estimated
molecular weights of a number of the other products is given.*

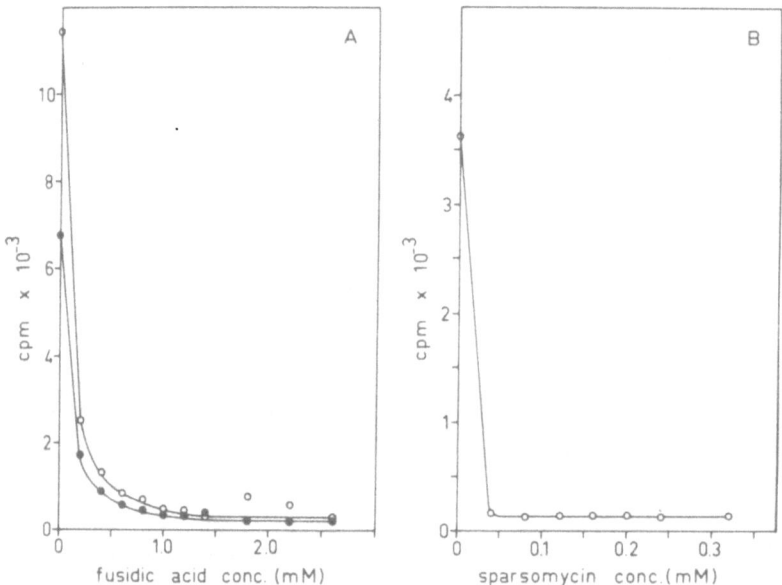

Fig.11. Inhibition of translation by fusidic acid (A) or sparso-
mycin (B). Reaction mixtures (100 μl) contained 2,5 μg (—○—)
RNA 4, (—•—) RNA 1.

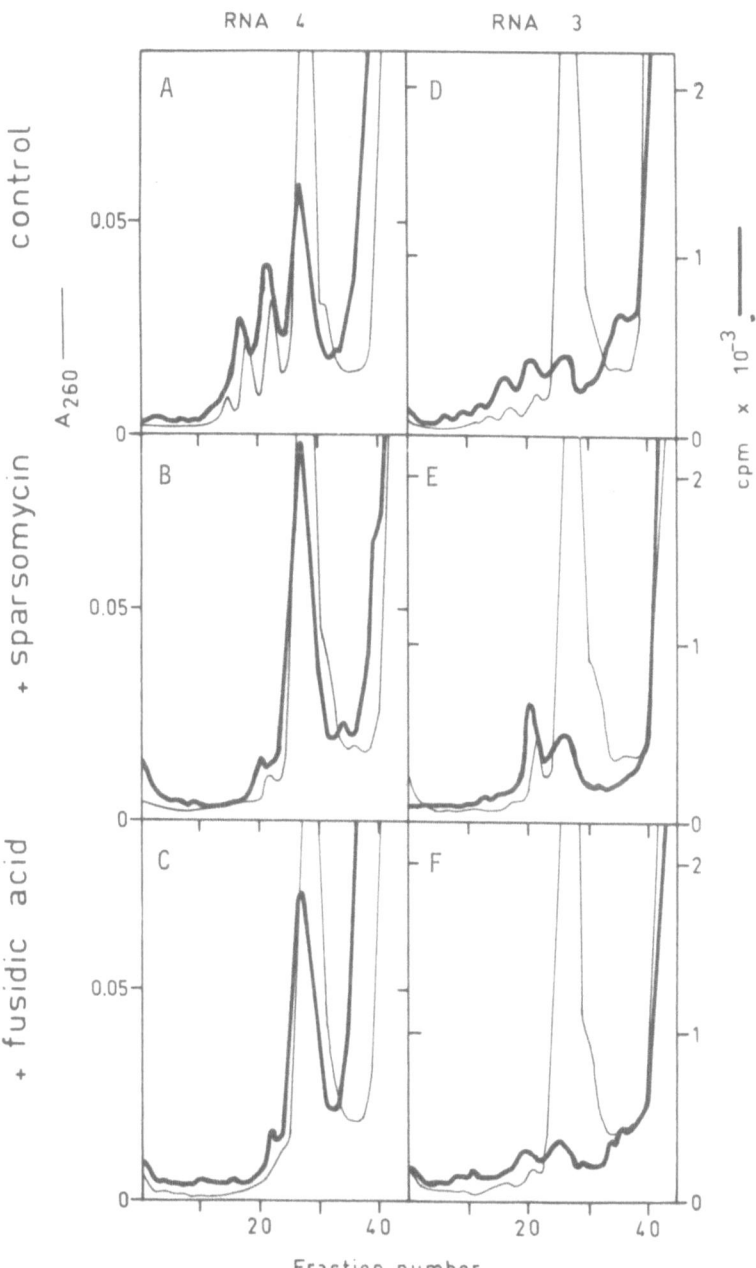

Fig.12. Sucrose density gradient profiles of polyribosomes directed by the (^{3}H) AMV-RNAs in a wheat germ cell-free system. Polyribosomes containing RNA 4 (A, B and C), RNA 3 (D, E and F), RNA 2 (G, H and

(continued next page)

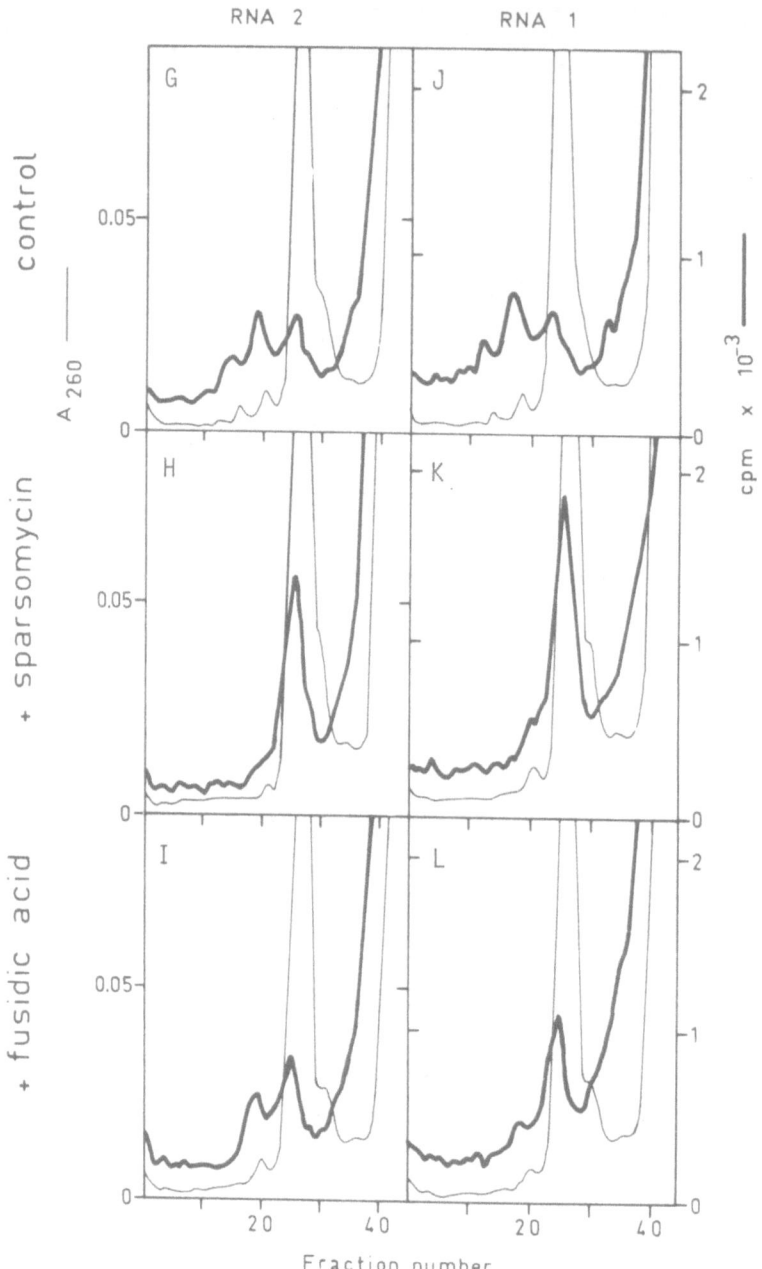

I) and RNA 1 (J,K and L), in the absence of inhibitors (A, D, G and J), in the presence of 0.2 mM sparsomycin (B, E, H and K), and in the presence of 1 mM fusidic acid (C, F, I and L).

senger. The small peak of disomes is probably due to a low level of
elongation, which permits occasionally a second ribosome to bind to
the initiation site. With RNA 3 we found that in the presence of both
inhibitors (E, F) disomes were formed, the amount of disomes is even
higher in the presence of sparsomycin than in the absence of any inhibitor (D). The figures obtained with RNA 2 are less clear cut. In
the presence of fusidic acid (I) disomes are formed, although less
than in the control experiment (G),the amount of disomes is too much
to be neglected. The experiment was therefore repeated at a higher
concentration of fusidic acid. Figure 13A shows that in the presence
of 1.5 mM fusidic acid only monosomes are found under the direction
of RNA 2. However at this concentration of fusidic acid other processes might have been disturbed too, since under these conditions
RNA 1 did not bind to the ribosomes (Fig.13B). The results obtained
with RNA 2 could be taken as evidence for two open initiation sites
on this RNA, which have a different sensitivity for fusidic acid.
This is in contrast with the results obtained with sparsomycin, here
only monosomes were formed (H). From (J, K and L) it is apparent that
on RNA 1 only one initiation site is available for the ribosomes.
We are fully aware of the fact that the results may have been influenced by a degradation of the disomes to monosomes by nucleolytic
activity, however, this is contradicted by the reproducibility of
the patterns obtained, even when the incubation time was varied.

These results suggest that,with the exception of RNA 3, all
AMV-RNAs have only one open initiation site.

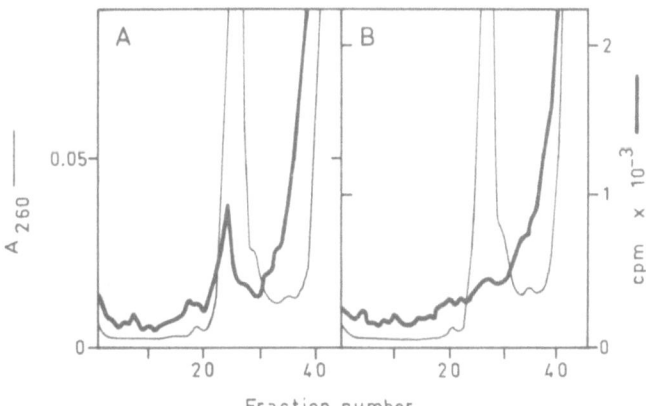

*Fig.13. Sucrose density gradient profile of polyribosomes directed
by RNA 2 (A) and RNA 1 (B) in a wheat germ cell-free system in the
presence of 1.5 mM fusidic acid.*

INFLUENCE OF $m^7G^{5'}p$ ON THE TRANSLATION

Since it is known (Pinck, 1975) that all AMV-RNAs contain a $m^7G^{5'}ppp^5Gp$ at their 5' terminus we tried to block the translation of the cistrons located near the 5' terminus with $m^7G^{5'}p$ (Hickey et al., 1976). It has already been shown by Roman et al.(1976) that the formation of initiation complexes with AMV Ta-RNA (identical to RNA 4) can be inhibited for about 75% by $m^7G^{5'}p$.

At low concentration of the messengers (0.5 µg) the translation of all AMV-RNAs was equally sensitive to the presence of $m^7G^{5'}p$ (results not shown). At higher RNA conc. (Fig. 14) the translation of RNA 1 and RNA 3 was much more inhibited by $m^7G^{5'}p$ than that of RNA 4 and RNA 2. The sensitivity of the translation of different

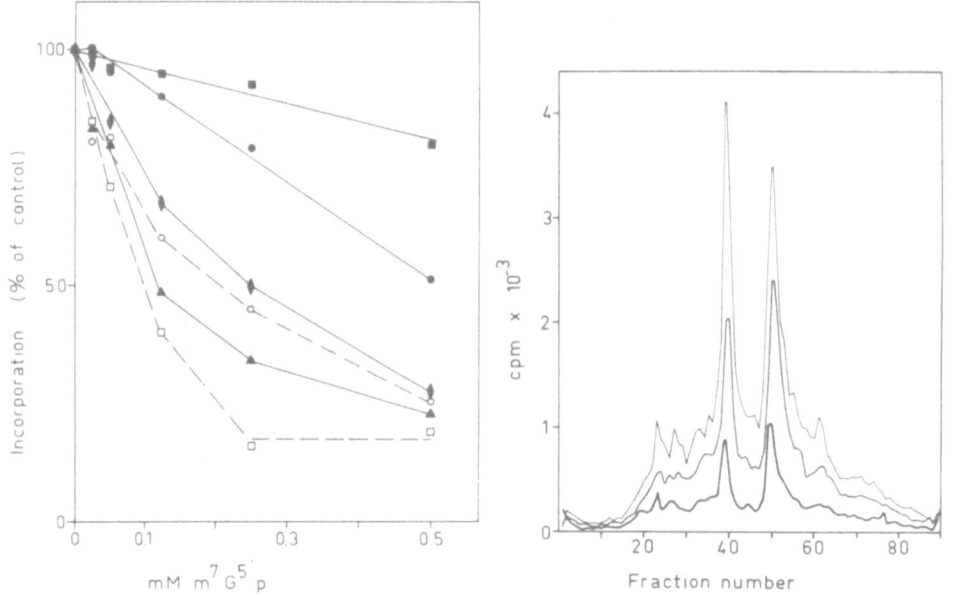

Fig.14. Inhibition of translation by increasing concentrations of $m^7G^{5'}p$. The incubation mixture (100 µl) contained 3 µg RNA. Translation in the presence of the nucleotide is expressed as % of the control. Control incorporation with AMV-RNA 1 (———) 36,534 cpm, AMV-RNA 2 (———) 52,671 cpm, AMV-RNA 3 (———) 16,467 cpm, AMV-RNA 4 (———) 63,044 cpm, TMV-RNA (—o—) 33,081 cpm, and BMV-RNA 4 (—o—) 27,108 cpm.

Fig.15. Comparison of the gel electrophoretic patterns of the products directed by AMV-RNA 3 in the presence of increasing amounts of $m^7G^{5'}p$. (———) no $m^7G^{5'}p$, (———) 0.125 m$\underline{M}$ $m^7G^{5'}p$ and (———) 0.5 m$\underline{M}$ $m^7G^{5'}p$.

RNAs for $m^7G^{5'}p$ might be inversely related to their affinity for
either the ribosomes or the initiation factors.

Comparison of the products directed in a cell-free system of
wheat germ by RNA 3 in the presence and the absence of $m^7G^{5'}p$ showed
that the formation of both products was inhibited although that of
coat protein slightly less than that of the 35,000 daltons protein
(Fig.15). Processing of RNA 3 followed by capping of the RNA 4 like
molecule is not likely, since the ratio of coat to 35,000 daltons
protein directed by RNA 3 is not influenced by the presence of S
adenosylhomocysteine (Van Vloten-Doting et al., 1976). Furthermore,
the translation of AMV-RNA 4 is rather resistent against $m^7G^{5'}p$
(Fig.14). If we assume that $m^7G^{5'}p$ only interferes with the initiation it would implicate that initiation at an internal site is also
inhibited by $m^7G^{5'}p$. Alternatively $m^7G^{5'}p$ may block other stages
of the protein synthesis.

THE 5' TERMINAL SEQUENCE OF AMV-RNA 4

Earlier studies of Verhoef et al. (1967, 1968, 1971) and Verhoef and Bosch (1971) with E. coli ribosomes, aimed at the identification of initiation sites on plant viral RNAs, showed that in the
absence of protein synthesis only one E. coli ribosome attaches to
AMV-RNA 4. For these experiments preparations of AMV-RNA 4 were
used which were contaminated for about 15% with the larger AMV-RNAs.
The monosomal complexes could accomodate besides fMet-tRNA three
other aminoacyl-tRNAs: Val-tRNA, Phe-tRNA, and Ile-tRNA. Moreover,
it was demonstrated that polypeptide synthesis directed by RNA 4
could be initiated on these complexes with either Phe-tRNA or Ile-
tRNA when the aminoacyl moieties were acetylated. After simultaneous
binding of acPhe-tRNA and Ile-tRNA the formation of the dipeptide
acPhe-Ile could be demonstrated. It was concluded that the E. coli
ribosome binding site on RNA 4 contains adjacent codons for Phe and

$m^7G^{5'}ppp^5G$-U-A-U-U-A-A-U-A-A-U-G   *Initiation site of BMV-
RNA 4 (Kaesberg, 1975;
Dasgupta et al., 1975)*

$m^7G^{5'}ppp^5G$-U-A-U-U-$\overset{U}{\underset{C}{-}}$-A-U-A-A-U-G   *Codons for Val, Phe, Ile,
and Met*

val   phe   ile   met

Fig.16. Initiation site of AMV-RNA 4 ?????

Ile and perhaps in close proximity the codons for Val and fMet. When writing successive the respective codons for Val, Phe and Ile and fMet (Fig.16) we were struck by the similarity with the ribosome binding site of BMV-RNA 4 (Kaesberg, 1975; Dasgupta et al., 1975).

This prompted us to repeat the experiments with highly purified AMV-RNA 4. In figure 17 the binding of Ile-tRNA to E.coli ribosomes programmed with AMV-RNA 4 and BMV-RNA 4 were compared. Significant binding was only observed in the presence of AMV-RNA 4. Also Val-tRNA and Phe-tRNA could only be bound to E.coli ribosomes in the presence of AMV-RNA 4. Interpretation of the negative results obtained with BMV-RNA 4 has to await the demonstration of a faithful translation of this RNA in the E.coli system.

Figure 18A shows that also purified preparations of AMV-RNA 4 can direct the incorporation of acPhe. The incorporation reaches a maximum after 15 min of incubation, thereafter there is a slight decrease in the incorporated label, probably due to proteolytic activity present in the cell-free extract. The experiments were carried out in the presence of all 20 unlabeled amino acids. Since methionine was present, the formation of fMet-tRNA is to be expected. To diminish the latter leucovorin was omitted from the incubation mixture. In the absence of leucovorin there is still a low level of formylation of methionine.

In figure 18B the gel electrophoretic pattern of the products initiated with acPhe and fMet are compared. The reaction mixture contained   ac(^{14}C)Phe-tRNA and   (^{3}H)Met and (^{3}H)Ile. From the figure can be seen that there is a difference in electrophoretic mo-

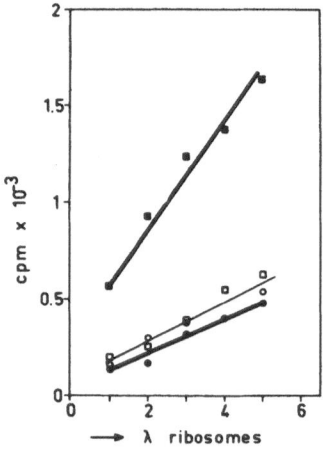

Fig.17. Comparison of the binding of Ile-tRNA to RNA 4 of AMV (——•——) and BMV (—•——) in the E.coli system. Reaction mixtures (100 µl) contained 20 µg RNA and 1 nmol (^{14}C)Ile-tRNA (11 mCi/mmol). (—◦—◦) level of binding in the absence of added RNA.

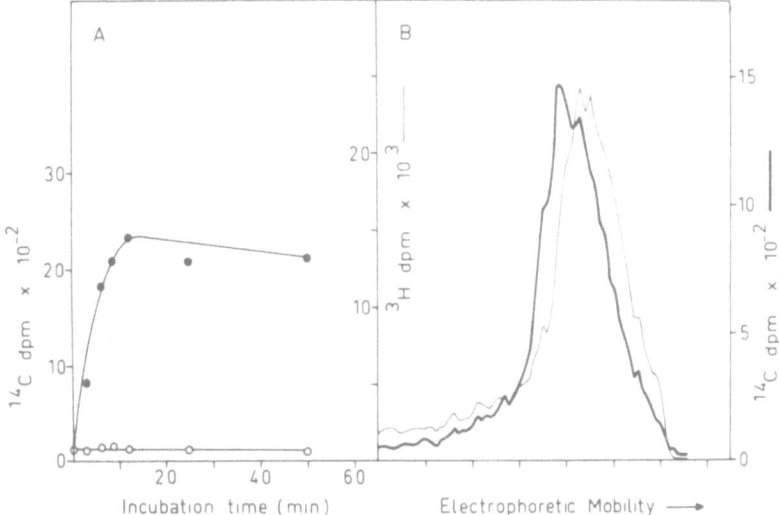

Fig.18A. Incorporation of ac(^{14}C)Phe as a function of time in the presence (—•—) and absence (—○—) of RNA 4. Reaction mixtures (100 µl) contained 15 µg RNA 4.17 pmoles ac(^{14}C)Phe-tRNA, while leucovorin was omitted. The reaction was stopped by the addition of 1 ml 7% TCA, radioactivity was determined on glass fibre filters after treatment with alkali.

Fig.18B. Gel electrophoretic pattern of products synthesized in the E.coli system under the direction of RNA 4. The reaction mixture (1000 µl) contained 127 pmoles ac(^{14}C)Phe-tRNA (495 mCi/mmol), 5 µCi (^{3}H)Met (8 Ci/mmol), 5 µCi (^{3}H)Ile (26 Ci/mmol), 200 µg RNA 4 and no leucovorin. The mixture was incubated for 20 min at 37° and treated with alkali after which the TCA-precipitable material was applied to the gel.

bility (corresponding to an apparent difference in molecular weight of about 1200 daltons) between the (^{14}C)- and the (^{3}H)-containing peaks. It has been shown that authentic AMV coat protein comigrates with the latter one (results not shown). These results suggest that acPhe can be incorporated at the N-terminal position of a product which is probably coat protein with a few additional amino acids at the N-terminus.

From these data we have to conclude that the 5' terminal sequence of AMV-RNA 4 and BMV-RNA 4 differ much more than depicted in Fig.16.

To investigate whether in wheat germ cell-free system the translation of the coat protein gene starts with a leader sequence, we compared the kinetics of the incorporation of Met and Phe. If there is a leader sequence of about 12 amino acids we expect the incorporation of Phe to precede or to be simultaneous with that

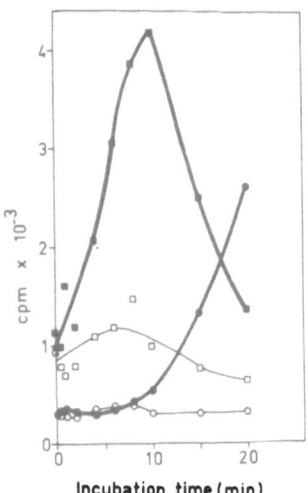

Fig.19. Comparison of the time of incorporation of penylalanine and methionine under the direction of AMV-RNA 4 in the wheat germ cell-free system. The reaction mixture (100 μl) contained 1 μg RNA, 0.125 μCi (¹⁴C)phenylalanine (495 mCi/mmol) (——•——), or 0.125 μCi (³⁵S)methionine (50 Ci/mmol) (——•——). The amount of (¹⁴C)Phe and (³⁵S)Met added was chosen sothat, considering the pool of amino acids, 1 pmol of amino acid incorporated corresponds to 400 cpm. Level of incorporation in the absence of AMV-RNA 4 is indicated by the thin lines. The incorporation experiment was carried out at 20° instead of 30° and the reaction was stopped by the addition of 4 ml of cold incubation buffer, radioactivity was determined after filtration over Millipore filters.

of Met, if there is no leader sequence the incorporation of Phe (position 66 in the authentic coat protein (Van Beynum, 1975; Kraal et al., 1976; Van Beynum et al., to be published))will start later than the incorporation of Met. The experiment was performed at 20° to slow down the protein synthesis. From figure 19 it is evident that the incorporation of Phe starts at least 12 min later than that of Met. From this picture two more things can be calculated: The rate of protein synthesis at 20° is about 5 amino acids per min (at 30° between 10 and 15) and the first Met is removed before the peptide chain is 50 amino acids long.

DISCUSSION

The results presented above leave us with a large number of questions. One of these questions concerns the specificity of the inhibition of translation by $m^7G^{5'}p$. Our results suggest that also the translation of internal cistrons might be inhibited.

Then there is the difference between the eukaryotic systems and the prokaryotic system. It is not so much surprising that after a start with formylmethionine the N-terminus is not processed to acSer, what did surprise us is the possibility to start with acPhe. Why does the E.coli ribosome, which evidently can translate AMV-RNA 4 with fidelity sometimes start on another triplet than AUG ahead of the initiation site ?

In analogy with the known initiation site for coat protein on BMV-RNA 4, we had expected the initiation codon on AMV-RNA 4 to be quite close to the 5' end of the RNA chain. However, from the difference in molecular weight between the E.coli products initiated with acPhe and fMet we can calculate that the distance between the 5' end and the initiation site is at least 40 nucleotides.

It is rather difficult to obtain well defined translation products of RNA 1 and 2. This could be due to some failure of the in vitro systems, or in the case of polysomes the translation of the larger RNAs could have been inhibited by some viral product. It is also possible that the sequence and structure of these RNAs is designed for a low efficiency of translation. The latter two explanations are in accordance with the finding that also in vivo (in protoplasts infected with cowpea chlorotic mottle virus, a virus with a coat protein independent tripartite genome) only very small amounts of translation products corresponding to the largest RNAs are found (unpublished results from F.Sakai, J.R.O.Dawson, J.W. Watts and J.B.Bancroft).

The results obtained up to now with RNA 1 and 2 suggest that both RNAs are monocistronic messengers, since the largest translation product found in different systems could represent a complete translation product,and only one ribosome could be bound to each RNA. However, analysis of the RNA content of polyribosomes isolated from tobacco infected with AMV revealed that apart from the four virion RNAs, two additional RNAs were present. The possibility that these two RNAs bear a similar relationship to RNA 1 and 2 respectively, as does RNA 4 to RNA 3 has been envisaged (Bol et al., 1976).If this assumption is correct, we can conclude from our ribosome binding experiments that the initiation site for the second cistron is blocked on the larger RNAs. Furthermore, in this case RNA 1 and 2 should both contain the genetic information for at least two proteins; consequently on each RNA there should be at least two complementation groups of temperature sensitive mutants. This is presently under investigation.

The major problem is connected with the translation of RNA 3. Mohier et al. (1975) could not detect the formation of material which reacted with anti-AMV-serum under the direction of AMV-RNA 3 in the cell-free systems of Krebs ascites cells or reticulocytes. However,our results suggest that on RNA 3 there are two initiation sites and both in wheat germ and in oocytes coat protein is readily translated from RNA 3. This is in contrast to the situation with BMV, where hardly any coat protein is translated from RNA 3 (Shih and Kaesberg, 1973).

The situation is the reverse from what we had expected, since
we know that in vivo coat protein can only be translated from a
mixture of AMV RNA 1, 2 and 3 in the presence of coat protein (Bol
et al., 1971), while the genome of BMV is infectious as such (Lane
and Kaesberg, 1971). There are two possible explanations : either
the regulation of translation in tobacco and bean cells is different
from that of wheat germ extract or oocytes, or the presence of
RNA 1 and/or RNA 2 prevents the translation of coat protein from
RNA 3 and this is reversed by coat protein itself. At the moment
we are trying to establish whether or not a mixture of RNA 1, 2 and
3 will yield coat protein in wheat germ and in oocytes.

ACKNOWLEDGEMENTS

We are indebted to Dr.M.Pranger (State University of Utrecht)
for a gift of sparsomycin, to Drs. J.W.Watts and G.van Dieijen for
permission to cite results prior to publication and to Dr.P.H.van
Knippenberg for critical reading of the manuscript.
This work was sponsored in part by the Netherlands Foundation
for Chemical Research (S.O.N.) with financial aid from the Netherlands Organization for the Advancement of Pure Research (Z.W.O.),
and in part by the "actions concertées" of the Belgian state.
G.M. is "chercheur qualifié" of the Belgian F.N.R.S.

REFERENCES

Bol, J.F., Van Vloten-Doting, L., and Jaspars, E.M.J. (1971)
      Virology 46, 73-85.
Bol, J.F., Bakhuizen, C.E.G.C., and Rutgers, T. (1976) Virology,
      in press.
Dasgupta, R., Shih, D.S., Saris, C., and Kaesberg, P. (1975)
      Nature 256, 624-628.
Dingjan-Versteegh, A., Van Vloten-Doting, L., and Jaspars, E.M.J.
      (1972) Virology 49, 716-722.
Haenni, A.L., and Chapeville, F. (1966) Biochim.Biophys.Acta 114,
      135-148.
Heller, G. (1966) Ph.D.Thesis, University of Göttingen.
Heijtink, R.A. (1974) Ph.D.Thesis, University of Leiden.
Hickey, E.D., Weber, L.A., and Baglioni, C. (1976) Proc.Nat.Acad.
      Sci. U.S.A. 73, 19-23.
Jackson, A.D., and Larkins, B.A. (1976) Plant Physiology 57, 5-10.
Kaesberg, P. (1975) INSERM 47, 205-210.
Knowland, J. (1974) Genetics 78, 383-394.
Kraal, B., Van Beynum, G.M.A., de Graaf, J.M., Castel, A., and
      Bosch, L. (1976) Virology 73, in press.
Lane, L.C., and Kaesberg, P. (1971) Nature New Biol.232, 40-43.
Mohier, E., Hirth., L., Le Meur, M. and Gerlinger, P. (1975)
      Virology 68, 349-359.

Nirenberg, M.W., and Matthaei, J.H. (1961) Proc.Nat.Acad.Sci.U.S.A.
    47, 1588-1602.
Pinck, L. (1975) FEBS Letters 59, 24-28.
Pestka, S. (1974).In "Methods in Enzymology XXX Nucleic Acids and
    Protein Synthesis. Part F.(Eds.K.Moldave and L.Grossman)Acad.
    Press,New York, London.
Roman, R., Brooker, J.D., Seal, S.N., and Marcus, A. (1976)
    Nature 260, 359-360.
Shih, D.S., and Kaesberg, P. (1973) Proc.Nat.Acad.Sci.U.S.A.70,
    1799-1803.
Thang, M.N., Dondon, L., Thang, D.C., Mohier, E., Hirth, L.,
    Le Meur, M.A.,and Gerlinger, P. (1975) INSERM 47, 225-232.
Van Beynum, G.M.A. (1975) Ph.D.Thesis, University of Leiden.
Van Vloten-Doting, L., Rutgers, T., Neeleman, L., and Bosch, L.
    (1975) INSERM 47, 233-242.
Van Vloten-Doting, L. (1976) Ann.Microbiol.(Inst.Pasteur) 1976,
    127A, 119-129.
Van Vloten-Doting, L., and Jaspars, E.M.J. (1976) in"Comprehensive
    Virology" (H.Fraenkel-Conrat and R.R.Wagner, eds.) Plenum Press
    New York, in press.
Verhoef, N.J., Reinecke, C.J., and Bosch, L. (1967) Biochim.Biophys.
    Acta 149, 305-307.
Verhoef, N.J., Kraal, B., and Bosch, L. (1968) Biochim.Biophys.
    Acta 155, 456-464.
Verhoef, N.J., Lupker, J.H., Cornelissen, M.C.E., and Bosch, L.
    (1971) Virology 45, 85-90.
Verhoef, N.J., and Bosch, L. (1971) Virology 45, 75-84.
Voorma, H.O., Gout, P.W., Van Duin, J., Hoogendam, B. W., and
    Bosch, L.(1965) Biochim.Biophys.Acta 95, 446-460.